Springer Optimization and Its Applications

Volume 237

Aims and Scope

Optimization has continued to expand in all directions at an astonishing rate. New algorithmic and theoretical techniques are continually developing and the diffusion into other disciplines is proceeding at a rapid pace, with a spot light on machine learning, artificial intelligence, and quantum computing. Our knowledge of all aspects of the field has grown even more profound. At the same time, one of the most striking trends in optimization is the constantly increasing emphasis on the interdisciplinary nature of the field. Optimization has been a basic tool in areas not limited to applied mathematics, engineering, medicine, economics, computer science, operations research, and other sciences.

The series **Springer Optimization and Its Applications (SOIA)** aims to publish state-of-the-art expository works (monographs, contributed volumes, textbooks, handbooks) that focus on theory, methods, and applications of optimization. Topics covered include, but are not limited to, nonlinear optimization, combinatorial optimization, continuous optimization, stochastic optimization, Bayesian optimization, optimal control, discrete optimization, multi-objective optimization, and more. New to the series portfolio include Works at the intersection of optimization and machine learning, artificial intelligence, and quantum computing.

Volumes from this series are indexed by Web of Science, zbMATH, Mathematical Reviews, and SCOPUS.

Chongyang Liu · Song Wang · Zhaohua Gong

Nonlinear Fractional Optimal Control

Theory and Numerical Computation

Springer

Chongyang Liu
School of Mathematics and Information Science
Shandong Technology and Business University
Yantai, Shandong, China

Song Wang
School of Electrical Engineering, Computing, and Mathematical Sciences
Curtin University
Bentley, WA, Australia

Zhaohua Gong
School of Mathematics and Information Science
Shandong Technology and Business University
Yantai, Shandong, China

ISSN 1931-6828 ISSN 1931-6836 (electronic)
Springer Optimization and Its Applications
ISBN 978-981-92-0273-7 ISBN 978-981-92-0274-4 (eBook)
https://doi.org/10.1007/978-981-92-0274-4

This Springer imprint is published by the registered company Springer Nature Singapore Pte Ltd.
The registered company address is: 152 Beach Road, #21-01/04 Gateway East, Singapore 189721, Singapore

Preface

Fractional calculus, a concept first introduced in a famous correspondence between G. W. Leibniz and G. L'Hôpital in 1695, has since gained significant importance in various fields. It has been recognised as an effective tool for modelling phenomena with memory effects and fractal structures, known as fractional models. These models present numerous theoretical and computational challenges for engineers and mathematicians, as conventional theories and computational tools struggle to handle them, particularly those involving differential equations with fractional-order derivatives.

Optimal control, on the other hand, focuses on finding the optimal control law for a given system to achieve a specific criterion. This concept has found applications in diverse fields, such as aerospace, process control, robotics, bioengineering, economics, finance, and management science. While the optimal control for a simple system can be derived using Pontryagin's maximum principle or analytically solving the corresponding Hamilton-Jacobi-Bellman equation, most real-world optimal control problems are too complex to allow analytical solutions, let alone those involving fractional differential equations. Consequently, numerical computational algorithms have become indispensable in determining optimal control laws for practical applications. This has garnered increasing interest from engineers and mathematicians.

This book presents the latest advances in theory and numerical computational algorithms for solving fractional optimal control problems. The book is organised as follows: Chap. 1 introduces fractional-order systems, optimal control, and their literature reviews. Chapters 2–4 review preliminary results on fractional integrals and derivatives, fractional-order systems, and constrained mathematical programming, respectively. In Chap. 5, we establish the first-order necessary optimality conditions and propose a numerical algorithm for fractional optimal control problems involving fractional ordinary differential systems with Caputo derivatives. Chapter 6 develops a numerical method for solving a class of fractional optimal control problems involving fractional time-delay systems. Chapter 7 presents a numerical solution approach for free final-time delay fractional optimal control problems. Chapter 8 develops a gradient-based optimization for solving robust fractional optimal control problems with uncertain parameters. Chapters 9 and 10 present numerical solution techniques for fractional optimal control problems with Atangana-Baleanu and conformable derivatives, respectively.

The book is self-contained and unified in presentation, providing the necessary background material for further exploration of the subject and the rich research literature. By introducing fresh novel methods and innovative tools, we aim to provide a state-of-the-art and comprehensive systematic treatment of fractional optimal control problems. This book can serve as a reference text for researchers in the control and optimization area, as well as a textbook for senior undergraduate or graduate students. Many examples and problems collected may also be of interest to practising scientists and engineers, enhancing their understanding of the numerical procedures for solving fractional optimal control problems.

We are delighted to express our gratitude to our colleagues and research collaborators for their invaluable contributions to the completion of this book. Specifically, we thank Profs. K. L. Teo, Y. H. Wu, W. Li, V. Rehbock, C. J. Yu, L. Wang, and H. T. Cheong for their thoughtful comments and encouragement. We also extend our deepest appreciation to Dr. X. P. Yi for writing the MATLAB codes of examples in Chap. 4. Additionally, we want to express our gratitude to the dedicated staff of Springer, particularly Daniel Wang and Kalai Shahethya, for their exceptional collaboration. Lastly, we acknowledge the financial support from the National Natural Science Foundation of China (Grants 12271307, 11771008, and 11201267) and the Shandong Province Natural Science Foundation of China (Grants ZR2023MA054, ZR2019MA031, and ZR2017MA005).

Yantai, China — Chongyang Liu
Bentley, Australia — Song Wang
Yantai, China — Zhaohua Gong
September 2025

Competing Interests The authors have no competing interests to declare that are relevant to the content of this manuscript.

Contents

1 Introduction ... 1
1.1 Fractional-Order System ... 1
1.2 Optimal Control ... 3
1.2.1 Integer-Order Optimal Control ... 3
1.2.2 Fractional-Order Optimal Control ... 4
1.3 Outline of the Book ... 5

2 Fractional Integrals and Derivatives ... 9
2.1 Introduction ... 9
2.2 Riemann-Liouville Fractional Integral and Derivative ... 10
2.3 Caputo Fractional Derivative ... 13
2.4 Atangana-Baleanu Fractional Integral and Derivative ... 16
2.5 Conformable Fractional Integral and Derivative ... 19

3 Fractional-Order Systems ... 23
3.1 Introduction ... 23
3.2 Basic Theory of Fractional Differential Equations ... 24
3.2.1 Fractional Ordinary Differential Equations ... 24
3.2.2 Fractional Delay Differential Equations ... 28
3.3 Numerical Solutions of Fractional Ordinary Differential Equations ... 31
3.3.1 Second-Order Approach ... 31
3.3.2 Third-Order Approach ... 41
3.4 Conclusions ... 53

4 Constrained Mathematical Programming ... 55
4.1 Introduction ... 55
4.2 Basic Concepts ... 56
4.3 Optimality Conditions ... 57
4.4 Penalty Methods ... 59
4.5 Sequential Quadratic Programming Algorithm ... 63

5 Fractional Optimal Control Problems 69
5.1 Introduction 69
5.2 Problem Statement 70
5.3 Necessary Optimality Conditions and Exact Penalization 71
5.3.1 Necessary Optimality Conditions 71
5.3.2 Exact Penalty Method 75
5.3.3 An Illustrative Example 81
5.4 Numerical Solution Method 83
5.4.1 Discretization Scheme 83
5.4.2 Solution Strategy 87
5.4.3 Numerical Examples 91
5.5 Conclusions 94

6 Delay Fractional Optimal Control Problems 95
6.1 Introduction 95
6.2 Problem Formulation 96
6.3 Numerical Solution Algorithm 97
6.3.1 Numerical Integration 97
6.3.2 Gradient Computation 108
6.4 Numerical Examples 114
6.5 Conclusions 121

7 Free Final-Time Delay Fractional Optimal Control Problems 123
7.1 Introduction 123
7.2 Problem Formulation 124
7.3 Numerical Solution Approach 127
7.3.1 Numerical Integration 127
7.3.2 Gradient Computation 131
7.4 Numerical Examples 137
7.5 Conclusions 141

8 Robust Fractional Optimal Control Problems 143
8.1 Introduction 143
8.2 Problem Formulation 144
8.3 Problem Transformation 145
8.3.1 Computing Sensitivity 145
8.3.2 Discretization Scheme 147
8.4 Gradient Computation 152
8.5 Numerical Examples 157
8.6 Conclusions 163

9 Fractional Optimal Control Problems with Atangana-Baleanu Derivatives 165
9.1 Introduction 165
9.2 Problem Statement 166
9.3 Numerical Solution Method 167
9.3.1 Discretization Scheme 167

9.3.2 Optimization Technique 170
9.4 Numerical Examples 174
9.5 Conclusions 178

10 Fractional Optimal Control Problems with Conformable Derivatives 181
10.1 Introduction 181
10.2 Problem Formulation 182
10.3 Numerical Solution Method 183
10.3.1 Discretization Scheme 183
10.3.2 Optimization Algorithm 186
10.4 Numerical Examples 189
10.5 Conclusions 195

References 197

Chapter 1
Introduction

1.1 Fractional-Order System

A fractional-order system is a dynamical system that can be modeled by fractional differential equations containing derivatives of arbitrary (non-integer) order [265]. Its origin can be traced back to 1695 when G.W. Leibniz and G. L'Hôpital communicated with each other regarding the possibility of a derivative to be fractional values, e.g., 1/2 [264]. In the past several decades, fractional-order systems have played very important roles in various fields such as physics, chemistry, mechanics, electrochemistry, biology, economics, signal and image processing, control theory and environmental science [24, 203, 292].

Compared with the classical integer-order systems, fractional-order systems evolve in a much more complex way. Thus, theoretical analysis for fractional systems have attracted increasing attention from many researchers [134, 242, 275]. The existence of solutions to various nonlinear fractional systems based on the Riemann-Liouville fractional derivative [22, 321, 341], the Caputo fractional derivative [30, 55, 107], the Atangana-Baleanu fractional derivative [133, 273] and the conformable fractional derivative [28] has been discussed. For system stability, the earliest study can be traced back to [208]. Then, various methods for the stability of linear fractional systems have been proposed; see, for example, frequency domain methods [40, 116, 251], linear matrix inequalities methods [220, 268] and conversion method [302]. In the past few decades, many tools have also been developed for the stability of nonlinear fractional-order systems, such as generalized Gronwall-Bellman inequality [233], the fractional-order comparison principle [319], the Mittag-Leffler stabilities [172, 335] and Lyapunov functions [8, 78, 173, 219]. Note that time-delays are common in practical engineering systems [261]. Hence, the stability analysis of fractional-order systems in the presence of time-delay has been one of the challenging issues [162]. The solution of the system has been directly examined for analyzing stability of fractional time-delay systems in [230]. Lyapunov-based methods have been used to

C. Liu et al., *Nonlinear Fractional Optimal Control*, Springer Optimization
and Its Applications 237, https://doi.org/10.1007/978-981-92-0274-4_1

check the stability of fractional time-delay systems in [21, 185]. The characteristic polynomials are exploited to check the stability of fractional time-delay systems in [301, 317]. In addition to stability, controllability and observability are two other important topics for fractional-order systems. The former is the ability that, under an admissible control over a finite time interval, the state of the system can be driven from an arbitrary initial point to another arbitrary end point. The latter means an ability of the system that the initial state will be uniquely recovered by the measured output and known or unknown input in a finite time interval. For fractional discrete-time systems, controllability and observability are investigated in terms of linear [106] and nonlinear [224] systems, respectively. The controllability was studied for fractional systems with state delay [231, 343] and input delay [117, 285]. Furthermore, the observability and/or controllability tests for fractional systems with impulsive effects are discussed in [45, 96, 310]. It should be noted that there exist other theoretical results on fractional-order systems, we refer the interested reader to [161, 229, 344, 345] for details.

As is well-known, deriving an analytical solution to a fractional-order system is usually difficult. Therefore, numerical methods for finding approximations to the solutions of fractional-order systems are vital and practically useful [52, 156]. In the last few decades, various efficient numerical techniques such as the finite difference method [168], Adomian decomposition method [148], radial basis function [122], and operational matrix method [153] have been proposed to solve fractional-order systems. Recently, spectral methods, whose advantages lie in their capacity to accomplish with high accurate outcomes, have been utilized for solution of various types of fractional-order systems [83, 142, 149, 241]. In addition, since solutions of fractional-order systems may contain some terms with fractional powers, classical polynomials are not reasonably well-suited for solving these problems. Consequently, numerous wavelets and wavelet operational matrices of fractional-order integrals and derivatives, which give rise to a system of algebraic equations, are investigated for solving fractional-order systems in [58, 59, 118, 166, 235]. For fractional time-delay systems, many classical solution methods have been modified and extended such as the fractional Adams method [33], decomposition and iterative method [281], Runge-Kutta method [316], finite difference method [218], predictor-corrector method [68] and spectral method [337]. More recently, basis functions or wavelet approximation methods have been developed for solving fractional time-delay systems [14, 227, 271, 284, 300], where shifted Jacobi polynomials [227], block pulse functions [300], polynomial spline functions [284], modified Chebyshev wavelet [271] and Haar wavelet [14] are used. As a special type of fractional time-delay systems, fractional-order systems with pantograph delay originated from the study of an electric locomotive in [237]. For fractional-order systems with pantograph delay, a variety of numerical solution methods have also been proposed such as Hermit wavelet method [270], spectral methods [234, 328], fractional-order Bernoulli wavelet method [255], fractional-order Boubaker polynomials [250], Taylor collocation methods [16, 312], Legendre multiwavelet collocation method [333], variational iteration method [56], collocation methods [35, 226] and generalized differential transform scheme [238]. For further details on numerical solution methods, we refer the reader to [112, 158, 338]. In this book, we shall focus on optimal control of nonlinear fractional-order systems.

1.2 Optimal Control

Optimal control is a subject where it is desired to determine the control law for a dynamical system such that an optimal performance index is achieved [154]. Over the years, optimal controls have found applications in many different fields, including aerospace, process control, robotics, bioengineering, economics, finance, and management science [137, 188, 294, 305, 307, 320].

1.2.1 Integer-Order Optimal Control

The theory of integer-order optimal control is an outcome of the classical calculus of variations, with a history stretching back to 1638 when G. Galilei posed two famous catenary and brachistochrone problems [99]. The generalization of the calculus of variations for optimal control theory has developed rapidly in the late 1950s. The decisive breakthroughs include: dynamic programming principle developed by R. Bellman, which makes use of the principle of optimality to solve discrete-time optimal control problems [29]; Pontryagin maximum principle formulated by L.S. Pontryagin, which is a set of first order necessary conditions of optimality for optimal control problems [245]; and linear quadratic regulator and linear quadratic Gaussian theory provided by R.E. Kalman, which are used to design optimal feedback controls [136]. Subsequently, the existence of optimal controls for various integer-order optimal control problems has been widely investigated in [49, 50, 87, 267, 291]. The optimality conditions for integer-order optimal control problems have also been discussed in [63, 77, 186, 187, 213, 244].

Practical problems, however, are usually too complex to be solved analytically using either Pontryagin maximum principle or dynamic programming. Thus, many numerical solution techniques have been developed. Among these, iterative dynamic programming (IDP) technique, which was initially developed by R. Luus in [192], is a computational variation of the dynamic programming principle. IDP reduces the state and control quantization to an arbitrarily small amount by first searching over a relatively coarse but large set of system inputs and states using dynamic programming, and then successively generating a denser and narrower search range centered about the previous result. This successive reduction in search range is then repeated, eventually allowing for an arbitrarily small variable quantization. The IDP technique has been successfully applied to solve a wide range of integer-order optimal control problems [174, 193, 195–199, 214, 215]. Indirect methods solve the optimal control problem by deriving the necessary conditions based on Pontryagin's maximum principle. The first step of this method is to formulate an appropriate two-point boundary value problem (TPBVP), and the second step is to solve the TPBVP numerically. Relative references on these methods are, for example, [17, 85, 100, 145, 146, 164, 209–211]. Direct methods are based on discretization of state and/or control variables over time and then solving the resulting discretized problem using a nonlinear programming technique. Based on the discretization of the state and/or control, direct methods can be categorized into three different approaches. The first approach is based on state and control variable parameterization [76, 79, 80, 86, 92, 93, 144, 266]. The second approach is control parameterization [105, 120, 155, 175, 189, 303]. The third approach is based on state parameterization only [129, 135, 287]. For more integer-order optimal control computation, we refer the reader to survey articles and books [32, 44, 94, 243, 259, 263, 276]. In this book, our attention is centred on fractional-order optimal control problems.

1.2.2 Fractional-Order Optimal Control

A fractional optimal control problem (FOCP) is an optimal control problem in which the dynamic constraints or the performance index or both of them include at least one fractional derivative term [67]. FOCPs arise in many fields such as mathematics, engineering, biology, economics, finance, and management [41, 57, 131, 240, 279].

As a generalization of integer-order optimal control problems, FOCPs have attracted increasing attention from many researchers in recent decades [67, 229, 311]. In the aspect of control theory, the existence of optimal controls for different kinds of FOCPs is investigated in [127, 139, 140, 150, 256]. The necessary optimality conditions in the form of Pontryagin's maximum principle for various types of FOCPs are also established in [26, 42, 95, 109, 138, 141, 200, 221, 222, 228]. It should, however, be noted that most of the fractional-order systems are highly nonlinear and thus it is often impossible to obtain the analytical solutions of FOCPs in general. For FOCPs with one state and one control, various numerical solution approaches have been developed; see, for example, [70, 84, 286, 306]. There are also numerical schemes for solving multi-dimensional FOCPs such as those in [12, 13, 36, 81, 132, 178, 181, 225, 269, 274, 298, 299, 331]. More specifically, Bernstein polynomials [12], shifted Legendre orthogonal polynomials [36] and Bell polynomials [269] are employed to approximate the state and control functions. A variational iteration method for solving the quadratic FOCPs is discussed in [13]. A solution approach using Grünwald-Letnikov approximation is discussed for solving FOCPs in [132]. Gradient-based optimization algorithms are developed for solving FOCPs in [178, 181, 225, 331]. Direct transcription methods based on fractional integral approximation formulas for solving FOCPs are provided in [274]. Fractional pseudospectral methods are investigated in the development of numerical methods for solving FOCPs in [81, 298, 299]. For free terminal time FOCPs, necessary optimality conditions for FOCPs involving nonlinear systems with both integer-order and fractional-orders are derived, and then an iterative algorithm is developed in [246]. For FOCPs with Atangana-Baleanu derivative, B-spline polynomials [297], shifted Legendre polynomials [130] and the extended Chebyshev cardinal wavelets [119] are exploited to develop numerical solution methods. For FOCPs with time-delay, a spectral collocation method based on Müntz-Legendre polynomials is employed in [125]. Legendre orthonormal polynomials [37], block-pulse functions [124, 204–206], Bernoulli wavelets [254], Bernstein polynomials [272] and Chelyshkov wavelets [223] are also used to develop numerical methods for solving FOCPs with time-delays. Moreover, a Legendre-Gauss collocation method is developed to solve FOCPs with time-varying delay in [257]. For more details on fractional-order optimal control, we refer the reader to survey papers [3, 5, 38, 54, 66]. In this book, our attention is focused on developing novel theory and numerical solution methods for fractional optimal control problems involving fractional-order systems with and without delays, fractional-order system with Atangana-Baleanu derivatives and fractional-order system with conformable derivatives, free final-time delay fractional optimal control problem and robust fractional optimal control problem. Note that these classes of problems have many practical applications across a wide range of disciplines such as those mentioned above.

1.3 Outline of the Book

The book contains ten chapters. This chapter briefly introduces the fractional-order system, optimal control, and their literature reviews. Besides this short introduction, there are nine major chapters, which are briefly summarized as follows.

To make the book self-contained, we present, in Chap. 2, a brief introduction to some basic results of several representative fractional integrals and derivatives, i.e., Riemann-Liouville fractional integral and derivative, Caputo fractional derivative, Atangana-Baleanu fractional integral and derivative, and conformable fractional integral and derivative. In Chap. 3, we state some basic theory of fractional-order systems including fractional ordinary differential equations and fractional delay differential equations. Two numerical methods for solving fractional ordinary differential equations are also developed. They are supported by rigorous convergence analysis. Engineers and practitioners should be able to follow the results in the subsequent chapters with the aid of Chaps. 2 and 3.

In Chap. 4, we review some results in constrained mathematical programming. This is important because after the application of an appropriate discretization scheme, a fractional optimal control problem is reduced to a series of finite-dimensional optimization problems, each of which is essentially a mathematical programming problem.

Starting from Chap. 5, we focus our attention on theory and numerical computation of nonlinear fractional optimal control problems. We start our discussions from fractional optimal control problems involving fractional ordinary systems with Caputo derivatives. The first-order necessary optimality conditions of this class of fractional optimal control problems are derived by using calculus of variation. Based on the necessary optimality conditions, we prove that a low-order power penalty method is exact for the fractional optimal control problem by showing that an optimal solution to the penalized problem satisfies the optimality conditions of the original one when the penalty parameter is sufficiently large. Furthermore, we propose a discretization scheme for the fractional optimal control problem, which results in a series of finite-dimensional optimization problems. The gradients of the discretized cost function with respect to the discretized controls are also derived. On this basis, a gradient-based optimization algorithm is developed for solving the resulting optimal control problem. Three example problems are solved to verify the necessary optimality conditions and to illustrate the effectiveness of the proposed optimization algorithm.

In Chap. 6, we aim to develop numerical method for solving a class of fractional optimal control problems involving fractional time-delay systems and subject to state inequality constraints. First, we propose a numerical integration scheme for the fractional time-delay system and prove that the convergence rate of the numerical solution to the exact one is 2nd-order based on Taylor expansion and linear interpolation. This gives rise to a discrete-time optimal control problem. Then, we derive the gradient formulas of the cost and constraint functions with respect to the decision variables and present a gradient computation procedure. With this procedure, the discrete-time optimal control problem can be solved as mathematical programming problems.

In Chap. 7, we consider a class of delay fractional optimal control problems with free terminal time. For this class problems, we show that the well-known time-scaling transformation for mapping the free time horizon into a fixed time interval yields a new fractional-order system with variable time-delay. Then, we propose an explicit numerical scheme for solving the resulting fractional time-varying delay system. Furthermore, we derive gradient formulas of the cost and constraint functions with respect to decision variables. On this basis, a gradient-based optimization approach is developed to solve the resulting optimization problem. Two example problems are solved to demonstrate the effectiveness of our proposed solution approach.

In Chap. 8, we focus on robust fractional optimal control problem with uncertain parameters, where the system sensitivity with respect to the uncertain parameters is explicitly included in the cost functional. For this problem, we prove that the system sensitivity can be expressed as the solution of an auxiliary fractional system. Then, we propose a numerical scheme for discretizing both the original and auxiliary fractional systems, resulting in a finite-dimensional optimization problem. Furthermore, a numerical solution algorithm based on gradients of the cost function is developed for the resulting optimization problem. Numerical results of two example problems are given to demonstrate the validity and applicability of the proposed algorithm.

In Chap. 9, we consider a class of fractional optimal control problems with Atangana-Baleanu-Caputo derivatives subject to canonical constraints of equality and inequality. To solve this problem, we present a discretization scheme based on the trapezoidal rule and a numerical integration technique. Then, the gradient formulas of the cost and constraint functions with respect to the decision variables are derived. Furthermore, a gradient-based optimization algorithm for solving the discretized optimal control problem is developed. The applicability and effectiveness of the proposed algorithm are verified through three non-trivial example problems.

In Chap. 10, numerical solution of a class of fractional optimal control problems with conformable derivatives is developed. We first state a nonlinear fractional optimal control problem subject to general equality and inequality constraints. Based on a numerical integration scheme, we then present a discretization method for the governing fractional-order system as well as the cost and constraint functionals, which yields a discretized approximate problem. We further derive the gradients of the cost and constraint functions with respect to the discretized controls. On this basis, a numerical optimization approach to solve the approximate problem is developed. Finally, three non-trivial example problems are solved to illustrate the applicability and effectiveness of the developed approach.

Chapter 2
Fractional Integrals and Derivatives

2.1 Introduction

In this chapter, we will introduce some basic results of several representative fractional integrals and derivatives, i.e., Riemann-Liouville fractional integral and derivative, Caputo fractional derivative, Atangana-Baleanu fractional integral and derivative, and conformable fractional integral and derivative. For more detailed discussions on these integrals and derivatives, we refer the reader to [2, 48, 53, 73, 121, 128, 134, 163, 242, 275].

The Riemann-Liouville fractional integral and derivative in next section, which are named after B. Riemann's work in 1847 [262] and J. Liouville's work in 1832 [176], generalize the ideas of the conventional integer-order integration and differentiation. The Riemann-Liouville fractional derivative played an important role in the development of theory of fractional derivatives and integrals and for its applications in pure mathematics, e.g., solution of integer-order differential equations, definitions of new function classes, and summation of series. The Caputo fractional derivative in Sect. 2.3, which was proposed by M. Caputo in 1967 [47], is very useful for modeling phenomena which take account of interactions between present and past information. These phenomena usually display nonlocal behaviors in derivatives. Note that the kernel functions reflecting the nonlocality of fractional derivatives can be either singular or non-singular. The most well-known fractional derivatives equipped with singular kernels are Riemann-Liouville and Caputo derivatives, while one of the most famous fractional derivatives with non-singular kernels is Atangana-Baleanu fractional derivative [20] to be discussed in Sect. 2.4. Furthermore, some challenges arise with these fractional derivatives in practical applications because they generally do not possess the properties in classic calculus such as integer-order derivatives. To overcome these challenges, a fractional derivative called conformable fractional derivative proposed in [147] is presented in Sect. 2.5. The above fractional integrals and derivatives will serve as the main modeling tools in the fractional-order optimal control problems of this book.

C. Liu et al., *Nonlinear Fractional Optimal Control*, Springer Optimization and Its Applications 237, https://doi.org/10.1007/978-981-92-0274-4_2

Before we start the main tasks, we introduce some function spaces which will be used in later discussions.

Definition 2.1 Let $I := [a, b]$ with $-\infty < a < b < \infty$, $0 < \mu \leq 1$, $k \in \mathbb{N}_0 := \{0, 1, 2, \ldots\}$, and $p \geq 1$.

$$L_p(I) := \left\{ f : I \to \mathbb{R} |\ f \text{ is measurable on } I \text{ and } \int_a^b |f(t)|^p dt < \infty \right\}, \quad (2.1)$$

$$L_\infty(I) := \{ f : I \to \mathbb{R} |\ f \text{ is measurable and essentially bounded on } I \}, \quad (2.2)$$

$$H_\mu(I) := \{ f : I \to \mathbb{R} |\ \exists c > 0, \forall x, y \in I, |f(x) - f(y)| \leq c|x - y|^\mu \}, \quad (2.3)$$

$$C^k(I) := \{ f : I \to \mathbb{R} |\ f \text{ has a continuous } k\text{th derivative } \}, \quad (2.4)$$

$$C(I) := C^0(I). \quad (2.5)$$

In other words, $L_p(I)$, $1 \leq p < \infty$, is the usual space of all p-power Lebesgue integrable functions on I, $H_\mu(I)$ is the set of all Hölder continuous functions of order μ, whereas $C(I)$ is the set of all continuous functions from I to $\mathbb{R}$.

Definition 2.2 The set of functions $f : I \to \mathbb{R}$ with an absolutely continuous $(n-1)$st derivative is denoted by $AC^n(I)$, i.e., the functions f for which there exists (almost everywhere) a function $g \in L_1(I)$ such that

$$f^{(n-1)}(x) = f^{(n-1)}(a) + \int_a^x g(t)dt. \quad (2.6)$$

In this case, we call g the nth derivative of f.

This definition implies that a function $f \in AC^1(I)$ possesses (almost everywhere) a derivative $f' \in L_1(I)$. However, this implication cannot be reversed in general.

2.2 Riemann-Liouville Fractional Integral and Derivative

We begin with the definition of the Riemann-Liouville fractional integral.

Definition 2.3 For $f \in L_1(I)$ and $\alpha > 0$, the left Riemann-Liouville fractional integral of order α starting from a, denoted as ${}_a\mathrm{I}_x^\alpha f$, is defined by

$${}_a\mathrm{I}_x^\alpha f(x) = \frac{1}{\Gamma(\alpha)} \int_a^x (x-s)^{\alpha-1} f(s)ds, \quad x > a, \quad (2.7)$$

and the right Riemann-Liouville fractional integral of order α terminating at b, denoted as ${}_x\mathrm{I}_b^\alpha f$, is defined by

$${}_x\mathrm{I}_b^\alpha f(x) = \frac{1}{\Gamma(\alpha)} \int_x^b (s-x)^{\alpha-1} f(s)ds, \quad x < b, \quad (2.8)$$

where $\Gamma : (0, \infty) \to \mathbb{R}$ is the Gamma function defined by

$$\Gamma(z) = \int_0^\infty t^{z-1} e^{-t} dt. \tag{2.9}$$

When $\alpha = k \in \mathbb{N} := \{1, 2, \ldots\}$, the definitions (2.7) and (2.8) coincide with the ordinary k-fold integrals. Furthermore, in case $\alpha = 0$, we adopt the convention ${}_a\mathrm{I}_x^0 f(x) = {}_x\mathrm{I}_b^0 f(x) = f(x)$. In fact, the integral operators ${}_a\mathrm{I}_x^\alpha$ and ${}_x\mathrm{I}_b^\alpha$ inherit certain properties from integer-order integrals. For example,

$${}_a\mathrm{I}_x^\alpha (x-a)^\gamma = \frac{\Gamma(\gamma+1)}{\Gamma(\gamma+\alpha+1)}(x-a)^{\gamma+\alpha}, \quad \gamma > -1,\ x > a,$$

and

$${}_x\mathrm{I}_b^\alpha (b-x)^\gamma = \frac{\Gamma(\gamma+1)}{\Gamma(\gamma+\alpha+1)}(b-x)^{\gamma+\alpha}, \quad \gamma > -1,\ x < b.$$

Example 2.1 For $\alpha > 0$, the Riemann-Liouville fractional integral ${}_0\mathrm{I}_x^\alpha e^{\lambda x}$, $\lambda \in \mathbb{R}$, is

$$\begin{aligned} {}_0\mathrm{I}_x^\alpha e^{\lambda x} &= {}_0\mathrm{I}_x^\alpha \sum_{k=0}^\infty \frac{(\lambda x)^k}{\Gamma(k+1)} = \sum_{k=0}^\infty \lambda^k \frac{{}_0\mathrm{I}_x^\alpha x^k}{\Gamma(k+1)} = \sum_{k=0}^\infty \lambda^k \frac{x^{k+\alpha}}{\Gamma(k+\alpha+1)} \\ &= x^\alpha \sum_{k=0}^\infty \frac{(\lambda x)^k}{\Gamma(k+\alpha+1)} = x^\alpha \mathbb{E}_{1,\alpha+1}(\lambda x), \end{aligned}$$

where $\mathbb{E}_{\alpha,\beta}$ is the two-parameter Mittag-Leffler function defined by

$$\mathbb{E}_{\alpha,\beta}(z) = \sum_{k=0}^\infty \frac{z^k}{\Gamma(k\alpha+\beta)}. \tag{2.10}$$

The following theorems summarize some important properties of ${}_a\mathrm{I}_x^\alpha$. Note that similar results hold for ${}_x\mathrm{I}_b^\alpha$.

Theorem 2.1 (Commutative Semigroup Property). *Let $\alpha, \beta \geq 0$ and $f \in L_1(I)$. Then,*

$${}_a\mathrm{I}_x^\alpha \left({}_a\mathrm{I}_x^\beta f(x)\right) = {}_a\mathrm{I}_x^\beta \left({}_a\mathrm{I}_x^\alpha f(x)\right) = {}_a\mathrm{I}_x^{\alpha+\beta} f(x) \tag{2.11}$$

holds almost everywhere on I.

Theorem 2.2 (Fractional Integration by Parts). *For $f \in L_p(I)$, $g \in L_q(I)$, $p, q \geq 1$ with $p^{-1} + q^{-1} \leq 1 + \alpha$ ($p, q \neq 1$, when $p^{-1} + q^{-1} = 1 + \alpha$), the following identity holds:*

$$\int_a^b g(x)\, {}_a\mathrm{I}_x^\alpha f(x) dx = \int_a^b f(x)\, {}_x\mathrm{I}_b^\alpha g(x) dx. \tag{2.12}$$

Theorem 2.3 (Interchange of Limit and Integration). *Let* $\alpha > 0$*, and* $\{f_n\}_{n=1}^{\infty}$ *be a uniformly convergent sequence of continuous functions on* I*. Then,*

$$ {}_a\mathrm{I}_x^\alpha \left(\lim_{n\to\infty} f_n(x)\right) = \lim_{n\to\infty} \left({}_a\mathrm{I}_x^\alpha f_n(x)\right), \quad x > a. \tag{2.13} $$

In particular, the sequence of functions $\{{}_a\mathrm{I}_x^\alpha f_n\}_{n=1}^{\infty}$ *is uniformly convergent.*

Theorem 2.4 (Continuity on Fractional Order). *Let* $1 \le p < \infty$*, and* $\{\alpha_n\}_{n=1}^{\infty}$ *be a convergent nonnegative sequence with limit* α*. Then, for every* $f \in L_p(I)$*,*

$$ \lim_{n\to\infty} {}_aI_x^{\alpha_n} f(x) = {}_a\mathrm{I}_x^\alpha f(x), \quad x > a, \tag{2.14} $$

where the convergence is in the sense of the $L_p(I)$ *norm.*

Based on the Riemann-Liouville fractional integral, we define the Riemann-Liouville fractional derivative as follows.

Definition 2.4 For $f \in L_1(I)$, and $n - 1 < \alpha \le n$, $n \in \mathbb{N}$, the left Riemann-Liouville fractional derivative of order α starting from a, denoted as ${}_a^R\mathrm{D}_x^\alpha f$, is defined by

$$ {}_a^R\mathrm{D}_x^\alpha f(x) = \frac{1}{\Gamma(n-\alpha)} \frac{d^n}{dx^n} \int_a^x (x-s)^{n-\alpha-1} f(s)ds, \quad x > a, \tag{2.15} $$

and the right Riemann-Liouville fractional derivative of order α terminating at b, denoted as ${}_x^R\mathrm{D}_b^\alpha f$, is defined by

$$ {}_x^R\mathrm{D}_b^\alpha f(x) = \frac{(-1)^n}{\Gamma(n-\alpha)} \frac{d^n}{dx^n} \int_x^b (s-x)^{n-\alpha-1} f(s)ds, \quad x < b, \tag{2.16} $$

where $\Gamma(\cdot)$ is the Gamma function as defined in (2.9).

For $\alpha = 0$, we set ${}_a^R\mathrm{D}_x^\alpha f(x) = {}_x^R\mathrm{D}_b^\alpha f(x) = f(x)$ as usual. In particular, if $\alpha = k \in \mathbb{N}$, ${}_a^R\mathrm{D}_x^\alpha f(x) = f^{(k)}(x)$ and ${}_x^R\mathrm{D}_b^\alpha f(x) = (-1)^k f^{(k)}(x)$. However, for $\alpha \notin \mathbb{N}_0$, ${}_a^R\mathrm{D}_x^\alpha f$ and ${}_x^R\mathrm{D}_b^\alpha f$ are inherently nonlocal, which influences their analytic properties. For example, the Riemann-Liouville fractional derivatives of a constant are, in general, not identically zero:

$$ {}_a^R\mathrm{D}_x^\alpha 1 = \frac{(x-a)^{-\alpha}}{\Gamma(1-\alpha)}, \quad {}_x^R\mathrm{D}_b^\alpha 1 = \frac{(b-x)^{-\alpha}}{\Gamma(1-\alpha)}. \tag{2.17} $$

On the other hand, for $j = 1, 2, \ldots, n$,

$$ {}_a^R\mathrm{D}_x^\alpha (x-a)^{\alpha-j} \equiv 0, \quad {}_x^R\mathrm{D}_b^\alpha (b-x)^{\alpha-j} \equiv 0. $$

Example 2.2 For $\alpha > 0$, the Riemann-Liouville fractional derivative ${}_0^R\mathrm{D}_x^\alpha e^{\lambda x}$, $\lambda \in \mathbb{R}$, is

$$
\begin{aligned}
{}_0^R\mathrm{D}_x^\alpha e^{\lambda x} &= {}_0^R\mathrm{D}_x^\alpha \left[\sum_{k=0}^{\infty} \frac{(\lambda x)^k}{\Gamma(k+1)} \right] = \sum_{k=0}^{\infty} \lambda^k \frac{{}_0^R\mathrm{D}_x^\alpha x^k}{\Gamma(k+1)} = \sum_{k=0}^{\infty} \frac{\lambda^k x^{k-\alpha}}{\Gamma(k+1-\alpha)} \\
&= x^{-\alpha} \sum_{k=0}^{\infty} \frac{(\lambda x)^k}{\Gamma(k+1-\alpha)} = x^{-\alpha} \mathbb{E}_{1,1-\alpha}(\lambda x),
\end{aligned}
$$

where $\mathbb{E}_{1,1-\alpha}(\cdot)$ is as defined in (2.10).

The following two theorems collect some important properties for ${}_a^R\mathrm{D}_x^\alpha$. Here, Theorem 2.5 shows that ${}_a^R\mathrm{D}_x^\alpha$ is the left inverse of ${}_a\mathrm{I}_x^\alpha$ on $L_1(I)$, but generally not a right inverse, and Theorem 2.6 gives the fractional version of integration by parts formula.

Theorem 2.5 *Let $\alpha > 0$, $n-1 < \alpha \le n$, $n \in \mathbb{N}$.*
(i) *For any $f \in L_1(I)$, ${}_a^R\mathrm{D}_x^\alpha({}_a\mathrm{I}_x^\alpha f(x)) = f(x)$ almost everywhere.*
(ii) *If ${}_a\mathrm{I}_x^{n-\alpha} f \in AC^n(I)$, then*

$$
{}_a\mathrm{I}_x^\alpha \left({}_a^R\mathrm{D}_x^\alpha f(x) \right) = f(x) - \sum_{j=0}^{n-1} \left[\frac{(x-a)^{\alpha-j-1}}{\Gamma(\alpha-j)} \lim_{z \to a^+} {}_a^R\mathrm{D}_z^{\alpha-j-1} f(z) \right]. \tag{2.18}
$$

Specifically, for $0 < \alpha \le 1$, we have

$$
{}_a\mathrm{I}_x^\alpha \left({}_a^R\mathrm{D}_x^\alpha f(x) \right) = f(x) - \frac{(x-a)^{\alpha-1}}{\Gamma(\alpha)} \lim_{z \to a^+} {}_a\mathrm{I}_z^{1-\alpha} f(z). \tag{2.19}
$$

Theorem 2.6 *Let $\alpha > 0$, $p, q \ge 1$ with $p^{-1} + q^{-1} \le 1 + \alpha$ ($p, q \ne 1$, when $p^{-1} + q^{-1} = 1 + \alpha$). Then,*

$$
\int_a^b g(x)\, {}_a^R\mathrm{D}_x^\alpha f(x) dx = \int_a^b f(x)\, {}_x^R\mathrm{D}_b^\alpha g(x) dx, \tag{2.20}
$$

provided that $f \in {}_a\mathrm{I}_x^\alpha(L_p(I))$ and $g \in {}_x\mathrm{I}_b^\alpha(L_q(I))$.

2.3 Caputo Fractional Derivative

Now, we turn to Caputo fractional derivative, which is one of the most popular fractional derivatives in fractional calculus and applications due to its ability to handle initial conditions in a physically transparent way.

Definition 2.5 For $n-1 < \alpha \le n$, $n \in \mathbb{N}$, and $f \in AC^n(I)$, the left Caputo fractional derivative of order α starting from a, denoted as ${}_a^C\mathrm{D}_x^\alpha f$, is defined by

$$ {}_a^C\mathrm{D}_x^\alpha f(x) = \frac{1}{\Gamma(n-\alpha)} \int_a^x (x-s)^{n-\alpha-1} f^{(n)}(s) ds, \quad x > a, \tag{2.21} $$

and the right Caputo fractional derivative of order α terminating at b, denoted as ${}_x^C\mathrm{D}_b^\alpha f$, is defined by

$$ {}_x^C\mathrm{D}_b^\alpha f(x) = \frac{(-1)^n}{\Gamma(n-\alpha)} \int_x^b (s-x)^{n-\alpha-1} f^{(n)}(s) ds, \quad x < b. \tag{2.22} $$

For $\alpha = k \in \mathbb{N}$, we have ${}_a^C\mathrm{D}_x^\alpha f(x) = f^{(k)}(x)$, which is same as the Riemann-Liouville fractional derivative. However, the Caputo fractional derivative is more restrictive than the Riemann-Liouville one since the definition requires $f \in AC^n(I)$. Furthermore, ${}_a^R\mathrm{D}_x^\alpha f(x) \neq {}_a^C\mathrm{D}_x^\alpha f(x)$ in general, even when both fractional derivatives are well defined. For example, when $f(x) \equiv 1$, ${}_a^C\mathrm{D}_x^\alpha f(x) = 0$, but ${}_a^R\mathrm{D}_x^\alpha f(x)$ is nonzero by (2.17).

Example 2.3 For $n-1 < \alpha \leq n, n \in \mathbb{N}$, the Caputo fractional derivative ${}_0^C\mathrm{D}_x^\alpha e^{\lambda x}$, $\lambda \in \mathbb{R}$, is

$$ \begin{aligned} {}_0^C\mathrm{D}_x^\alpha e^{\lambda x} &= {}_0^C\mathrm{D}_x^\alpha \left[\sum_{k=0}^{\infty} \frac{\lambda^k x^k}{k!} \right] = \sum_{k=0}^{\infty} \lambda^k \frac{{}_0^C\mathrm{D}_x^\alpha x^k}{k!} = \sum_{k=n}^{\infty} \lambda^k \frac{x^{k-\alpha}}{\Gamma(k-\alpha+1)} \\ &= \lambda^n x^{n-\alpha} \sum_{k=0}^{\infty} \frac{(\lambda x)^k}{\Gamma(k+n-\alpha+1)} = \lambda^n x^{n-\alpha} \mathbb{E}_{1,n-\alpha+1}(\lambda x), \end{aligned} $$

where $\mathbb{E}_{1,n-\alpha+1}(\cdot)$ is as defined in (2.10).

This result is completely different from that in Example 2.2, except the case $\alpha = 1$. Nonetheless, the Riemann-Liouville and Caputo fractional derivatives are closely related to each other.

Theorem 2.7 *Let $n-1 < \alpha \leq n$, $n \in \mathbb{N}$, and $f \in AC^n(I)$. Then,*

$$ {}_a^R\mathrm{D}_x^\alpha f(x) = {}_a^C\mathrm{D}_x^\alpha f(x) + \sum_{j=0}^{n-1} \frac{(x-a)^{j-\alpha}}{\Gamma(j-\alpha+1)} f^{(j)}(a), \quad x > a, \tag{2.23} $$

and

$$ {}_x^R\mathrm{D}_b^\alpha f(x) = {}_x^C\mathrm{D}_b^\alpha f(x) + \sum_{j=0}^{n-1} \frac{(-1)^j (b-x)^{j-\alpha}}{\Gamma(j-\alpha+1)} f^{(j)}(b), \quad x < b. \tag{2.24} $$

Theorem 2.8 *Let $n-1 < \alpha \leq n$, $n \in \mathbb{N}$, and $f \in AC^n(I)$. Then,*

$$ {}_a^C\mathrm{D}_x^\alpha f(x) = {}_a^R\mathrm{D}_x^\alpha \left(f(x) - \sum_{k=0}^{n-1} \frac{(x-a)^k}{k!} f^{(k)}(a) \right), \quad x > a, \tag{2.25} $$

and

$$ {}_{x}^{C}\mathrm{D}_{b}^{\alpha} f(x) = {}_{x}^{R}\mathrm{D}_{b}^{\alpha} \left(f(x) - \sum_{k=0}^{n-1} \frac{(-1)^k (b-x)^k}{k!} f^{(k)}(b) \right), \quad x < b. \tag{2.26} $$

Theorem 2.8 indicates that ${}_{a}^{C}\mathrm{D}_{x}^{\alpha} f$ and ${}_{x}^{C}\mathrm{D}_{b}^{\alpha} f$ can be regarded as ${}_{a}^{R}\mathrm{D}_{x}^{\alpha} f$ and ${}_{x}^{R}\mathrm{D}_{b}^{\alpha} f$ with initial corrections, respectively. In fact,

$$ {}_{a}^{C}\mathrm{D}_{x}^{\alpha} f(x) = {}_{a}^{R}\mathrm{D}_{x}^{\alpha} f(x), \quad x > a, \ \text{if } f^{(j)}(a) = 0, j = 0, 1, \ldots, n-1, \tag{2.27} $$

and

$$ {}_{x}^{C}\mathrm{D}_{b}^{\alpha} f(x) = {}_{x}^{R}\mathrm{D}_{b}^{\alpha} f(x), \quad x < b, \ \text{if } f^{(j)}(b) = 0, j = 0, 1, \ldots, n-1. \tag{2.28} $$

For the Caputo fractional derivatives, we have similar results as for the Riemann-Liouville fractional derivatives in Theorems 2.5 and 2.6.

Theorem 2.9 *Let $\alpha > 0$, $n-1 < \alpha < n$, $n \in \mathbb{N}$. Then, the following statements hold.*
(i) *If $f \in L_1(I)$ with ${}_{a}\mathrm{I}_{x}^{\alpha-n+1} f \in AC^1(I)$ and ${}_{a}\mathrm{I}_{x}^{\alpha-n+1} f(a) = 0$, then*

$$ {}_{a}^{C}\mathrm{D}_{x}^{\alpha} \left({}_{a}\mathrm{I}_{x}^{\alpha} f(x) \right) = f(x), \quad x \in I, \tag{2.29} $$

almost everywhere.
(ii) *If $f \in AC^n(I)$, then*

$$ {}_{a}\mathrm{I}_{x}^{\alpha} \left({}_{a}^{C}\mathrm{D}_{x}^{\alpha} f(x) \right) = f(x) - \sum_{k=0}^{n-1} \frac{(x-a)^k}{k!} f^{(k)}(a), \quad x \in I, \tag{2.30} $$

almost everywhere. Moreover, if $f \in C^n(I)$, then the identity (2.30) holds for all $x \in I$.

Theorem 2.10 (Integration by Parts). *Let $\alpha > 0$, $n-1 < \alpha \leq n$, $n \in \mathbb{N}$. Then,*

$$ \begin{aligned} &\int_a^b g(x)\, {}_{a}^{C}\mathrm{D}_{x}^{\alpha} f(x) dx \\ &= \int_a^b f(x)\, {}_{x}^{R}\mathrm{D}_{b}^{\alpha} g(x) dx + \sum_{j=0}^{n-1} \left[{}_{x}^{R}\mathrm{D}_{b}^{\alpha+j-n} g(x)\, {}_{x}^{R}\mathrm{D}_{b}^{n-1-j} f(x) \right]_a^b, \end{aligned} \tag{2.31} $$

and

$$\int_a^b g(x)\,{}_x^C\mathrm{D}_b^\alpha f(x)dx$$
$$= \int_a^b f(x)\,{}_a^R\mathrm{D}_x^\alpha g(x)dx - \sum_{j=0}^{n-1} \left[{}_a^R\mathrm{D}_x^{\alpha+j-n} g(x)\,{}_a^R\mathrm{D}_x^{n-1-j} f(x) \right]_a^b, \tag{2.32}$$

where ${}_a^R\mathrm{D}_x^k g(x) = {}_a\mathrm{I}_x^{-k} g(x)$ *and* ${}_x^R\mathrm{D}_b^k g(x) = {}_x\mathrm{I}_b^{-k} g(x)$*, if* $k < 0$.

The following theorem generalizes the classical integer-order mean value theorem.

Theorem 2.11 *Let* $n - 1 < \alpha < n$, $n \in \mathbb{N}$, *and* $f \in C^n(I)$ *with* ${}_a^C\mathrm{D}_x^\alpha f \in C(I)$. *Then, there exists some* $\xi \in (a, b)$ *such that*

$$\frac{f(b) - \sum_{k=0}^{n-1} \frac{(b-a)^k}{k!} f^{(k)}(a)}{(b-a)^\alpha} = \frac{{}_a^C\mathrm{D}_x^\alpha f(\xi)}{\Gamma(\alpha+1)}. \tag{2.33}$$

2.4 Atangana-Baleanu Fractional Integral and Derivative

The Atangana-Baleanu fractional derivative is a nonlocal fractional one with non-singular Mittag-Leffler kernel, which is connected with a variety of applications [9, 20, 102, 126].

Definition 2.6 For $0 < \alpha < 1$ and $f \in H_1(I)$, the left Atangana-Baleanu fractional derivative of order α in Caputo sense starting from a, denoted as ${}^{ABC}_{\ a}\mathrm{D}_x^\alpha f$, is defined by

$${}^{ABC}_{\ a}\mathrm{D}_x^\alpha f(x) = \frac{\mathbb{B}(\alpha)}{1-\alpha} \int_a^x \mathbb{E}_\alpha\Big[-\alpha \frac{(x-s)^\alpha}{1-\alpha} \Big] f'(s)ds, \quad x > a, \tag{2.34}$$

and the right Atangana-Baleanu fractional derivative of order α in Caputo sense terminating at b, denoted as ${}^{ABC}_{\ x}\mathrm{D}_b^\alpha f$, is defined by

$${}^{ABC}_{\ x}\mathrm{D}_b^\alpha f(x) = -\frac{\mathbb{B}(\alpha)}{1-\alpha} \int_x^b \mathbb{E}_\alpha\Big[-\alpha \frac{(s-x)^\alpha}{1-\alpha} \Big] f'(s)ds, \quad x < b, \tag{2.35}$$

where $\mathbb{B}(\alpha)$ is a normalized function with $\mathbb{B}(0) = \mathbb{B}(1) = 1$, and $\mathbb{E}_\alpha(\cdot)$ is the one-parameter Mittag-Leffler function defined by

$$\mathbb{E}_\alpha(z) = \sum_{k=0}^{\infty} \frac{z^k}{\Gamma(k\alpha+1)}. \tag{2.36}$$

Definition 2.7 For $0 < \alpha < 1$ and $f \in L_1(I)$, the left Atangana-Baleanu fractional derivative of order α in Riemann-Liouville sense starting from a, denoted as ${}^{ABR}_{\ a}\mathrm{D}_x^\alpha f$, is defined by

$$ {}^{ABR}_{a}\mathrm{D}^{\alpha}_{x} f(x) = \frac{\mathbb{B}(\alpha)}{1-\alpha} \frac{d}{dx} \int_a^x \mathbb{E}_{\alpha}\Big[-\alpha \frac{(x-s)^{\alpha}}{1-\alpha}\Big] f(s) ds, \quad x > a, \tag{2.37} $$

and the right Atangana-Baleanu fractional derivative of order α in Riemann-Liouville sense terminating at b, denoted as ${}^{ABR}_{x}\mathrm{D}^{\alpha}_{b} f$, is defined by

$$ {}^{ABR}_{x}\mathrm{D}^{\alpha}_{b} f(x) = -\frac{\mathbb{B}(\alpha)}{1-\alpha} \frac{d}{dx} \int_x^b \mathbb{E}_{\alpha}\Big[-\alpha \frac{(s-x)^{\alpha}}{1-\alpha}\Big] f(s) ds, \quad x < b. \tag{2.38} $$

Example 2.4 For $f(x) \equiv 1$, the left Atangana-Baleanu fractional derivative of order α in Riemann-Liouville sense is

$$ \begin{aligned} {}^{ABR}_{a}\mathrm{D}^{\alpha}_{x} 1 &= \frac{\mathbb{B}(\alpha)}{1-\alpha} \frac{d}{dx} \int_a^x \mathbb{E}_{\alpha}\Big[-\alpha \frac{(x-s)^{\alpha}}{1-\alpha}\Big] ds \\ &= \frac{\mathbb{B}(\alpha)}{1-\alpha} \frac{d}{dx} \sum_{k=0}^{\infty} \Big[\frac{(\frac{-\alpha}{1-\alpha})^k}{\Gamma(k\alpha+1)} \int_a^x (x-s)^{k\alpha} ds \Big] \\ &= \frac{\mathbb{B}(\alpha)}{1-\alpha} \frac{d}{dx} \Big\{ (x-a) \sum_{k=0}^{\infty} \frac{\Big[-\alpha \frac{(x-a)^{\alpha}}{1-\alpha}\Big]^k}{\Gamma(k\alpha+2)} \Big\} \\ &= \frac{\mathbb{B}(\alpha)}{1-\alpha} \mathbb{E}_{\alpha}\Big[-\alpha \frac{(x-a)^{\alpha}}{1-\alpha}\Big]. \end{aligned} $$

The Atangana-Baleanu Caputo and the Atangana-Baleanu Riemann-Liouville fractional derivatives are related as shown in the following theorem.

Theorem 2.12 *For $0 < \alpha < 1$ and $f \in H_1(I)$, the following identities hold:*

$$ {}^{ABC}_{a}\mathrm{D}^{\alpha}_{x} f(x) = {}^{ABR}_{a}\mathrm{D}^{\alpha}_{x} f(x) - \frac{\mathbb{B}(\alpha)}{1-\alpha} f(a) \mathbb{E}_{\alpha}\Big[-\alpha \frac{(x-a)^{\alpha}}{1-\alpha}\Big], \quad x > a, \tag{2.39} $$

and

$$ {}^{ABC}_{x}\mathrm{D}^{\alpha}_{b} f(x) = {}^{ABR}_{x}\mathrm{D}^{\alpha}_{b} f(x) - \frac{\mathbb{B}(\alpha)}{1-\alpha} f(b) \mathbb{E}_{\alpha}\Big[-\alpha \frac{(b-x)^{\alpha}}{1-\alpha}\Big], \quad x < b. \tag{2.40} $$

For the Atangana-Baleanu fractional derivatives, the associated Atangana-Baleanu fractional integrals are defined as

Definition 2.8 For $\alpha > 0$ and $f \in L_1(I)$, the left Atangana-Baleanu fractional integral of order α starting from a, denoted as ${}^{AB}_{a}\mathrm{I}^{\alpha}_{x} f$, is defined by

$$ {}^{AB}_{a}\mathrm{I}^{\alpha}_{x} f(x) = \frac{1-\alpha}{\mathbb{B}(\alpha)} f(x) + \frac{\alpha}{\mathbb{B}(\alpha)\Gamma(\alpha)} \int_a^x \frac{f(s)}{(x-s)^{1-\alpha}} ds, \quad x > a, \tag{2.41} $$

and the right Atangana-Baleanu fractional integral of order α terminating at b, denoted as ${}^{AB}_{x}\mathrm{I}^{\alpha}_{b} f$, is defined by

$$
{}^{AB}_{\ x}\mathrm{I}^{\alpha}_{b} f(x) = \frac{1-\alpha}{\mathbb{B}(\alpha)} f(x) + \frac{\alpha}{\mathbb{B}(\alpha)\Gamma(\alpha)} \int_x^b \frac{f(s)}{(s-x)^{1-\alpha}} ds, \quad x < b. \tag{2.42}
$$

Note that the Atangana-Baleanu fractional integrals recover function $f(x)$ if $\alpha = 0$ and are the usual ordinary integrals if $\alpha = 1$. Furthermore, we state some important properties of Atangana-Baleanu fractional derivatives and integrals as follows.

Theorem 2.13 *Let* $0 < \alpha < 1$.
(i) *For* $f \in L_1(I)$,

$$
{}^{AB}_{\ a}\mathrm{I}^{\alpha}_{x}\big({}^{ABR}_{\ \ a}\mathrm{D}^{\alpha}_{x} f(x)\big) = {}^{ABR}_{\ \ a}\mathrm{D}^{\alpha}_{x}\big({}^{AB}_{\ a}\mathrm{I}^{\alpha}_{x} f(x)\big) = f(x), \quad x > a, \tag{2.43}
$$

$$
{}^{AB}_{\ x}\mathrm{I}^{\alpha}_{b}\big({}^{ABR}_{\ \ x}\mathrm{D}^{\alpha}_{b} f(x)\big) = {}^{ABR}_{\ \ x}\mathrm{D}^{\alpha}_{b}\big({}^{AB}_{\ x}\mathrm{I}^{\alpha}_{b} f(x)\big) = f(x), \quad x < b. \tag{2.44}
$$

(ii) *For* $f \in H_1(I)$,

$$
{}^{ABC}_{\ \ a}\mathrm{D}^{\alpha}_{x} f(x) = -\,{}^{ABC}_{\ \ x}\mathrm{D}^{\alpha}_{b} f(x), \quad a < x < b, \tag{2.45}
$$

$$
{}^{AB}_{\ a}\mathrm{I}^{\alpha}_{x}\big({}^{ABC}_{\ \ a}\mathrm{D}^{\alpha}_{x} f(x)\big) = f(x) - f(a), \quad x > a, \tag{2.46}
$$

$$
{}^{AB}_{\ x}\mathrm{I}^{\alpha}_{b}\big({}^{ABC}_{\ \ x}\mathrm{D}^{\alpha}_{b} f(x)\big) = f(x) - f(b), \quad x < b. \tag{2.47}
$$

As a result, ${}^{ABR}_{\ \ a}\mathrm{D}^{\alpha}_{x}$ is both a left and right inverse of ${}^{AB}_{\ a}\mathrm{I}^{\alpha}_{x}$ on $L_1(I)$. In addition, the right Atangana-Baleanu fractional derivative in Caputo sense ${}^{ABC}_{\ \ x}\mathrm{D}^{\alpha}_{b}$ is equivalent to the left Atangana-Baleanu fractional derivative in Caputo sense ${}^{ABC}_{\ \ a}\mathrm{D}^{\alpha}_{x}$ on $H_1(I)$. Next, we give the following integration by parts for Atangana-Baleanu fractional derivatives and integrals.

Theorem 2.14 (Integration by Parts). *Let* $\alpha > 0,\ p, q \geq 1$ *with* $p^{-1} + q^{-1} \leq 1 + \alpha$ $(p, q \neq 1$, *when* $p^{-1} + q^{-1} = 1 + \alpha)$. *Then, for any* $f(x) \in L_p(I)$ *and* $g(x) \in L_q(I)$, *we have*

$$
\int_a^b f(x)\, {}^{ABR}_{\ \ a}\mathrm{D}^{\alpha}_{x} g(x)dx = \int_a^b g(x)\, {}^{ABR}_{\ \ x}\mathrm{D}^{\alpha}_{b} f(x)dx, \tag{2.48}
$$

$$
\int_a^b f(x)\, {}^{AB}_{\ a}\mathrm{I}^{\alpha}_{x} g(x)dx = \int_a^b g(x)\, {}^{AB}_{\ x}\mathrm{I}^{\alpha}_{b} f(x)dx, \tag{2.49}
$$

$$
\begin{aligned}
\int_a^b f(x)\, {}^{ABC}_{\ \ a}\mathrm{D}^{\alpha}_{x} g(x)dx &= \int_a^b g(x)\, {}^{ABR}_{\ \ x}\mathrm{D}^{\alpha}_{b} f(x)dx \\
&\quad + \frac{\mathbb{B}(\alpha)}{1-\alpha} g(x) \tilde{\mathbb{E}}^{1}_{\alpha,1,\frac{-\alpha}{1-\alpha},b} f(x)\Big|_a^b,
\end{aligned} \tag{2.50}
$$

$$
\begin{aligned}
\int_a^b f(x)\, {}^{ABC}_{\ \ x}\mathrm{D}^{\alpha}_{b} g(x)dx &= \int_a^b g(x)\, {}^{ABR}_{\ \ a}\mathrm{D}^{\alpha}_{x} f(x)dx \\
&\quad - \frac{\mathbb{B}(\alpha)}{1-\alpha} g(x) \tilde{\mathbb{E}}^{1}_{\alpha,1,\frac{-\alpha}{1-\alpha},a} f(x)\Big|_a^b,
\end{aligned} \tag{2.51}
$$

where $\tilde{\mathbb{E}}^{\beta}_{\gamma,\mu,\omega,a}$ *is the left generalized fractional integral defined by*

$$\tilde{\mathbb{E}}^{\beta}_{\gamma,\mu,\omega,a}y(t) = \int_a^t (t-\tau)^{\mu-1}\bar{\mathbb{E}}^{\beta}_{\gamma,\mu}[\omega(t-\tau)^{\gamma}]y(\tau)d\tau, \tag{2.52}$$

and $\tilde{\mathbb{E}}^{\beta}_{\gamma,\mu,\omega,b}$ *is the right generalized fractional integral defined by*

$$\tilde{\mathbb{E}}^{\beta}_{\gamma,\mu,\omega,b}y(t) = \int_t^b (\tau-t)^{\mu-1}\bar{\mathbb{E}}^{\beta}_{\gamma,\mu}[\omega(\tau-t)^{\gamma}]y(\tau)d\tau. \tag{2.53}$$

In (2.52) and (2.53), $\bar{\mathbb{E}}^{\beta}_{\gamma,\mu}(z) = \sum_{k=0}^{\infty} \frac{(\beta)_k z^k}{\Gamma(\gamma k+\mu)k!}$, $\beta > 0$, *is the generalized two-parameter Mittag-Leffler function, and* $(\beta)_k$ *is the Pochhammer symbol with* $(\beta)_k = \frac{\Gamma(\beta+k)}{\Gamma(\beta)}$.

2.5 Conformable Fractional Integral and Derivative

The conformable fractional derivative is a well-behaved local fractional derivative depending on the basic limit definition of the derivative.

Definition 2.9 For function $f : I \to \mathbb{R}$, and $0 < \alpha \leq 1$, the left conformable fractional derivative of order α starting from a is defined by

$$\frac{d_a^{\alpha}}{dx_a^{\alpha}} f(x) = f_a^{(\alpha)}(x) = \lim_{\epsilon \to 0} \frac{f(x+\epsilon(x-a)^{1-\alpha}) - f(x)}{\epsilon}, \quad x > a, \tag{2.54}$$

and the right conformable fractional derivative of order α terminating at b is defined by

$$\frac{{}_b d^{\alpha}}{{}_b dx^{\alpha}} f(x) = {}_b f^{(\alpha)}(x) = -\lim_{\epsilon \to 0} \frac{f(x+\epsilon(b-x)^{1-\alpha}) - f(x)}{\epsilon}, \quad x < b. \tag{2.55}$$

If limit (2.54) exists, then we say that f is left α-differentiable. In addition, if $f_a^{(\alpha)}(x)$ exists for $x \in (a, b)$, then

$$f_a^{(\alpha)}(a) = \lim_{x \to a^+} f_a^{(\alpha)}(x) \tag{2.56}$$

and

$$f_a^{(\alpha)}(b) = \lim_{x \to b^-} f_a^{(\alpha)}(x). \tag{2.57}$$

If limit (2.55) exists, and likewise the right α-differentiable, ${}_b f^{(\alpha)}(a)$ and ${}_b f^{(\alpha)}(b)$ can be defined. It should be noted that if f is differentiable, then $f_a^{(\alpha)}(x) = (x-a)^{1-\alpha} f'(x)$ and ${}_b f^{(\alpha)}(x) = -(b-x)^{1-\alpha} f'(x)$. For example,

Example 2.5 The conformable fractional derivatives $\frac{d_a^\alpha e^{\lambda x}}{dx_a^\alpha}$, and $\frac{{}_b d^\alpha e^{\lambda x}}{{}_b dx^\alpha}$, $\lambda \in \mathbb{R}$, are

$$\frac{d_a^\alpha e^{\lambda x}}{dx_a^\alpha} = \lim_{\epsilon \to 0} \frac{e^{\lambda(x+\epsilon(x-a)^{1-\alpha})} - e^{\lambda x}}{\epsilon} = \lambda(x-a)^{1-\alpha} e^{\lambda x}, \quad x \geq a, \tag{2.58}$$

$$\frac{{}_b d^\alpha e^{\lambda x}}{{}_b dx^\alpha} = -\lim_{\epsilon \to 0} \frac{e^{\lambda(x+\epsilon(b-x)^{1-\alpha})} - e^{\lambda x}}{\epsilon} = -\lambda(b-x)^{1-\alpha} e^{\lambda x}, \quad x \leq b. \tag{2.59}$$

The most remarkable consequence of definitions (2.54) and (2.55) is that the conformable fractional derivatives satisfy the simple fractional version of chain and product rules.

Theorem 2.15 *Let $0 < \alpha \leq 1$, and f and g be α-differentiable functions. Then,*
(i)

$$(fg)_a^{(\alpha)}(x) = f_a^{(\alpha)}(x)g(x) + f(x)g_a^{(\alpha)}(x) \tag{2.60}$$

and

$$_b(fg)^{(\alpha)}(x) = {}_bf^{(\alpha)}(x)g(x) + f(x)\,{}_bg^{(\alpha)}(x). \tag{2.61}$$

(ii)

$$\left(\frac{f}{g}\right)_a^{(\alpha)}(x) = \frac{f_a^{(\alpha)}(x)g(x) - f(x)g_a^{(\alpha)}(x)}{g^2(x)}, \tag{2.62}$$

and

$$_b\left(\frac{f}{g}\right)^{(\alpha)}(x) = \frac{{}_bf^{(\alpha)}(x)g(x) - f(x)\,{}_bg^{(\alpha)}(x)}{g^2(x)}. \tag{2.63}$$

(iii) *If $g(x) \geq a$, then*

$$(f \circ g)_a^{(\alpha)}(x) = f_a^{(\alpha)}(g(x))g_a^{(\alpha)}(x)(g(x)-a)^{\alpha-1}. \tag{2.64}$$

(iv) *If $g(x) \leq b$, then*

$$_b(f \circ g)_a^{(\alpha)}(x) = {}_bf^{(\alpha)}(g(x))\,{}_bg^{(\alpha)}(x)(b-g(x))^{\alpha-1}. \tag{2.65}$$

(v) *If $g(x) < a$, then*

$$(f \circ g)_a^{(\alpha)}(x) = -\,{}_af^{(\alpha)}(g(x))g_a^{(\alpha)}(x)(a-g(x))^{\alpha-1}. \tag{2.66}$$

(vi) *If $g(x) > b$, then*

$$_b(f \circ g)_a^{(\alpha)}(x) = -f_b^{(\alpha)}(g(x))\,{}_bg^{(\alpha)}(x)(g(x)-b)^{\alpha-1}. \tag{2.67}$$

The corresponding conformable fractional integrals are defined as follows.

Definition 2.10 For $f \in L_1(I)$, and $0 < \alpha \leq 1$, the left conformable fractional integral of order α starting from a, denoted as $I_a^\alpha f(x)$, is defined by

$$I_a^\alpha f(x) = \int_a^x f(s) d_a^\alpha s = \int_a^x \frac{f(s)}{(s-a)^{1-\alpha}} ds, \quad x > a, \tag{2.68}$$

and the right conformable fractional integral of order α terminating at b, denoted as ${}_bI^\alpha f(x)$, is defined by

$${}_bI^\alpha f(x) = \int_x^b f(s)\, {}_bd^\alpha s = \int_x^b \frac{f(s)}{(b-s)^{1-\alpha}} ds, \quad x < b. \tag{2.69}$$

For $\alpha = 1$, the conformable fractional integrals (2.68) and (2.69) reduce to ordinary first-order integrals. In addition, the following important properties of conformable fractional derivatives and integrals are very useful.

Theorem 2.16 *Let $f \in C(I)$, and $0 < \alpha \leq 1$. Then,*

$$\frac{d_a^\alpha}{dx_a^\alpha}(I_a^\alpha f(x)) = f(x) \tag{2.70}$$

and

$$\frac{{}_bd^\alpha}{{}_bdx^\alpha}({}_bI^\alpha f(x)) = f(x), \tag{2.71}$$

for all $x \in I$.

Theorem 2.17 (Fundamental Theorem of Conformable Fractional Calculus). *Let $f \in C^1(I)$ and $0 < \alpha \leq 1$. Then,*

$$I_a^\alpha (f_a^{(\alpha)}(x)) = f(x) - f(a) \tag{2.72}$$

and

$${}_bI^\alpha ({}_bf^\alpha(x)) = f(x) - f(b), \tag{2.73}$$

for all $x \in I$.

Theorem 2.18 (Integration by Parts). *Let $0 < \alpha \leq 1$ and $f, g : I \to \mathbb{R}$ be two functions such that fg is differentiable. Then,*

$$\int_a^b f(x) g_a^{(\alpha)}(x) d_a^\alpha x = f(x)g(x)\Big|_a^b - \int_a^b g(x) f_a^{(\alpha)}(x) d_a^\alpha x \tag{2.74}$$

and

$$\int_a^b f(x)\, {}_bg^{(\alpha)}(x)\, {}_bd^\alpha x = -f(x)g(x)\Big|_a^b - \int_a^b g(x)\, {}_bf^{(\alpha)}(x)\, {}_bd^\alpha x. \tag{2.75}$$

In particular, if f and g are differentiable, then

$$\int_a^b f(x)g_a^{(\alpha)}(x)d_a^\alpha x = f(x)g(x)\big|_a^b + \int_a^b g(x)\,{}_bf^{(\alpha)}(x)\,{}_bd^\alpha x. \tag{2.76}$$

Theorem 2.19 (Mean Value Theorem). *Let $f, g : I \to \mathbb{R}$ be functions such that*
(i) *f and g are continuous on I;*
(ii) *f and g are α-differentiable for some $0 < \alpha \leq 1$.*
Then, there exist some $\xi \in (a, b)$ such that

$$\frac{f_a^{(\alpha)}(\xi)}{g_a^{(\alpha)}(\xi)} = \frac{f(b) - f(a)}{g(b) - g(a)}. \tag{2.77}$$

Theorem 2.20 (Chain Rule for Functions of Several Variables). *Let $f : \mathbb{R}^n \to \mathbb{R}$, $n \in \mathbb{N}$, be a differentiable function in all its arguments and $y_1, \ldots, y_n : \mathbb{R} \to \mathbb{R}$ be α-differentiable functions. Then,*

$$\frac{d_a^\alpha}{dx_a^\alpha} f(y_1(x), \ldots, y_n(x)) = \frac{\partial f}{\partial y_1} y_{1a}^{(\alpha)} + \frac{\partial f}{\partial y_2} y_{2a}^{(\alpha)} + \cdots + \frac{\partial f}{\partial y_n} y_{na}^{(\alpha)} \tag{2.78}$$

and

$$\frac{{}_bd^\alpha}{{}_bdx^\alpha} f(y_1(x), \ldots, y_n(x)) = \frac{\partial f}{\partial y_1}\,{}_by_1^{(\alpha)} + \frac{\partial f}{\partial y_2}\,{}_by_2^{(\alpha)} + \cdots + \frac{\partial f}{\partial y_n}\,{}_by_n^{(\alpha)}. \tag{2.79}$$

Chapter 3
Fractional-Order Systems

3.1 Introduction

In this chapter, we will review some basic theory of several fractional-order systems, which will be involved in the fractional optimal control problems in this book. These fractional systems include ordinary differential equations containing the Riemann-Liouville, Caputo, Atangana-Baleanu and conformable fractional derivatives, and delay differential equations with Caputo fractional derivatives. Some basic concepts and theorems for these fractional-order systems are briefly introduced. Furthermore, we will present two classes of numerical methods for solving fractional ordinary differential equations (FODEs).

Fractional differential equations are non-integer order differential equations. They provide a powerful tool in capturing memory effects of different substances and the nature of the inheritance. In next section, we will introduce the existence, uniqueness, and regularity of solutions to various fractional ordinary differential equations and to fractional delay differential equations. These qualitative theories of problem to be treated here may be found in [65, 73, 133, 134, 160, 308, 330, 342]. For a more rigorous and general treatment, we refer the reader to [1, 110, 151, 216, 242, 275, 329, 345]. For numerical methods, two main strategies for solving FODEs have been extensively studied in the literature. The first strategy involves a direct approximation of the fractional derivatives in FODEs, e.g., [90, 168, 239, 327, 340], while the second strategy focuses on solving the equivalent Volterra integral equations; see, for example, [71, 74, 75, 152]. In Sect. 3.3, we develop two time-stepping algorithms for numerical solutions of nonlinear FODEs by using the second strategy. In Sect. 3.3.1, based on a one-point quadrature rule, we propose an explicit second-order numerical integration scheme. In Sect. 3.3.2, based on a third-order Taylor expansion (for the first two mesh points by a second-order Taylor expansion), we present corresponding implicit nonlinear algebraic equations in the unknowns, which can be iteratively solved by Newton's method. A convergence analysis shows

C. Liu et al., *Nonlinear Fractional Optimal Control*, Springer Optimization and Its Applications 237, https://doi.org/10.1007/978-981-92-0274-4_3

that convergence rate of this scheme is third order. Numerical examples are also provided to demonstrate the effectiveness of the above two numerical methods.

The main references of the numerical solution methods are [171, 332].

3.2 Basic Theory of Fractional Differential Equations

3.2.1 Fractional Ordinary Differential Equations

To begin with, we consider the following FODEs in Riemann-Liouville sense:

$$ {}_0^R\mathrm{D}_t^{\alpha} x(t) = f(t, x(t)), \quad t \in (0, T], \tag{3.1}$$

with the initial conditions

$$\lim_{t\to 0+} {}_0^R\mathrm{D}_t^{\alpha-k} x(t) = b_k,\ k = 1, \ldots, m-1,\ \lim_{t\to 0+} {}_0\mathrm{I}_t^{\alpha-m} x(t) = b_m, \tag{3.2}$$

where α, satisfying $\alpha \in (m-1, m)$ with $m \in \mathbb{N}$, is the fractional order, and b_k, $k = 1, \ldots, m$, are given real numbers.

For the fractional system (3.1) with (3.2), the existence and uniqueness theorem is given as follows.

Theorem 3.1 (Local Existence and Uniqueness). *Let $K > 0$, and $h^* \in (0, T]$. Define*

$$G := \{(t, x) \in \mathbb{R}^2 \mid 0 \le t \le h^*, x \in \mathbb{R}\ \text{for}\ t = 0,\ \text{and}$$
$$|t^{m-\alpha}x - \sum_{k=1}^{m} b_k t^{m-k}/\Gamma(\alpha - k + 1)| < K\ \text{otherwise}\},$$

and assume that the function $f : G \to \mathbb{R}$ is continuous and bounded in G and that it is Lipschitz continuous in the second variable, i.e., there exists a constant $L > 0$ such that for all (t, x_1) and $(t, x_2) \in G$, we have

$$|f(t, x_1) - f(t, x_2)| < L|x_1 - x_2|. \tag{3.3}$$

Then, the fractional system (3.1) with (3.2) has a unique continuous solution $x(\cdot)$ on $[0, h]$, where

$$h := \min\left\{h^*, \tilde{h}, \left(\frac{\Gamma(\alpha+1)K}{M}\right)^{1/m}\right\} \tag{3.4}$$

with $M := \sup_{(t,y)\in G} |f(t, y)|$ *and* $\tilde{h}$ *being an arbitrary positive number satisfying the constraint*

$$\tilde{h} < \left(\frac{\Gamma(2\alpha - m + 1)}{\Gamma(\alpha - m + 1)L}\right)^{1/\alpha}.$$

Theorem 3.2 (Global Existence and Uniqueness). *Let* $\tilde{G} \subset \mathbb{R}$ *be an open set, and* $f : (0, T] \times \tilde{G} \to \mathbb{R}$ *be a function such that* $f(t, x) \in L_1((0, T])$ *for any* $x \in \tilde{G}$ *and be Lipschitz continuous in* x*, i.e., there exists a constant* $L > 0$ *such that for all* (t, x_1) *and* $(t, x_2) \in [0, T] \times \tilde{G}$*, inequality (3.3) holds. Then, there exists a unique solution* $x(\cdot) \in L_1([0, T])$ *to the fractional system (3.1) with (3.2).*

The results are very similar to the known classical results for first-order equations. Moreover, fractional system (3.1) with (3.2) can be transformed into an equivalent Volterra integral equation as indicated in the following theorem.

Theorem 3.3 *Let* $x(\cdot) \in L_1([0, T])$ *with* ${}_0\mathrm{I}_t^{\alpha-m} x \in AC^n([0, T])$ *and* $t \mapsto f(t, x(t)) \in L_1((0, T])$*. Then,* $x(\cdot)$ *is a solution of fractional system (3.1) with (3.2) if and only if it is a solution of the Volterra integral equation:*

$$x(t) = \sum_{k=1}^{m} \frac{b_k t^{\alpha-k}}{\Gamma(\alpha - k + 1)} + \frac{1}{\Gamma(\alpha)} \int_0^t (t - s)^{\alpha-1} f(s, x(s))ds, \quad t \in [0, T]. \tag{3.5}$$

Next, we give a generalized Gronwall's inequality, which appears often in the study of fractional system.

Theorem 3.4 *Let* $\alpha > 0$*,* $a(\cdot)$ *be a nonnegative local integrable function on* $(0, T]$*, and* $b(\cdot)$ *be a nonnegative, nondecreasing continuous function on* $[0, T]$*. Furthermore, let* $y(\cdot)$ *be nonnegative and locally integrable on* $(0, T]$ *with*

$$y(t) \le a(t) + b(t) \int_0^t (t - s)^{\alpha-1} y(s)ds, \quad t \in (0, T]. \tag{3.6}$$

Then,

$$y(t) \le a(t) + \int_0^t \left[\sum_{k=1}^{\infty} \frac{(b(t))^k}{\Gamma(k\alpha)} (t - s)^{k\alpha-1} a(s)\right] ds, \quad t \in (0, T]. \tag{3.7}$$

In particular, the following assertions hold:
(i) *If* $a(t) \equiv a_1 t^{-\beta}$ *with* $\beta \in (0, 1)$ *and* $a_1 > 0$*, and* $b(t) \equiv b_1 > 0$*, then*

$$y(t) \le a_1 \Gamma(1 - \beta) \mathbb{E}_{\alpha,1-\beta}(b_1 t^{\alpha}) t^{-\beta}, \quad t \in (0, T], \tag{3.8}$$

where $\mathbb{E}_{\alpha,1-\beta}(\cdot)$ *is as defined in (2.10).*
(ii) *If* $a(t)$ *is nondecreasing on* $[0, T]$ *and* $b(t) \equiv b_1$*, then*

$$y(t) \le a(t) \mathbb{E}_{\alpha}(b_1 t^{\alpha}), \quad t \in (0, T], \tag{3.9}$$

where $\mathbb{E}_\alpha(\cdot)$ is as defined in (2.36).

As an application of Theorem 3.4, the following dependence of the solution on the fractional order and the initial condition can be confirmed.

Theorem 3.5 *Let $\alpha > 0$, $\delta > 0$ such that $0 < \alpha - \delta < \alpha \le 1$, and function f be continuous and fulfill a Lipschitz condition with respect to the second variable. For $t \in [0, h] \subset [0, T)$, assume that $x(t)$ and $y(t)$ are the solutions of fractional systems*

$$\begin{cases} {}_0^R\mathrm{D}_t^\alpha x(t) = f(t, x(t)), \quad t \in (0, T], \\ \lim\limits_{t\to 0+} {}_0\mathrm{I}_t^{\alpha-1} x(t) = \eta, \end{cases} \tag{3.10}$$

and

$$\begin{cases} {}_0^R\mathrm{D}_t^{\alpha-\delta} y(t) = f(t, y(t)), \quad t \in (0, T], \\ \lim\limits_{t\to 0+} {}_0\mathrm{I}_t^{\alpha-\delta-1} y(t) = \tilde{\eta}, \end{cases} \tag{3.11}$$

respectively. Then, for $t \in (0, h]$, the following inequality holds:

$$|x(t) - y(t)| \le A(t) + \int_0^t \left[\sum_{n=1}^{\infty} \left(\frac{L}{\Gamma(\alpha)} \Gamma(\alpha - \delta) \right)^n \frac{(t-s)^{n(\alpha-\delta)-1}}{\Gamma(n(\alpha-\delta))} A(s) \right] ds,$$

where h is as defined in (3.4); $L > 0$ is the Lipschitz constant;

$$\begin{aligned} A(t) = &\left| \frac{\tilde{\eta}}{\Gamma(\alpha-\delta)} t^{\alpha-\delta-1} - \frac{\eta}{\Gamma(\alpha)} t^{\alpha-1} \right| + \left| \frac{t^{\alpha-\delta}}{(\alpha-\delta)\Gamma(\alpha)} - \frac{t^\alpha}{\Gamma(\alpha+1)} \right| M \\ &+ \left| \frac{t^{\alpha-\delta}}{\alpha-\delta} \left[\frac{1}{\Gamma(\alpha-\delta)} - \frac{1}{\Gamma(\alpha)} \right] \right| M; \end{aligned}$$

and

$$M := \max_{0 \le t \le h} |f(t, x)|.$$

Now, we turn to the following FODEs in Caputo sense, which have been widely used in modeling and simulating real-world problems:

$${}_0^C\mathrm{D}_t^\alpha x(t) = f(t, x(t)), \quad t \in (0, T], \tag{3.12}$$

with the initial conditions

$$x^{(k)}(0) = b_k, \quad k = 0, 1, \ldots, m-1, \tag{3.13}$$

where α, satisfying $\alpha \in (m-1, m]$ with $m \in \mathbb{N}$, is the fractional order and b_k, $k = 0, \ldots, m-1$, are given real numbers. For the fractional system (3.12) with (3.13), we have the following results:

Theorem 3.6 (Local Existence and Uniqueness). *Let $K > 0$ and $h^* \in (0, T]$. Define $G := \{(t, x)|\ t \in [0, h^*], |x - \sum_{k=0}^{m-1} t^k b_k/k!| \le K\}$, and let function $f : G \to \mathbb{R}$ be continuous and be Lipschitz continuous in the second variable. Furthermore, define $M := \sup_{(t,z)\in G} |f(t, z)|$ and*

$$h := \begin{cases} h^*, & \text{if } M = 0, \\ \min\{h^*, (K\Gamma(\alpha + 1)/M)^{1/\alpha}\}, & \text{otherwise}. \end{cases} \tag{3.14}$$

Then, the fractional system (3.12) with (3.13) admits a unique continuous solution $x(\cdot)$ on $[0, h]$.

Theorem 3.7 *Under the assumptions in Theorem 3.6, $x(\cdot)$ is a continuous solution of the fractional system (3.12) with (3.13) if and only if it is a solution of the Volterra integral equation:*

$$x(t) = \sum_{k=0}^{m-1} \frac{b_k t^k}{k!} + \frac{1}{\Gamma(\alpha)} \int_0^t (t - s)^{\alpha-1} f(s, x(s))ds, \quad t \in [0, T], \tag{3.15}$$

with $m - 1 < \alpha < m$.

Theorem 3.8 (Global Existence and Uniqueness) *Let $f : (0, T] \times \mathbb{R} \to \mathbb{R}$ be a continuous function and be Lipschitz continuous in x for all $t \in [0, T]$. Then, there exists a unique continuous solution $x(\cdot)$ on $[0, T]$ to the fractional system (3.12) with (3.13).*

Theorem 3.9 (Dependence on Initial Values) *Let $x(\cdot)$ be the solution of the fractional system (3.12) with (3.13), and $y(\cdot)$ be the solution of the following fractional system:*

$$\begin{cases} {}_0^C\mathrm{D}_t^\alpha y(t) = f(t, y(t)), & t \in (0, T], \\ y^{(k)}(0) = c_k, & k = 0, \ldots, m - 1. \end{cases} \tag{3.16}$$

Moreover, let $\epsilon := \max\{|b^k - c^k| : k = 0, \ldots, m - 1\}$. Then, if ϵ is sufficiently small, there exists some $h \in (0, T]$ such that both the functions $x(\cdot)$ and $y(\cdot)$ are defined on $[0, h]$, and

$$\sup_{0\le t\le h} |x(t) - y(t)| = \mathcal{O}\big(\max_{k\in\{0,\ldots,m-1\}} \{|b^k - c^k|\}\big). \tag{3.17}$$

Theorem 3.10 (Dependence on Fractional Order) *Let $x(\cdot)$ be the solution of the fractional system (3.12) with (3.13), and $y(\cdot)$ be the solution of the following fractional system:*

$$\begin{cases} {}_0^C\mathrm{D}_t^{\tilde\alpha} y(t) = f(t, y(t)), & t \in (0, T], \\ y^{(k)}(0) = b_k, & k = 0, \ldots, \tilde m - 1, \end{cases} \tag{3.18}$$

where $\tilde\alpha > \alpha$ and $\tilde m - 1 < \tilde\alpha \le \tilde m$, $\tilde m \in \mathbb{N}$. Moreover, let $\epsilon := \tilde\alpha - \alpha$ and

$$\epsilon^* := \begin{cases} 0, & \textit{if } m = \tilde{m}, \\ \max\limits_{m \le k \le \tilde{m}-1} \{|b_k|\}, & \textit{otherwise.} \end{cases} \tag{3.19}$$

Then, if ϵ *and* ϵ^* *are sufficiently small, there exists some* $h \in (0, T]$ *such that both the functions* $x(\cdot)$ *and* $y(\cdot)$ *are defined on* $[0, h]$*, and*

$$\sup_{0 \le t \le h} |x(t) - y(t)| = \mathcal{O}(\tilde{\alpha} - \alpha) + \mathcal{O}\big(\max\{0, \max\{|b_k| : m \le k \le \tilde{m} - 1\}\}\big). \tag{3.20}$$

Finally, we briefly introduce the existence and uniqueness of solutions to FODEs with Atangana-Baleanu-Caputo and conformable fractional derivatives as follows.

Theorem 3.11 *Let* $0 < \alpha < 1$*, and* $f \in C((0, T] \times \mathbb{R})$ *be Lipschitz continuous in the second variable with constant* $L > 0$ *satisfying*

$$L\Big(\frac{1-\alpha}{\mathbb{B}(\alpha)} + \frac{T^\alpha}{\mathbb{B}(\alpha)\Gamma(\alpha)}\Big) < 1. \tag{3.21}$$

Then, if $f(0, x(0)) = 0$*, the fractional system*

$$\begin{cases} {}^{ABC}_{\ 0}\mathrm{D}_t^\alpha x(t) = f(t, x(t)), & t \in (0, T], \\ x(0) = x^0 \end{cases} \tag{3.22}$$

has a unique solution on $[0, T]$*, where* x^0 *is a given real number.*

Theorem 3.12 *Let* $0 < \alpha < 1$*, and* $f \in C((0, T] \times \mathbb{R})$ *be Lipschitz continuous in the second variable. Then, the fractional system*

$$\begin{cases} x_0^{(\alpha)}(t) = f(t, x(t)), & t \in (0, T], \\ x(0) = x^0, \end{cases} \tag{3.23}$$

admits a unique solution on $[0, T]$*, where* $x_0^{(\alpha)}(t)$ *is the conformable fractional derivative of* x *starting from* $t = 0$*, and* x^0 *is a given real number.*

3.2.2 *Fractional Delay Differential Equations*

Consider the following fractional differential equations with a constant delay:

$${}^C_0\mathrm{D}_t^\alpha \boldsymbol{x}(t) = \boldsymbol{f}(t, \boldsymbol{x}(t), \boldsymbol{x}(t-\tau)), \quad t \in (0, T], \tag{3.24}$$

with the initial condition

$$x(t) = \phi(t), \quad t \in [-\tau, 0], \tag{3.25}$$

where $\tau > 0$ is a given delay; $x(t) \in \mathbb{R}^n$; $\alpha \in (0, 1)$ is the fractional order; $f : (0, T] \times \mathbb{R}^n \times \mathbb{R}^n \to \mathbb{R}^n$ and $\phi : [-\tau, 0] \to \mathbb{R}^n$ are given continuous functions; and ${}_0^C\mathrm{D}_t^\alpha x(t) = ({}_0^C\mathrm{D}_t^\alpha x_1(t), \dots, {}_0^C\mathrm{D}_t^\alpha x_n(t))^\top$ with ${}_0^C\mathrm{D}_t^\alpha x_i(t)$, $i = 1, \dots, n$, being the Caputo fractional derivatives of $x_i(t)$.

A continuous function $\psi(\cdot|\phi)$ on $[-\tau, T]$ is called a *solution* of the fractional delay system (3.24) with (3.25) over the interval $[-\tau, T]$ if

$$\begin{cases} {}_0^C\mathrm{D}_t^\alpha \psi(t|\phi) = f(t, \psi(t|\phi), \psi(t-\tau|\phi)), & t \in (0, T], \\ \psi(t|\phi) = \phi(t), & t \in [-\tau, 0]. \end{cases} \tag{3.26}$$

The following theorem indicates that the fractional delay system (3.24) with (3.25) is equivalent to a integral equation with initial condition (3.25).

Theorem 3.13 *The continuous function $\psi(\cdot|\phi)$ on $[-\tau, T]$ is a solution of the fractional delay system (3.24) with (3.25) over the interval $[-\tau, T]$ if and only if it is a solution of the delay integral equation*

$$x(t) = \phi(0) + \frac{1}{\Gamma(\alpha)} \int_0^t (t-s)^{\alpha-1} f(s, x(s), x(s-\tau))ds, \quad t \in (0, T], \tag{3.27}$$

with the initial condition

$$x(t) = \phi(t), \quad t \in [-\tau, 0]. \tag{3.28}$$

Now, we present the existence and uniqueness of global solutions to the fractional delay system (3.24) with (3.25).

Theorem 3.14 (Global Existence and Uniqueness) *Let function $f : (0, T] \times \mathbb{R}^n \times \mathbb{R}^n \to \mathbb{R}^n$ be continuous and be Lipschitz continuous in the second variable, i.e., there exists a nonnegative continuous function $L : (0, T] \times \mathbb{R}^n \to \mathbb{R}$ such that*

$$\|f(t, x, y) - f(t, \hat{x}, y)\| \le L(t, y)\|x - \hat{x}\|, \tag{3.29}$$

for all $t \in [0, T]$, x, y, $\hat{x} \in \mathbb{R}^n$, and $\phi : [-\tau, 0] \to \mathbb{R}^n$ be continuous, where $\|\cdot\|$ denotes the Euclidean norm. Then, the fractional system (3.24) with (3.25) has a unique global solution $\psi(\cdot|\phi)$ on the interval $[-\tau, T]$.

It is not difficult to extend Theorems 3.13 and 3.14 to fractional systems with multiple constant time delays. Next, we focus on the following fractional system with time-varying delays:

$${}_0^C\mathrm{D}_t^\alpha x(t) = f(t, x(t), x(t-h_1(t)), \dots, x(t-h_m(t))), \quad t \in (0, T], \tag{3.30}$$

with the initial condition

$$\boldsymbol{x}(t) = \boldsymbol{\phi}(t), \quad t \in [-r, 0], \tag{3.31}$$

where $\boldsymbol{x}(t) = (x_1(t), \dots, x_n(t))^\top \in \mathbb{R}^n$; $\boldsymbol{\alpha} = (\alpha_1, \dots, \alpha_n)^\top \in (0, 1)^n$; $r \geq 0$ is a given real number; $h_j(t) > 0$, $j = 1, \dots, m$, are continuous functions and $h_j(t) \leq r + t$ for all $t \geq 0$; $\boldsymbol{f} : (0, T] \times \mathbb{R}^{(m+1)n} \to \mathbb{R}^n$ and $\boldsymbol{\phi} : [-r, 0] \to \mathbb{R}^n$ are given functions; and ${}_0^C\mathrm{D}_t^{\boldsymbol{\alpha}} \boldsymbol{x}(t) = ({}_0^C\mathrm{D}_t^{\alpha_1} x_1(t), \dots, {}_0^C\mathrm{D}_t^{\alpha_n} x_n(t))^\top$ with ${}_0^C\mathrm{D}_t^{\alpha_i} x_i(t)$, $i = 1, \dots, n$, being the Caputo fractional derivatives of x_i.

Similarly, a continuous function $\boldsymbol{\psi}(\cdot|\boldsymbol{\phi})$ is called a *solution* of the fractional delay system (3.30) with (3.31) on interval $[-r, T]$ if

$$\begin{cases} {}_0^C\mathrm{D}_t^{\alpha} \boldsymbol{\psi}(t|\boldsymbol{\phi}) = \boldsymbol{f}(t, \boldsymbol{\psi}(t), \boldsymbol{\psi}(t - h_1(t)), \dots, \boldsymbol{\psi}(t - h_m(t))), & t \in (0, T], \\ \boldsymbol{\psi}(t|\boldsymbol{\phi}) = \boldsymbol{\phi}(t), & t \in [-r, 0]. \end{cases} \tag{3.32}$$

For the fractional delay system (3.30) with (3.31), we have the following theorems, which are similar to Theorems 3.13 and 3.14.

Theorem 3.15 *The continuous function $\boldsymbol{\psi}(\cdot|\boldsymbol{\phi})$ on $[-r, T]$ is a solution of the fractional delay system (3.30) with (3.31) on $[-r, T]$ if and only if it is a solution of the delay integral system:*

$$x_i(t) = \phi_i(0) + \frac{1}{\Gamma(\alpha_i)} \int_0^t (t - s)^{\alpha_i - 1} f_i(s, \boldsymbol{x}(s), \boldsymbol{x}(s - h_1(s)), \dots, \boldsymbol{x}(s - h_m(s)))ds, \tag{3.33}$$

with $t \in (0, T]$, $i = 1, \dots, n$, and satisfies the initial condition

$$\boldsymbol{x}(t) = \boldsymbol{\phi}(t), \quad t \in [-r, 0]. \tag{3.34}$$

Theorem 3.16 (Global Existence and Uniqueness) *Let function $\boldsymbol{f} : (0, T] \times \mathbb{R}^{(m+1)n} \to \mathbb{R}^n$ be continuous and be Lipschitz continuous in the second variable, i.e., there exists a constant $L > 0$ such that*

$$\| f_i(t, \boldsymbol{x}) - f_i(t, \boldsymbol{y})\| \leq L \|\boldsymbol{x} - \boldsymbol{y}\|, \tag{3.35}$$

for all $t \in [0, T]$, $\boldsymbol{x}$, $\boldsymbol{y} \in \mathbb{R}^{(m+1)n}$, $i = 1, \dots, n$, and function $\boldsymbol{\phi} : [-r, 0] \to \mathbb{R}^n$ be continuous. Then, the fractional system (3.30) with (3.31) has a unique global solution $\boldsymbol{\psi}(\cdot|\boldsymbol{\phi})$ on the interval $[-r, T]$.

In fact, the interval $[-r, T]$ in Theorem 3.16 does not depend on the parameters in (3.30). Thus, under the assumptions that $\boldsymbol{f}$ is continuous and fulfills a Lipschitz condition with respect to the second variable for $t \in [0, \infty)$, the assertion of Theorem 3.16 holds on the half real line, that is, the fractional delay system (3.30) with (3.31) has a unique global solution on $[-r, \infty)$. Furthermore, if a condition about growth in time of the function $\boldsymbol{f}$ at the origin is added, then the following exponential boundedness of solutions can be established.

Theorem 3.17 (Exponential Boundedness). *Let $r \geq 0$ and $h_j(t) > 0$, $j = 1, \ldots, m$, be continuous functions and $h_j(t) \leq r + t$ for all $t \geq 0$. Consider the fractional delay system*

$$\begin{cases} {}_0^C\mathrm{D}_t^{\boldsymbol{\alpha}} \boldsymbol{x}(t) = \boldsymbol{f}(t, \boldsymbol{x}(t), \boldsymbol{x}(t - h_1(t)), \ldots, \boldsymbol{x}(t - h_m(t))), & t > 0, \\ \boldsymbol{x}(t) = \phi(t), & t \in [-r, 0]. \end{cases} \tag{3.36}$$

Suppose that $\boldsymbol{f}$ is continuous and fulfills a Lipschitz condition (3.35) on $[0, \infty)$, ϕ is continuous, and there exists a constant $\gamma > \max\limits_{i \in \{1,\ldots,n\}} (mL)^{1/\alpha_i}$ such that

$$\max_{i \in \{1,\ldots,n\}} \sup_{t \geq 0} \frac{\int_0^t (t-s)^{\alpha_i - 1} |f_i(s, 0, \ldots, 0)| ds}{\exp(\gamma t)} < \infty. \tag{3.37}$$

Then, the fractional system (3.36) has a unique global solution $\psi(\cdot|\phi)$ on $[-r, \infty)$ and

$$\|\psi(t|\phi)\| < M \exp(\gamma t), \quad t \geq 0, \tag{3.38}$$

where M is some positive constant which depends on $\psi(\cdot|\phi)$.

3.3 Numerical Solutions of Fractional Ordinary Differential Equations

3.3.1 Second-Order Approach

We consider the following FODE with Caputo fractional derivative:

$$\begin{cases} {}_0^C\mathrm{D}_t^{\alpha} x(t) = f(t, x(t)), & t \in (0, T], \\ x(0) = x_0, \end{cases} \tag{3.39}$$

where $\alpha \in (0, 1)$ is the fractional order and $f : (0, T] \times \mathbb{R} \to \mathbb{R}$ is a twice continuously differentiable function. Higher-order fractional initial value problems can be transformed into a system of fractional initial value problems of the form (3.39) and any efficient numerical method developed for (3.39) can be extended to a vector-valued initial value problem. For FODE with the Riemann-Liouville fractional derivative can be readily transformed into (3.39) as demonstrated in [91].

Numerical Scheme

It follows from Theorem 3.7 that (3.39) is equivalent to the following Volterra integral equation:

$$x(t) = x_0 + \frac{1}{\Gamma(\alpha)} \int_0^t (t-\tau)^{\alpha-1} f(\tau, x(\tau))\, d\tau, \quad \alpha \in (0, 1). \tag{3.40}$$

For a given positive integer N, we first divide $(0, T]$ into N sub-intervals with mesh points $t_i = ih$ for $i = 0, 1, \ldots, N$, where $h = T/N$. Thus, (3.40) can be written as

$$x(t_i) = x_0 + \frac{1}{\Gamma(\alpha)} \sum_{j=1}^{i} \int_{(j-1)h}^{jh} (ih-\tau)^{\alpha-1} f(\tau, x(\tau)) d\tau. \tag{3.41}$$

We now consider approximation of the integral on right-hand side of (3.41). For $j = 1, 2, \ldots, i$, we apply Taylor's theorem to $f(\tau, x(\tau))$ at τ_{ij} to get

$$f(\tau, x(\tau)) = f(\tau_{ij}, x(\tau_{ij})) + K_{ij}(\tau - \tau_{ij}) + c_{ij}(\tau - \tau_{ij})^2, \tag{3.42}$$

where c_{ij} is a constant representing the second derivative of f at a point between τ_{ij} and τ, and

$$K_{ij} = f_\tau(\tau_{ij}, x(\tau_{ij})) + f_x(\tau_{ij}, x(\tau_{ij}))x'_\tau(\tau_{ij}).$$

Therefore, substituting the right-hand side of (3.42) into the integrand of the last term in (3.41), we have for any $j = 1, \ldots, i$,

$$\begin{aligned}
&\frac{1}{\Gamma(\alpha)} \int_{(j-1)h}^{jh} (ih-\tau)^{\alpha-1} f(\tau, x(\tau)) d\tau \\
&= \frac{h^\alpha}{\Gamma(\alpha+1)} f(\tau_{ij}, x(\tau_{ij}))[(i-j+1)^\alpha - (i-j)^\alpha] \\
&+ \frac{K_{ij}}{\Gamma(\alpha)} \int_{(j-1)h}^{jh} (ih-\tau)^{\alpha-1}(\tau - \tau_{ij}) d\tau + R_{ij},
\end{aligned} \tag{3.43}$$

where $R_{ij} = \dfrac{1}{\Gamma(\alpha)} \displaystyle\int_{(j-1)h}^{jh} (ih-\tau)^{\alpha-1} c_{ij}(\tau - \tau_{ij})^2 d\tau.$

From (3.43), it is clear that τ_{ij} should be chosen such that the second term becomes zero so that the truncation error in (3.43) is just R_{ij}. The choice of τ_{ij} is given in the following theorem.

Theorem 3.18 *For $i = 1, 2, \ldots, N$ and $j = 1, 2 \ldots, i$, the unique solution to*

$$\int_{(j-1)h}^{jh} (ih-\tau)^{\alpha-1}(\tau - \tau_{ij}) d\tau = 0 \tag{3.44}$$

is given by

$$\tau_{ij} = h\frac{[(i-j+1)^{\alpha+1} - (i-j)^{\alpha+1}] + (\alpha+1)[(i-j+1)^{\alpha}(j-1) - (i-j)^{\alpha} j]}{(\alpha+1)[(i-j+1)^{\alpha} - (i-j)^{\alpha}]}. \tag{3.45}$$

Furthermore, $(j-1)h < \tau_{ij} < jh$.

Proof By using ordinary integration by parts for the left-hand side of (3.44), we obtain

$$\begin{aligned}
&\int_{(j-1)h}^{jh} (ih-\tau)^{\alpha-1}(\tau-\tau_{ij})d\tau \\
&= -\frac{[(i-j)^{\alpha} j - (i-j+1)^{\alpha}(j-1)]h^{\alpha+1}}{\alpha} - \frac{[(i-j+1)^{\alpha} - (i-j)^{\alpha}]h^{\alpha}}{\alpha}\tau_{ij} \\
&\quad + \frac{(i-j+1)^{\alpha+1} - (i-j)^{\alpha+1}}{\alpha(\alpha+1)}h^{\alpha+1} = 0.
\end{aligned}$$

From the above, we have

$$\begin{aligned}
&[(i-j+1)^{\alpha} - (i-j)^{\alpha}]\tau_{ij} \\
&= \frac{(i-j+1)^{\alpha+1} - (i-j)^{\alpha+1}}{\alpha+1}h + [(i-j+1)^{\alpha}(j-1) - (i-j)^{\alpha} j]h.
\end{aligned}$$

Solving this for τ_{ij}, we obtain (3.45).

We now show that $(j-1)h < \tau_{ij} < jh$. From (3.45),

$$\frac{\tau_{ij}}{h} - j = \frac{(i-j+1)^{\alpha+1} - (i-j)^{\alpha+1} - (\alpha+1)(i-j+1)^{\alpha}}{(\alpha+1)[(i-j+1)^{\alpha} - (i-j)^{\alpha}]}.$$

To prove $\tau_{ij} - jh < 0$, we only need to show that

$$(i-j+1)^{\alpha+1} - (i-j)^{\alpha+1} - (\alpha+1)(i-j+1)^{\alpha} < 0.$$

Using the mean value theorem, we have

$$(i-j+1)^{\alpha+1} - (i-j)^{\alpha+1} - (\alpha+1)(i-j+1)^{\alpha} = (\alpha+1)\xi^{\alpha} - (\alpha+1)(i-j+1)^{\alpha} < 0,$$

since $\xi \in (i-j, i-j+1)$. Thus, we conclude that $\tau_{ij} - jh < 0$. In a similar way, we can prove that $\tau_{ij} - (j-1)h > 0$. □

Combining Theorem 3.18 and (3.41), we have the following representation for $x(t_i)$.

$$x(t_i) = x_0 + \frac{h^{\alpha}}{\Gamma(\alpha+1)}\sum_{j=1}^{i} f(\tau_{ij}, x(\tau_{ij}))[(i-j+1)^{\alpha} - (i-j)^{\alpha}] + R_i \tag{3.46}$$

for $i = 1, 2, \ldots, N$, where τ_{ij} is as given in (3.45) for $j = 1, 2, \ldots, i$ and $R_i = \sum_{j=1}^{i} R_{ij}$. Omitting the remainder R_i in (3.46), we have an equation approximating (3.41) which has the truncation error R_i. An upper bound for R_i is given in the following theorem.

Theorem 3.19 *Suppose that $f(t, x(t))$ is twice continuously differentiable in t and x. Then, for any $i = 1, 2, .., N$, the following estimate holds:*

$$|R_i| \le Ch^2, \tag{3.47}$$

where C denotes a positive constant independent of h and α.

Proof From the definition of R_i and Theorem 3.18, we have for $j = 1, 2, \ldots, i$,

$$|R_{ij}| \le \frac{|c_{ij}|h^2}{\Gamma(\alpha)} \left\{ \frac{[(i-j+1)h]^\alpha}{\alpha} - \frac{[(i-j)h]^\alpha}{\alpha} \right\}.$$

Since f is twice continuously differentiable, c_{ij} is bounded on $[0, T]$. Let $c = \max_i(\max_{1\le j\le i} |c_{ij}|)$. Then,

$$\begin{aligned} |R_i| &\le \frac{ch^2}{\Gamma(\alpha)} \sum_{j=1}^{i} \left\{ \frac{[(i-j+1)h]^\alpha}{\alpha} - \frac{[(i-j)h]^\alpha}{\alpha} \right\} \\ &\le \frac{ch^2 T^\alpha}{\Gamma(1+\alpha)} = C_1 h^2, \end{aligned} \tag{3.48}$$

where $C_1 = \dfrac{cT^\alpha}{\Gamma(1+\alpha)}$. It follows from $\alpha \in (0, 1)$ and [159] that $2^{\alpha-1} \le \Gamma(1+\alpha) \le 1$. Also, it is obvious that $T^\alpha < \max\{1, T\}$ for $\alpha \in (0, 1)$. Therefore, (3.47) holds true for a C independent of both h and α, which completes the proof. □

From (3.46), it is clear that to compute $x(t_i)$, we need to calculate $f(\tau_{ij}, x(\tau_{ij}))$, $j = 1, 2, \ldots, i$. However, $x(\tau_{ij})$, $j = 1, \ldots, i$, are not available directly from the scheme although the τ_{ij} are known. Thus, we use the following linear interpolation to approximate $x(\tau_{ij})$:

$$x(\tau_{ij}) = x(t_{j-1}) + \rho_{ij}(x(t_j) - x(t_{j-1})) + \mathcal{O}(h^2), \tag{3.49}$$

where

$$\rho_{ij} := \frac{\tau_{ij} - t_{j-1}}{h} \in (0, 1). \tag{3.50}$$

Then, we approximate $f(\tau_{ij}, x(\tau_{ij}))$ by

$$f(\tau_{ij}, x(\tau_{ij})) = f\left(\tau_{ij}, x(t_{j-1}) + \rho_{ij}(x(t_j) - x(t_{j-1}))\right) + \mathcal{O}(h^2) \tag{3.51}$$

Algorithm 3.1 Explicit one-step algorithm for solving (3.40).

Step 0. For a given positive integer N, let $t_i = ih$ for $i = 0, 1, \ldots, N$, where $h = T/N$.
Step 1. For $i = 1, 2, \ldots, N$, compute

$$x_i = \frac{x_0 + h_\alpha g_i(x_0, x_1, \ldots, x_{i-1}) + h_\alpha\big(f(\tau_{ii}, x_{i-1}) - \rho_{ii} f_x(\tau_{ii}, x_{i-1})x_{i-1}\big)}{1 - h_\alpha \rho_{ii} f_x(\tau_{ii}, x_{i-1})}, \tag{3.56}$$

where $\tau_{ij}, \rho_{ij}, h_\alpha$ and $g_i(x_0, \ldots, x_{i-1})$ are as defined in (3.45), (3.50), (3.53) and (3.54), respectively.

for $1 \le j \le i \le N$. Clearly, if we replace $f(\tau_{ij}, x(\tau_{ij}))$ in (3.46) with the above expression and omit the truncation error terms of order $\mathcal{O}(h^2)$, we define the following single-step time-stepping scheme for (3.40):

$$\begin{aligned} x_i &= x_0 + h_\alpha \sum_{j=1}^{i} f\left(\tau_{ij}, x_{j-1} + \rho_{ij}(x_j - x_{j-1})\right)[(i-j+1)^\alpha - (i-j)^\alpha] \\ &= x_0 + h_\alpha g_i(x_0, x_1, \ldots, x_{i-1}) + h_\alpha f(\tau_{ii}, x_{i-1} + \rho_{ii}(x_i - x_{i-1})) \end{aligned} \tag{3.52}$$

for $i = 1, 2, \ldots, N$, where τ_{ij} is as defined in (3.45);

$$h_\alpha = h^\alpha / \Gamma(1+\alpha); \tag{3.53}$$

and

$$g_i(x_0, x_1, \ldots, x_{i-1}) = \begin{cases} 0, & i = 1, \\ \displaystyle\sum_{j=1}^{i-1} f\left(\tau_{ij}, x_{j-1} + \rho_{ij}(x_j - x_{j-1})\right)[(i-j+1)^\alpha - (i-j)^\alpha], & i > 1. \end{cases} \tag{3.54}$$

The above scheme is implicit, as the last term on the right-hand side of (3.52) contains a nonlinear function of x_i. Furthermore, we may define an explicit single-step scheme by further approximating the last term in (3.52) by the following truncated Taylor's expansion:

$$\begin{aligned} &f(\tau_{ii}, x_{i-1} + \rho_{ii}(x_i - x_{i-1})) \\ &= f(\tau_{ii}, x_{i-1}) + f_x(\tau_{ii}, x_{i-1})\rho_{ii}(x_i - x_{i-1}) + \mathcal{O}(h^2). \end{aligned} \tag{3.55}$$

Combining (3.52) and (3.55), we propose Algorithm 3.1 for solving (3.40).

Convergence Analysis

Using linear interpolation theory and Taylor's theorem, we are able to prove that for any $i = 1, 2, \dots, N$, x_i generated by Algorithm 3.1 converges to $x(t_i)$ at the rate $\mathcal{O}(h^2)$ when $h \to 0^+$. This is formalized in the following theorem.

Theorem 3.20 *Let $x(t)$ be the exact solution of (3.40). If $f(t, x)$ is twice continuously differentiable in t and x, then there exists an $\bar{h} > 0$ such that $h_\alpha < \bar{h}$ implies*

$$|x(t_i) - x_i| \leq Ch^2, \quad i = 1, 2, \dots, N, \tag{3.57}$$

where $\{x_i\}$ is the sequence generated by Algorithm 3.1; C is a positive constant independent of h and α; and h_α is defined in (3.53).

Proof We will prove this theorem by mathematical induction and consider the following two cases.

Case 1. (3.57) holds for $i = 1$

By (3.46), we have

$$x(t_1) = x_0 + h_\alpha f(\tau_{11}, x(\tau_{11})) + R_1.$$

Furthermore, from (3.49) and (3.51), we have

$$x(t_1) = x_0 + h_\alpha \left[f(\tau_{11}, x_0) - f_x(\tau_{11}, x_0)\rho_{11}x_0\right] + h_\alpha f_x(\tau_{11}, x_0)\rho_{11}x(t_1) \\ + \mathcal{O}(h^2) + R_1,$$

where ρ_{11} is as defined in (3.50). Solving this equation for $x(t_1)$ and using (3.56) with $i = 1$ we have

$$x(t_1) = x_1 + \frac{\mathcal{O}(h^2) + R_1}{1 - h_\alpha f_x(\tau_{11}, x_0)\rho_{11}}.$$

Therefore, using (3.47), the previous equation yields the inequality

$$|x(t_1) - x_1| \leq \frac{Ch^2}{|1 - h_\alpha f_x(\tau_{11}, x_0)\rho_{11}|}. \tag{3.58}$$

Note that $\rho_{11} \in (0, 1)$ by (3.50) and $h_\alpha > 0$. So, if $f_x(\tau_{11}, x_0) \leq 0$, from (3.58), we see that (3.57) is satisfied for $i = 1$. However, when $f_x(\tau_{11}, x_0) > 0$, we need to choose an upper bound for h_α so that the denominator of (3.58) is bounded below by a positive constant. Clearly, for a given constant $\sigma \in (0, 1)$, if we choose $\bar{h}_1 := \dfrac{1 - \sigma}{\max\{\rho_{11} f_x(\tau_{11}, x_0), 1\}}$, then (3.57) is satisfied for $i = 1$, when $h_\alpha \leq \bar{h}_1$. To be more general, we choose

$$\bar{h} := \frac{1 - \sigma}{\max\{\max\limits_{1 \leq i \leq N} \rho_{ii} f_x(\tau_{ii}, x_{i-1}), 1\}}, \tag{3.59}$$

and when $h_\alpha \leq \bar{h}$, (3.57) is satisfied for $i = 1$.

Case 2. (3.57) *holds for* $i \geq 2$

Assume that

$$\max_{1 \leq j \leq i-1} |x(t_j) - x_j| \leq Ch^2 \tag{3.60}$$

when $h_\alpha \leq \bar{h}$. We now show that $\max_{1 \leq j \leq i} |x(t_j) - x_j| \leq Ch^2$ or, equivalently, $|x(t_i) - x_i| \leq Ch^2$.

Using (3.51) and (3.55), we have from (3.46) and (3.47),

$$x(t_i) = x_0 + A_{i-1} + B_i + \mathcal{O}(h^2), \tag{3.61}$$

where

$$\begin{aligned} A_{i-1} =& h_\alpha \sum_{j=1}^{i-1} f(\tau_{ij}, x(t_{j-1}) + \rho_{ij}(x(t_j) - x(t_{j-1})) \\ &+ \mathcal{O}(h^2))[(i-j+1)^\alpha - (i-j)^\alpha], \qquad (3.62)\\ B_i =& h_\alpha [f(\tau_{ii}, x(t_{i-1})) + f_x(\tau_{ii}, x(t_{i-1}))\rho_{ii}(x(t_i) - x(t_{i-1}))]. \qquad (3.63) \end{aligned}$$

Note that (3.56) can be rewritten in the following form:

$$x_i = x_0 + \tilde{A}_{i-1} + \tilde{B}_i, \tag{3.64}$$

where

$$\tilde{A}_{i-1} = h_\alpha \sum_{j=1}^{i-1} f(\tau_{ij}, x_{j-1} + \rho_{ij}(x_j - x_{j-1})[(i-j+1)^\alpha - (i-j)^\alpha], \tag{3.65}$$

$$\tilde{B}_i = h_\alpha \left[f(\tau_{ii}, x_{i-1}) + f_x(\tau_{ii}, x_{i-1})\rho_{ii}(x_i - x_{i-1}) \right]. \tag{3.66}$$

Subtracting (3.64) from (3.61) gives

$$x(t_i) - x_i = (A_{i-1} - \tilde{A}_{i-1}) + (B_i - \tilde{B}_i) + \mathcal{O}(h^2). \tag{3.67}$$

Let us first estimate $B_i - \tilde{B}_i$. From (3.63) and (3.66), we have

$$\begin{aligned} B_i - \tilde{B}_i =& h_\alpha [f(\tau_{ii}, x(t_{i-1})) - f(\tau_{ii}, x_{i-1})] \\ &+ h_\alpha \rho_{ii} [f_x(\tau_{ii}, x(t_{i-1}))x(t_i) - f_x(\tau_{ii}, x_{i-1})x_i] \\ &- h_\alpha \rho_{ii} [f_x(\tau_{ii}, x(t_{i-1}))x(t_{i-1}) - f_x(\tau_{ii}, x_{i-1})x_{i-1}]. \qquad (3.68) \end{aligned}$$

Note that f is twice continuously differentiable. Using a Taylor expansion, we get

$$f_x(\tau_{ii}, x(t_{i-1})) = f_x(\tau_{ii}, x_{i-1}) + r_i, \tag{3.69}$$

where

$$r_i = f_{xx}(\tau_{ii}, \xi)(x(t_{i-1}) - x_{i-1})$$

with ξ being a point between $x(t_{i-1})$ and x_{i-1}. Since $|x(t_{i-1}) - x_{i-1}| \le Ch^2$ by (3.60), we have

$$r_i = f_{xx}(\tau_{ii}, \xi) \cdot \mathcal{O}(h^2) = \mathcal{O}(h^2).$$

Similarly, we have

$$f(\tau_{ii}, x(t_{i-1})) - f(\tau_{ii}, x_{i-1}) = \mathcal{O}(h^2).$$

Using (3.68) and (3.69) together with the above estimate, we have

$$B_i - \tilde{B}_i = h_\alpha \rho_{ii} f_x(\tau_{ii}, x_{i-1})[x(t_i) - x_i] + \mathcal{O}(h^{2+\alpha}),$$

since $h_\alpha = h^\alpha/\Gamma(1+\alpha)$ and $|x_{i-1} - x(t_{i-1})| \le Ch^2$ from (3.60). Thus, from the above expression and (3.67), we get

$$x(t_i) - x_i = (A_{i-1} - \tilde{A}_{i-1}) + h_\alpha \rho_{ii} f_x(\tau_{ii}, x_{i-1})[x(t_i) - (x_i)] + \mathcal{O}(h^2).$$

This implies

$$x(t_i) - x_i = \frac{(A_{i-1} - \tilde{A}_{i-1}) + \mathcal{O}(h^2)}{1 - h_\alpha \rho_{ii} f_x(\tau_{ii}, x_{i-1})},$$

and so

$$|x(t_i) - x_i| \le \frac{|A_{i-1} - \tilde{A}_{i-1}|}{|1 - h_\alpha \rho_{ii} f_x(\tau_{ii}, x_{i-1})|} + \frac{\mathcal{O}(h^2)}{|1 - h_\alpha \rho_{ii} f_x(\tau_{ii}, x_{i-1})|}. \tag{3.70}$$

To estimate $|x(t_i) - x_i|$, we need to estimate $|A_{i-1} - \tilde{A}_{i-1}|$. To simplify our notation, let $x_{ij} = x_{j-1} + \rho_{ij}(x_j - x_{j-1})$. We also use either the right-hand side or left-hand side of (3.49) to represent the point $x(\tau_{ij})$. From the definitions of A_{i-1} and $\tilde{A}_{i-1}$ in (3.62) and (3.65), respectively, and using (3.49), we have

$$\begin{aligned} &|A_{i-1} - \tilde{A}_{i-1}| \\ &\le h_\alpha \sum_{j=1}^{i-1} \big|f(\tau_{ij}, x(\tau_{ij})) - f(\tau_{ij}, x_{ij})\big| [(i-j+1)^\alpha - (i-j)^\alpha], \end{aligned} \tag{3.71}$$

since z^α is an increasing function of z for $\alpha \in (0, 1)$. Because f is twice continuously differentiable, we have (recall that C is a generic positive constant independent of h)

$$\begin{aligned} |f(\tau_{ij}, x(\tau_{ij})) - f(\tau_{ij}, x_{ij})| \le& C(|x(t_{j-1}) - x_{j-1}| + |x(t_j) - x_j)| \\ &+ |x(t_{j-1}) - x_{j-1}|) + \mathcal{O}(h^2), \end{aligned}$$

since $\rho_{ij} \in (0, 1)$. Furthermore, from (3.60), we have

$$|f(\tau_{ij}, x(\tau_{ij})) - f(\tau_{ij}, x_{\tau_{ij}})| \leq Ch^2.$$

Replacing $|f(\tau_{ij}, x(\tau_{ij})) - f(\tau_{ij}, x_{\tau_{ij}})|$ in (3.71) with the above upper bound, we have

$$|A_{i-1} - \tilde{A}_{i-1}| \leq Ch^2.$$

Combining the above error bound with (3.70), we have

$$|x(t_i) - x_i| \leq \frac{Ch^2}{|1 - h_\alpha \rho_{ii} f_x(\tau_{ii}, x_{i-1})|}.$$

Therefore, when $h_\alpha \leq \bar{h}$ with $\bar{h}$ being as defined in (3.59), we have

$$|x(t_i) - x_i| \leq \frac{Ch^2}{|1 - h_\alpha \rho_{ii} f_x(\tau_{ii}, x_{i-1})|} \leq \frac{Ch^2}{\sigma}$$

for a given $\sigma > 0$.

It should be noted that the positive constant C used in the above proof is independent of h, but a function of $T^\alpha/\Gamma(1+\alpha)$. However, in the proof of Theorem 3.19, we have shown that $T^\alpha/\Gamma(1+\alpha)$ is bounded above by a positive constant independent of α. The proof is complete. □

Remark 3.1 In the above theorem, we have established an upper error bound of order $\mathcal{O}(h^2)$ for the numerical approximation to (3.39) generated by our proposed single-step Algorithm 3.1.

Remark 3.2 Algorithm 3.1 is a linearized form of our implicit method (3.52). This linearization does not affect the $\mathcal{O}(h^2)$-order convergence rate of (3.52).

Numerical Examples

Example 3.1 Consider the following FODE [71]:

$$\begin{aligned} {}_0^C\mathrm{D}_t^\alpha x(t) &= \frac{\Gamma(5)}{\Gamma(5-\alpha)} t^{4-\alpha} - x(t) + t^4, \quad t \in (0, 1], \\ x(0) &= 0. \end{aligned}$$

The exact solution is

$$x(t) = t^4.$$

By using Algorithm 3.1, we solve this example for various values of α and mesh sizes h. Here, the computed errors $E_{h_k} = \max_{1 \leq i \leq 1/h_k} |x_i - x(t_i)|$ for $h_k = 1/(2^k \times$

Table 3.1 Maximum errors and convergence rates for Example 3.1

h	Our results		Results from [71]		Our results		Results from [71]	
	α=0.1	Order	α=0.1	Order	α=0.5	Order	α=0.5	Order
1/10	4.06e–03	–	0.364	–	8.49e–03	–	0.0355	–
1/20	1.11e–03	1.86	0.170	1.10	2.16e–03	1.98	0.00879	2.01
1/40	3.02e–04	1.89	7.13e–02	1.26	5.43e–04	1.99	2.16e–03	2.03
1/80	8.08e–05	1.90	2.88e–02	1.31	1.37e–04	1.99	5.31e–04	2.02
1/160	2.14e–05	1.92	1.15e–02	1.32	3.43e–05	2.00	1.31e–04	2.02
1/320	5.65e–06	1.93	4.64e–03	1.31	8.58e–06	2.00	3.24e–05	2.02
1/640	1.47e–06	1.93	1.88e–03	1.30	2.15e–06	2.00	8.03e–06	2.01
1/1280	3.84e–07	1.94	–	–	5.38e–07	2.00	–	–

10), $k = 0, 1, \ldots, 6$, and the chosen values of α are listed in Table 3.1. To estimate the rates of convergence, we calculate $\log_2(E_{h_k}/E_{h_{k+1}})$ for $k = 0, 1, \ldots, 5$, and the computed rates of convergence are also listed in Table 3.1. From Table 3.1, we see that the computed rates of convergence are very close to the theoretical one in Theorem 3.20. In Table 3.1, we also compare our results with those obtained by method in [71]. The latter method is a single-step predictor-corrector method with a theoretical rate of convergence of order $\mathcal{O}(h^{\min(1+2\alpha,2)})$. The rates of convergence of our method are much higher than those method in [71] for $\alpha = 0.1$. From the table, we also see that the absolute errors from our method are much smaller than those in [71]. Clearly, our proposed method is superior to that in [71]. In addition, one can expect that a predictor-corrector method should be computationally more expensive than our method. From Table 3.1, we see that when α is close to zero, the computed rate of convergence is slightly worse than $\mathcal{O}(h^2)$. This may be because the constant C in (3.57) contains the term $1/\Gamma(1+\alpha)$, as noted in the previous section. However, when h decreases, the rate of convergence of our method increases.

Example 3.2 Consider the following nonlinear FODE:

$$
\begin{aligned}
{}_0^C\mathrm{D}_t^\alpha x(t) &= \frac{\Gamma(4+\alpha)}{6} t^3 + t^{4(3+\alpha)} - x^4(t), \quad t \in (0, 1], \\
x(0) &= 0.
\end{aligned}
$$

The exact solution is $x(t) = t^{3+\alpha}$.

This example is solved using Algorithm 3.1 for various values of h and α, and the computed errors and rates of convergence are listed in Table 3.2. From Table 3.2, we see that the computed order of convergence is greater than 2 when α is small.

Table 3.2 Maximum errors and convergence rates for Example 3.2

h	α=0.1	Order	α=0.3	Order	α=0.5	Order	α=0.9	Order
1/10	6.19e–02	–	4.27e–02	–	2.56e–02	–	3.70e–03	–
1/20	1.69e–02	1.87	1.03e–02	2.05	5.70e–03	2.17	6.90e–04	2.42
1/40	4.42e–03	1.93	2.368e–03	2.13	1.10e–03	2.37	1.88e–04	1.88
1/80	1.10e–03	2.00	5.055e–04	2.23	1.85e–04	2.57	5.92e–05	1.67
1/160	2.65e–04	2.05	1.025e–04	2.30	2.63e–05	2.82	2.11e–05	1.49
1/320	6.26e–05	2.08	2.00e–05	2.36	2.63e–06	3.32	6.19e–06	1.77
1/640	1.46e–05	2.10	3.77e–06	2.41	5.30e–07	2.31	1.68e–06	1.88
1/1280	3.41e–06	2.10	6.86e–07	2.46	1.55e–07	1.77	4.37e–07	1.94

3.3.2 Third-Order Approach

Consider the following general nonlinear FODEs with Caputo fractional derivatives:

$$\begin{cases} {}_{t_0}^{C}\mathrm{D}_t^{\boldsymbol{\alpha}}\boldsymbol{x}(t) = \boldsymbol{f}(t, \boldsymbol{x}(t)), & t \in (t_0, t_f], \\ \boldsymbol{x}(t_0) = \boldsymbol{x}^0, \end{cases} \tag{3.72}$$

where $\boldsymbol{\alpha} = (\alpha_1, \alpha_2, \ldots, \alpha_n)^\top \in (0, 1]^n$ is a given vector of fractional orders; $\boldsymbol{x}(t) = (x_1(t), x_2(t), \ldots, x_n(t))^\top \in \mathbb{R}^n$ is the state vector at time t; $\boldsymbol{f} : \mathbb{R} \times \mathbb{R}^n \to \mathbb{R}^n$ is a given thrice continuously differentiable function with respect to all its arguments; t_0 is a given initial time; t_f is a given terminal time; $\boldsymbol{x}^0 \in \mathbb{R}^n$ is a given initial state vector; and ${}_{t_0}^{C}\mathrm{D}_t^{\boldsymbol{\alpha}}\boldsymbol{x}(t) = \left({}_{t_0}^{C}\mathrm{D}_t^{\alpha_1}x_1(t), {}_{t_0}^{C}\mathrm{D}_t^{\alpha_2}x_2(t), \ldots, {}_{t_0}^{C}\mathrm{D}_t^{\alpha_n}x_n(t)\right)^\top$, with ${}_{t_0}^{C}\mathrm{D}_t^{\alpha_i}x_i(t)$ denoting the Caputo fractional derivative of order α_i of $x_i(t)$.

Numerical Scheme

It follows from Theorem 3.7 that (3.72) is equivalent to the following Volterra integral equations:

$$x_i(t) = x_i^0 + \frac{1}{\Gamma(\alpha_i)} \int_{t_0}^{t} (t - \tau)^{\alpha_i - 1} f_i(\tau, \boldsymbol{x}(\tau))d\tau, \quad t \in (t_0, t_f], \tag{3.73}$$

for $i = 1, 2, \ldots, n$, where $x_i^0 = x_i(t_0)$. For a given positive integer N, define a mesh on $[t_0, t_f]$ with $N + 1$ mesh points $t_q, q = 0, 1, \ldots, N$, satisfying $t_0 < t_1 < \cdots < t_N = t_f$. For any $i = 1, 2, \ldots, n$, and $q = 0, 1, \ldots, N$, (3.73) can be rewritten as

$$x_i(t_q) = x_i^0 + \frac{1}{\Gamma(\alpha_i)} \sum_{j=0}^{q-1} \int_{t_j}^{t_{j+1}} (t_q - \tau)^{\alpha_i - 1} f_i(\tau, \boldsymbol{x}(\tau)) d\tau. \tag{3.74}$$

We now consider the approximation of the integrand $f_i(\tau, \boldsymbol{x}(\tau))$ on the right-hand side of (3.74) in $(t_0, t_1]$ by the following Taylor expansion:

$$f_i(\tau, \boldsymbol{x}(\tau)) = a_{i0} + b_{i0}(\tau - t_1) + c_{i0}(\tau - t_1)^2, \tag{3.75}$$

where a_{i0} and b_{i0} are coefficients to be determined and $c_{i0}(\tau - t_1)^2$ is the remainder. To determine a_{i0} and b_{i0}, we omit the remainder and force the following equations to hold:

$$\begin{cases} f_i(t_0, \boldsymbol{x}^0) = a_{i0} + b_{i0}(t_0 - t_1), \\ f_i(t_1, \boldsymbol{x}(t_1)) = a_{i0}. \end{cases}$$

Solving these coupled equations gives

$$\begin{aligned} \Delta^0 f_i^1 &:= a_{i0} = f_i(t_1, \boldsymbol{x}(t_1)), \\ \Delta^1 f_i^1 &:= b_{i0} = \frac{f_i(t_1, \boldsymbol{x}(t_1)) - f_i(t_0, \boldsymbol{x}^0)}{h_1}, \end{aligned}$$

where $h_1 = t_1 - t_0$. Using these divided differences, (3.75) can be expressed as

$$f_i(\tau, \boldsymbol{x}(\tau)) = \Delta^0 f_i^1 + \Delta^1 f_i^1(\tau - t_1) + c_{i0}(\tau - t_1)^2. \tag{3.76}$$

Substituting the right-hand side of (3.76) into the first term of the sum in (3.74) yields

$$\begin{aligned} &\frac{1}{\Gamma(\alpha_i)} \int_{t_0}^{t_1} (t_q - \tau)^{\alpha_i - 1} \left[\Delta^0 f_i^1 + \Delta^1 f_i^1(\tau - t_1) + c_{i0}(\tau - t_1)^2 \right] d\tau \\ &= -\frac{I_{0,1}^i}{h_1} f_i(t_0, \boldsymbol{x}^0) + \left(I_{0,0}^i + \frac{I_{0,1}^i}{h_1} \right) f_i(t_1, \boldsymbol{x}(t_1)) + R_{i0}, \end{aligned} \tag{3.77}$$

where

$$I_{0,1}^i = \frac{-h_1(t_q - t_0)^{\alpha_i}}{\Gamma(\alpha_i + 1)} + \frac{(t_q - t_0)^{\alpha_i + 1} - (t_q - t_1)^{\alpha_i + 1}}{\Gamma(\alpha_i + 2)}, \tag{3.78}$$

$$I_{0,0}^i = \frac{(t_q - t_0)^{\alpha_i} - (t_q - t_1)^{\alpha_i}}{\Gamma(\alpha_i + 1)}, \tag{3.79}$$

$$R_{i0} = \frac{c_{i0}}{\Gamma(\alpha_i)} \int_{t_0}^{t_1} (t_q - \tau)^{\alpha_i - 1} (\tau - t_1)^2 d\tau.$$

For any $j = 1, 2, \ldots, q - 1$, and $q = 2, 3, \ldots, N$, we approximate $f_i(\tau, \boldsymbol{x}(\tau))$ on $(t_j, t_{j+1}]$ by the following third-order Taylor expansion:

$$f_i(\tau, \boldsymbol{x}(\tau)) = a_{ij} + b_{ij}(\tau - t_{j+1}) + c_{ij}(\tau - t_{j+1})^2 + d_{ij}(\tau - t_{j+1})^3, \tag{3.80}$$

where a_{ij}, b_{ij}, and c_{ij} are coefficients to be determined, and $d_{ij}(\tau - t_{j+1})^3$ is the remainder. Omitting the remainder, we can use the values of $f_i(\tau, \boldsymbol{x}(\tau))$ at points t_{j-1}, t_j and t_{j+1} to determine a_{ij}, b_{ij}, and c_{ij}, yielding the following divided differences:

$$\begin{aligned}
\Delta^0 f_i^{j+1} &:= a_{ij} = f_i(t_{j+1}, \boldsymbol{x}(t_{j+1})),\\
\Delta^1 f_i^{j+1} &:= b_{ij} = \frac{h_{j+1} f_i(t_{j-1}, \boldsymbol{x}(t_{j-1}))}{h_j(h_j + h_{j+1})} - \frac{(h_j + h_{j+1}) f_i(t_j, \boldsymbol{x}(t_j))}{h_j h_{j+1}}\\
&\quad + \frac{(h_j + 2h_{j+1}) f_i(t_{j+1}, \boldsymbol{x}(t_{j+1}))}{h_{j+1}(h_j + h_{j+1})},\\
\Delta^2 f_i^{j+1} &:= c_{ij} = \frac{f_i(t_{j-1}, \boldsymbol{x}(t_{j-1}))}{h_j(h_j + h_{j+1})} - \frac{f_i(t_j, \boldsymbol{x}(t_j))}{h_j h_{j+1}} + \frac{f_i(t_{j+1}, \boldsymbol{x}(t_{j+1}))}{h_{j+1}(h_j + h_{j+1})},
\end{aligned}$$

where $h_{j+1} = t_{j+1} - t_j$ and $h_j = t_j - t_{j-1}$. Using these differences, we rewrite (3.80) as

$$\begin{aligned}
f_i(\tau, \boldsymbol{x}(\tau)) &= \Delta^0 f_i^{j+1} + \Delta^1 f_i^{j+1}(\tau - t_{j+1}) + \Delta^2 f_i^{j+1}(\tau - t_{j+1})^2\\
&\quad + d_{ij}(\tau - t_{j+1})^3.
\end{aligned} \tag{3.81}$$

Substituting the right-hand side of (3.81) into the $(j+1)$th term of the sum in (3.74) gives

$$\begin{aligned}
&\frac{1}{\Gamma(\alpha_i)} \int_{t_j}^{t_{j+1}} (t_q - \tau)^{\alpha_i - 1} \big[\Delta^0 f_i^{j+1} + \Delta^1 f_i^{j+1}(\tau - t_{j+1}) + \Delta^2 f_i^{j+1}(\tau - t_{j+1})^2\\
&\quad + d_{ij}(\tau - t_{j+1})^3 \big] d\tau\\
&= \frac{h_{j+1} I_{j,1}^i + I_{j,2}^i}{h_j(h_j + h_{j+1})} f_i(t_{j-1}, \boldsymbol{x}(t_{j-1})) - \frac{(h_j + h_{j+1}) I_{j,1}^i + I_{j,2}^i}{h_j h_{j+1}} f_i(t_j, \boldsymbol{x}(t_j))\\
&\quad + \left[I_{j,0}^i + \frac{(h_j + 2h_{j+1}) I_{j,1}^i + I_{j,2}^i}{h_{j+1}(h_j + h_{j+1})} \right] f_i(t_{j+1}, \boldsymbol{x}(t_{j+1})) + R_{ij},
\end{aligned} \tag{3.82}$$

where

$$\begin{aligned}
I_{j,2}^i &= \frac{h_{j+1}^2 (t_q - t_j)^{\alpha_i}}{\Gamma(\alpha_i + 1)} - \frac{2h_{j+1}(t_q - t_j)^{\alpha_i + 1}}{\Gamma(\alpha_i + 2)}\\
&\quad - 2\frac{(t_q - t_{j+1})^{\alpha_i + 2} - (t_q - t_j)^{\alpha_i + 2}}{\Gamma(\alpha_i + 3)},
\end{aligned} \tag{3.83}$$

$$I_{j,1}^i = \frac{-h_{j+1}(t_q - t_j)^{\alpha_i}}{\Gamma(\alpha_i + 1)} - \frac{(t_q - t_{j+1})^{\alpha_i + 1} - (t_q - t_j)^{\alpha_i + 1}}{\Gamma(\alpha_i + 2)}, \tag{3.84}$$

$$I_{j,0}^{i} = \frac{(t_q - t_j)^{\alpha_i} - (t_q - t_{j+1})^{\alpha_i}}{\Gamma(\alpha_i + 1)}, \tag{3.85}$$

$$R_{ij} = \frac{d_{ij}}{\Gamma(\alpha_i)} \int_{t_j}^{t_{j+1}} (t_q - \tau)^{\alpha_i - 1} (\tau - t_{j+1})^3 d\tau.$$

Combining (3.74), (3.77), and (3.82), we obtain the following equations, which are equivalent to the equations appearing in (3.74),

$$\begin{aligned} x_i(t_q) = x_i^0 &- \frac{I_{0,1}^i}{h_1} f_i(t_0, \boldsymbol{x}^0) + \left(I_{0,0}^i + \frac{I_{0,1}^i}{h_1}\right) f_i(t_1, \boldsymbol{x}(t_1)) + \sum_{j=1}^{q-1} \left\{ \frac{h_{j+1} I_{j,1}^i + I_{j,2}^i}{h_j (h_j + h_{j+1})} \right. \\ &\times f_i(t_{j-1}, \boldsymbol{x}(t_{j-1})) - \frac{(h_j + h_{j+1}) I_{j,1}^i + I_{j,2}^i}{h_j h_{j+1}} f_i(t_j, \boldsymbol{x}(t_j)) \\ &\left. + \left[I_{j,0}^i + \frac{(h_j + 2h_{j+1}) I_{j,1}^i + I_{j,2}^i}{h_{j+1}(h_j + h_{j+1})} \right] f_i(t_{j+1}, \boldsymbol{x}(t_{j+1})) \right\} + R_i^q \end{aligned} \tag{3.86}$$

for $i = 1, 2, \ldots, n$, and $q = 1, 2, \ldots, N$, where $R_i^q = \sum_{j=0}^{q-1} R_{ij}$ is the cumulative truncation error up to the point t_q.

Omitting the truncation error R_i^q in (3.86), we define the following numerical scheme for (3.74):

$$\begin{aligned} x_i^q = x_i^0 &- \frac{I_{0,1}^i}{h_1} f_i(t_0, \boldsymbol{x}^0) + \left(I_{0,0}^i + \frac{I_{0,1}^i}{h_1}\right) f_i(t_1, \boldsymbol{x}^1) + \sum_{j=1}^{q-1} \left\{ \frac{h_{j+1} I_{j,1}^i + I_{j,2}^i}{h_j (h_j + h_{j+1})} \right. \\ &\times f_i(t_{j-1}, \boldsymbol{x}^{j-1}) - \frac{(h_j + h_{j+1}) I_{j,1}^i + I_{j,2}^i}{h_j h_{j+1}} f_i(t_j, \boldsymbol{x}^j) \\ &\left. + \left[I_{j,0}^i + \frac{(h_j + 2h_{j+1}) I_{j,1}^i + I_{j,2}^i}{h_{j+1}(h_j + h_{j+1})} \right] f_i(t_{j+1}, \boldsymbol{x}^{j+1}) \right\} \end{aligned} \tag{3.87}$$

for $i = 1, 2, \ldots, n$, and $q = 1, 2, \ldots, N$, where x_i^q represents an approximation of $x_i(t_q)$ for each feasible i and q.

We comment that (3.87) defines an implicit time-stepping scheme for the numerical solution of (3.72). Since (3.87) is implicit, we need to use a technique, such as a Newton's method, to solve the nonlinear algebraic equations at each point $t_q, q = 1, 2, \ldots, N$.

Convergence Analysis

Note that (3.87) is a nonlinear system in $\boldsymbol{x}^q$ that will be solved by Newton's iterative method. Thus, we need to show that the Jacobian matrix of this nonlinear system is invertible for any feasible q. For $i = 1, 2, \ldots, n$, and $q = 1, 2, \ldots, N$, let

$$\begin{aligned} F_i(\boldsymbol{x}^q) &= x_i^q - x_i^0 + \frac{I_{0,1}^i}{h_1} f_i(t_0, \boldsymbol{x}^0) - \left(I_{0,0}^i + \frac{I_{0,1}^i}{h_1}\right) f_i(t_1, \boldsymbol{x}^1) \\ &\quad - \sum_{j=1}^{q-1} \left\{ \frac{h_{j+1} I_{j,1}^i + I_{j,2}^i}{h_j(h_j + h_{j+1})} f_i(t_{j-1}, \boldsymbol{x}^{j-1}) - \frac{(h_j + h_{j+1}) I_{j,1}^i + I_{j,2}^i}{h_j h_{j+1}} f_i(t_j, \boldsymbol{x}^j) \right. \\ &\quad \left. + \left[I_{j,0}^i + \frac{(h_j + 2h_{j+1}) I_{j,1}^i + I_{j,2}^i}{h_{j+1}(h_j + h_{j+1})} \right] f_i(t_{j+1}, \boldsymbol{x}^{j+1}) \right\}. \end{aligned}$$

We then have the following theorem:

Theorem 3.21 *Let $\boldsymbol{F}(\boldsymbol{x}^q) = (F_1(\boldsymbol{x}^q), F_2(\boldsymbol{x}^q), \ldots, F_n(\boldsymbol{x}^q))^\top$ for $q \in \{1, \ldots, N\}$. Then, the Jacobian matrix of $\boldsymbol{F}(\boldsymbol{x}^q)$ is invertible when $h_q > 0$ is sufficiently small.*

Proof From direct computation, we see that the Jacobian matrix of $\boldsymbol{F}(\boldsymbol{x}^q)$, denoted as $\boldsymbol{F}'(\boldsymbol{x}^q)$, can be rewritten as

$$\boldsymbol{F}'(\boldsymbol{x}^q) = \boldsymbol{I}_n - [b_{il}^q]_{n\times n}$$

for $i = 1, 2, \ldots, n$, $l = 1, 2, \ldots, n$, and $q = 1, 2, \ldots, N$, where $\boldsymbol{I}_n$ is the $n \times n$ identity matrix and $[b_{il}^q]_{n\times n}$ is an $n \times n$ matrix with

$$b_{il}^q = \begin{cases} \left(I_{0,0}^i + \dfrac{I_{0,1}^i}{h_1}\right) \dfrac{\partial f_i}{\partial x_l}\bigg|_{(t_1, \boldsymbol{x}^1)}, & \text{if } q = 1, \\ \left[I_{q-1,0}^i + \dfrac{(h_{q-1} + 2h_q) I_{q-1,1}^i + I_{q-1,2}^i}{h_q(h_{q-1} + h_q)} \right] \dfrac{\partial f_i}{\partial x_l}\bigg|_{(t_q, \boldsymbol{x}^q)}, & \text{if } q = 2, \ldots, N. \end{cases} \tag{3.88}$$

Since $\boldsymbol{f}$ is thrice continuously differentiable, $\dfrac{\partial f_i}{\partial x_l}$ is bounded on $[t_0, t_f]$ for $i = 1, 2, \ldots, n$, and $l = 1, 2, \ldots, n$. Let

$$M = \max_{\substack{i \in \{1, \ldots, n\} \\ l \in \{1, \ldots, n\}}} \left\{ \left| \frac{\partial f_i}{\partial x_l} \right| \right\}. \tag{3.89}$$

Combining (3.78), (3.79), and (3.88), we obtain

$$|b_{il}^1| \leq \left| \left(I_{0,0}^i + \frac{I_{0,1}^i}{h_1}\right) \right| M = \frac{h_1^{\alpha_i} M}{\Gamma(\alpha_i + 2)} \tag{3.90}$$

for $i = 1, 2, \ldots, n$, and $l = 1, 2, \ldots, n$. Furthermore, combining (3.83)–(3.85) and (3.88), we obtain

$$|b_{il}^q| \leq \left| I_{q-1,0}^i + \frac{(h_{q-1} + 2h_q) I_{q-1,1}^i + I_{q-1,2}^i}{h_q(h_{q-1} + h_q)} \right| M < \frac{4 h_q^{\alpha_i} M}{\Gamma(\alpha_i + 1)} \tag{3.91}$$

for $i = 1, 2, \ldots, n$, $l = 1, 2, \ldots, n$, and $q = 2, 3, \ldots, N$.

For given constants $\sigma_i \in (0, 1)$, $i = 1, 2, \ldots, n$, define

$$\hat{h} = \min_{i \in \{1,\ldots,n\}} \left\{ \left(\frac{\Gamma(\alpha_i + 1)(1 - \sigma_i)}{4nM} \right)^{\frac{1}{\alpha_i}} \right\}. \tag{3.92}$$

Select $h_1 \leq \hat{h}$ such that for $i = 1, 2, \ldots, n$,

$$\begin{aligned} \sum_{l=1}^{n} |b_{il}^1| &\leq \frac{n h_1^{\alpha_i} M}{\Gamma(\alpha_i + 2)} \leq \frac{n \hat{h}^{\alpha_i} M}{\Gamma(\alpha_i + 2)} \leq \frac{\Gamma(\alpha_i + 1)(1 - \sigma_i)}{4\Gamma(\alpha_i + 2)} \\ &< 1 - \sigma_i \leq \max_{1 \leq \iota \leq n} \{1 - \sigma_\iota\} = 1 - \min_{1 \leq \iota \leq n} \{\sigma_\iota\}. \end{aligned} \tag{3.93}$$

Choose $h_q \leq \hat{h}$ such that for $i = 1, 2, \ldots, n$, and $q = 2, 3, \ldots, N$,

$$\sum_{l=1}^{n} |b_{il}^q| < \frac{4n h_q^{\alpha_i} M}{\Gamma(\alpha_i + 1)} \leq \frac{4n \hat{h}^{\alpha_i} M}{\Gamma(\alpha_i + 1)} = 1 - \sigma_i \leq \max_{1 \leq \iota \leq n} \{1 - \sigma_\iota\} = 1 - \min_{1 \leq \iota \leq n} \{\sigma_\iota\}. \tag{3.94}$$

For $i = 1, 2, \ldots, n$, and $q = 1, 2, \ldots, N$, combining (3.93) and (3.94), we obtain

$$0 < \min_{1 \leq \iota \leq n} \{\sigma_\iota\} < 1 - \sum_{l=1}^{n} |b_{il}^q| = 1 - |b_{ii}^q| - \sum_{\substack{l=1 \\ l \neq i}}^{n} |b_{il}^q| \leq |1 - b_{ii}^q| - \sum_{\substack{l=1 \\ l \neq i}}^{n} |b_{il}^q|. \tag{3.95}$$

It follows from (3.95) that $\boldsymbol{F}'(\boldsymbol{x}^q)$ is strictly diagonally dominant. According to the Levy-Desplanques theorem in [309], $\boldsymbol{F}'(\boldsymbol{x}^q)$ is non-singular. Therefore, $\boldsymbol{F}'(\boldsymbol{x}^q)$ is invertible, which completes the proof. □

Theorem 3.22 *Let $\boldsymbol{x}(t_q)$ be the exact solution of (3.74), and let $\boldsymbol{x}^q$ be the solution of (3.87) for $q = 1, 2, \ldots, N$. Then, there exists a positive constant C independent of $\boldsymbol{\alpha}$ such that when h_1 is chosen to satisfy $h_1 \leq h^{3/2}$, it holds that*

$$\|\boldsymbol{x}(t_q) - \boldsymbol{x}^q\|_\infty \leq Ch^3, \tag{3.96}$$

where $\|\cdot\|_\infty$ is the infinity norm; $h = \max_{\ell \in \{2,\ldots,N\}} \{h_\ell\}$ satisfies the condition $h \leq \hat{h}$ ($\hat{h}$ is as defined in (3.92)).

Proof The proof is carried out in three steps.

Step 1. The truncation error $\boldsymbol{R}^q$ is of order $\mathcal{O}(h^3)$

Let $\boldsymbol{R}^q = (R_1^q, R_2^q, \ldots, R_n^q)^\top$ for $q \in \{1, 2, \ldots, N\}$. From the definition of R_i^q given by (3.86) and the thrice continuous differentiability of $\boldsymbol{f}$, we deduce that for $i = 1, 2, \ldots, n$, and $q = 1, 2, \ldots, N$,

$$
\begin{aligned}
|R_i^q| &\le |R_{i0}| + \sum_{j=1}^{q-1} |R_{ij}| \\
&\le \frac{dh^3}{\Gamma(\alpha_i)} \sum_{j=0}^{q-1} \int_{t_j}^{t_{j+1}} (t_q - \tau)^{\alpha_i - 1} d\tau = \frac{dh^3}{\Gamma(\alpha_i)} \frac{(t_q - t_0)^{\alpha_i}}{\alpha_i} \le \frac{dh^3 t_f^{\alpha_i}}{\Gamma(\alpha_i + 1)},
\end{aligned}
$$

because $h_1 \le h^{3/2}$, where $d = \max_{\substack{i \in \{1,\ldots,n\} \\ j \in \{1,\ldots,q-1\}}} \{|c_{i0}|, |d_{ij}|\}$ and $h = \max_{\ell \in \{2,\ldots,N\}} \{h_\ell\}$.

Recall that $\alpha_i \in (0, 1]$ for $i = 1, 2, \ldots, n$. Thus, $2^{\alpha_i - 1} \le \Gamma(\alpha_i + 1) \le 1$. Also, it is obvious that $t_f^{\alpha_i} \le \max\{1, t_f\}$. Therefore, choosing $\hat{C} = 2d \max\{1, t_f\}$, we obtain

$$
\|\boldsymbol{R}^q\|_\infty = \mathcal{O}(h^3) \tag{3.97}
$$

for $q = 1, 2, \ldots, N$.

Step 2. The initial step of the mathematical induction process involving (3.96)

Subtracting (3.87) from (3.86) and applying the Lagrange mean value theorem, we obtain

$$
x_i(t_1) - x_i^1 = [u_{il}^1]_{1 \times n} (\boldsymbol{x}(t_1) - \boldsymbol{x}^1) + R_i^1, \tag{3.98}
$$

for $i = 1, 2, \ldots, n$ and $l = 1, 2, \ldots, n$, where u_{il}^1 is defined as

$$
u_{il}^1 = \left(I_{0,0}^i + \frac{I_{0,1}^i}{h_1} \right) \frac{\partial f_i}{\partial x_l} \bigg|_{(t_1, \xi_l^1)}. \tag{3.99}
$$

Here, for $l = 1, 2, \ldots, n$, $\xi_l^1 = x_l(t_1) + \lambda_l^1 (x_l^1 - x_l(t_1))$ with $\lambda_l^1 \in (0, 1)$.

Based on (3.98), we obtain

$$
\begin{aligned}
|x_i(t_1) - x_i^1| \le \left| [u_{il}^1]_{1 \times n} (\boldsymbol{x}(t_1) - \boldsymbol{x}^1) \right| + |R_i^1| &\le n \left\| [u_{il}^1]_{1 \times n} \right\|_\infty \left\| \boldsymbol{x}(t_1) - \boldsymbol{x}^1 \right\|_\infty \\
&+ \|\boldsymbol{R}^1\|_\infty
\end{aligned} \tag{3.100}
$$

for $i = 1, 2, \ldots, n$ and $l = 1, 2, \ldots, n$.

Similar to the derivation of (3.90), it follows that $\left\| [u_{il}^1]_{1 \times n} \right\|_\infty \le \dfrac{h_1^{\alpha_i} M}{\Gamma(\alpha_i + 2)}$, where M is as defined in (3.89). Thus, (3.100) can be rewritten as

$$
\begin{aligned}
|x_i(t_1) - x_i^1| &\le \frac{n h_1^{\alpha_i} M}{\Gamma(\alpha_i + 2)} \left\| \boldsymbol{x}(t_1) - \boldsymbol{x}^1 \right\|_\infty + \|\boldsymbol{R}^1\|_\infty \\
&\le \max_{1 \le i \le n} \left\{ \frac{n h_1^{\alpha_i} M}{\Gamma(\alpha_i + 2)} \right\} \left\| \boldsymbol{x}(t_1) - \boldsymbol{x}^1 \right\|_\infty + \|\boldsymbol{R}^1\|_\infty,
\end{aligned} \tag{3.101}
$$

for all feasible i, which implies that

$$\left(1-\max_{1\le i\le n}\left\{\frac{nh_1^{\alpha_i}M}{\Gamma(\alpha_i+2)}\right\}\right)\left\|\boldsymbol{x}(t_1)-\boldsymbol{x}^1\right\|_\infty \le \|\boldsymbol{R}^1\|_\infty. \tag{3.102}$$

According to the derivation of (3.93), if we select $h_1 \le \hat{h}$ (recall that $\hat{h}$ is as defined in (3.92)), then $1-\dfrac{nh_1^{\alpha_i}M}{\Gamma(\alpha_i+2)} > \min\limits_{1\le\iota\le n}\{\sigma_\iota\} > 0$ is satisfied for all feasible i. Thus, $1-\max\limits_{1\le i\le n}\left\{\dfrac{nh_1^{\alpha_i}M}{\Gamma(\alpha_i+2)}\right\} > \min\limits_{1\le\iota\le n}\{\sigma_\iota\} > 0$ holds. From (3.97) and (3.102), we obtain

$$\left\|\boldsymbol{x}(t_1)-\boldsymbol{x}^1\right\|_\infty \le \frac{\|\boldsymbol{R}^1\|_\infty}{1-\max\limits_{1\le i\le n}\left\{\dfrac{nh_1^{\alpha_i}M}{\Gamma(\alpha_i+2)}\right\}} < \frac{\|\boldsymbol{R}^1\|_\infty}{\min\limits_{1\le\iota\le n}\{\sigma_\iota\}} = \mathcal{O}(h^3). \tag{3.103}$$

Step 3. Inductive step for (3.96)

We now consider the case $q \ge 2$. In Step 2, we have shown that (3.103), i.e., the case $q = 1$ is valid. To apply the mathematical induction, we assume that

$$\|\boldsymbol{x}(t_j)-\boldsymbol{x}^j\|_\infty = \mathcal{O}(h^3), \tag{3.104}$$

for $j = 1, 2, \ldots, q-1$. Then, we need to show that $\|\boldsymbol{x}(t_q)-\boldsymbol{x}^q\|_\infty = \mathcal{O}(h^3)$.

Subtracting (3.87) from (3.86) and applying the Lagrange mean value theorem, we obtain

$$\begin{aligned} x_i(t_q)-x_i^q = {} & [u_{il}^1]_{1\times n}(\boldsymbol{x}(t_1)-\boldsymbol{x}^1) + \sum_{j=1}^{q-1}\Big\{[v_{il}^{j-1}]_{1\times n}(\boldsymbol{x}(t_{j-1})-\boldsymbol{x}^{j-1}) \\ & - [w_{il}^j]_{1\times n}(\boldsymbol{x}(t_j)-\boldsymbol{x}^j)\Big\} + \sum_{j=1}^{q-2}\Big\{[z_{il}^{j+1}]_{1\times n}(\boldsymbol{x}(t_{j+1})-\boldsymbol{x}^{j+1})\Big\} \\ & + [z_{il}^q]_{1\times n}(\boldsymbol{x}(t_q)-\boldsymbol{x}^q) + R_i^q \end{aligned} \tag{3.105}$$

for $i = 1, 2, \ldots, n$ and $l = 1, 2, \ldots, n$, where u_{il}^1 is as defined in (3.99), while v_{il}^{j-1}, w_{il}^j, and z_{il}^{j+1} are defined as

$$\begin{aligned} v_{il}^{j-1} &= \frac{h_{j+1}I_{j,1}^i + I_{j,2}^i}{h_j(h_j+h_{j+1})}\frac{\partial f_i}{\partial x_l}\bigg|_{(t_{j-1},\xi_l^{j-1})}, \\ w_{il}^j &= \frac{(h_j+h_{j+1})I_{j,1}^i + I_{j,2}^i}{h_jh_{j+1}}\frac{\partial f_i}{\partial x_l}\bigg|_{(t_j,\xi_l^j)}, \\ z_{il}^{j+1} &= \left[I_{j,0}^i + \frac{(h_j+2h_{j+1})I_{j,1}^i + I_{j,2}^i}{h_{j+1}(h_j+h_{j+1})}\right]\frac{\partial f_i}{\partial x_l}\bigg|_{(t_{j+1},\xi_l^{j+1})}. \end{aligned}$$

Here, for $l = 1, 2, \ldots, n$ and $k = 0, 1, \ldots, q$, $\xi_l^k = x_l(t_k) + \lambda_l^k(x_l^k - x_l(t_k))$ with $\lambda_l^k \in (0, 1)$.

Based on (3.105), we can deduce that

$$\begin{aligned}|x_i(t_q) - x_i^q| \leq & \Bigg| [u_{il}^1]_{1\times n}(\boldsymbol{x}(t_1) - \boldsymbol{x}^1) + \sum_{j=1}^{q-1}\Big\{[v_{il}^{j-1}]_{1\times n}(\boldsymbol{x}(t_{j-1}) - \boldsymbol{x}^{j-1}) \\ & - [w_{il}^j]_{1\times n}(\boldsymbol{x}(t_j) - \boldsymbol{x}^j)\Big\} + \sum_{j=1}^{q-2}\Big\{[z_{il}^{j+1}]_{1\times n}(\boldsymbol{x}(t_{j+1}) - \boldsymbol{x}^{j+1})\Big\}\Bigg| \\ & + n\big\|[z_{il}^q]_{1\times n}\|_\infty \|\boldsymbol{x}(t_q) - \boldsymbol{x}^q\|_\infty + \|\boldsymbol{R}^q\|_\infty \end{aligned} \tag{3.106}$$

for $i = 1, 2, \ldots, n$ and $l = 1, 2, \ldots, n$.

Similar to the derivation of (3.91), we can show that $\big\|[z_{il}^q]_{1\times n}\big\|_\infty < \dfrac{4h_q^{\alpha_i}M}{\Gamma(\alpha_i + 1)}$. Thus, (3.106) can be expressed as

$$\begin{aligned}|x_i(t_q) - x_i^q| < & \Bigg| [u_{il}^1]_{1\times n}(\boldsymbol{x}(t_1) - \boldsymbol{x}^1) + \sum_{j=1}^{q-1}\Big\{[v_{il}^{j-1}]_{1\times n}(\boldsymbol{x}(t_{j-1}) - \boldsymbol{x}^{j-1}) \\ & - [w_{il}^j]_{1\times n}(\boldsymbol{x}(t_j) - \boldsymbol{x}^j)\Big\} + \sum_{j=1}^{q-2}\Big\{[z_{il}^{j+1}]_{1\times n}(\boldsymbol{x}(t_{j+1}) - \boldsymbol{x}^{j+1})\Big\}\Bigg| \\ & + \max_{1\leq i\leq n}\left\{\frac{n4h_q^{\alpha_i}M}{\Gamma(\alpha_i + 1)}\right\}\|\boldsymbol{x}(t_q) - \boldsymbol{x}^q\|_\infty + \|\boldsymbol{R}^q\|_\infty\end{aligned}$$

for all feasible i and l, which implies that

$$\begin{aligned}&\left(1 - \max_{1\leq i\leq n}\left\{\frac{n4h_q^{\alpha_i}M}{\Gamma(\alpha_i + 1)}\right\}\right)\|\boldsymbol{x}(t_q) - \boldsymbol{x}^q\|_\infty \\ &< \Bigg|[u_{il}^1]_{1\times n}(\boldsymbol{x}(t_1) - \boldsymbol{x}^1) + \sum_{j=1}^{q-1}\Big\{[v_{il}^{j-1}]_{1\times n}(\boldsymbol{x}(t_{j-1}) - \boldsymbol{x}^{j-1}) - [w_{il}^j]_{1\times n}(\boldsymbol{x}(t_j) \\ &\quad - \boldsymbol{x}^j)\Big\} + \sum_{j=1}^{q-2}\Big\{[z_{il}^{j+1}]_{1\times n}(\boldsymbol{x}(t_{j+1}) - \boldsymbol{x}^{j+1})\Big\}\Bigg| + \|\boldsymbol{R}^q\|_\infty. \end{aligned} \tag{3.107}$$

According to the derivation of (3.94), if we select $h_q \leq \hat{h}$, then $1 - \dfrac{n4h_q^{\alpha_i}M}{\Gamma(\alpha_i + 1)} \geq \min\limits_{1\leq\iota\leq n}\{\sigma_\iota\} > 0$ is satisfied for all feasible i. Thus,

$$1 - \max_{1\leq i\leq n}\left\{\frac{n4h_q^{\alpha_i}M}{\Gamma(\alpha_i + 1)}\right\} \geq \min_{1\leq\iota\leq n}\{\sigma_\iota\} > 0$$

holds. From (3.97), (3.104), and (3.107) and the boundedness of u_{il}^1, v_{il}^{j-1}, w_{il}^j, and z_{il}^{j+1} for $i = 1, 2, \ldots, n, l = 1, 2, \ldots, n$, and $j = 1, 2, \ldots, q-1$, we obtain

$$\|\boldsymbol{x}(t_q) - \boldsymbol{x}^q\|_\infty = \frac{\mathcal{O}(h^3) + \|\boldsymbol{R}^q\|_\infty}{1 - \max\limits_{1\le i\le n}\left\{\dfrac{n4h_q^{\alpha_i}M}{\Gamma(\alpha_i+1)}\right\}} = \mathcal{O}(h^3).$$

This completes the proof. □

Remark 3.3 The choice of $h_1 \le h^{3/2}$ aligns with the Rannacher time-stepping technique as discussed in [258, 339]. This ensures that the overall accuracy of the proposed numerical method is not affected by the first time step.

Theorem 3.23 *Let $\hat{\boldsymbol{x}}^q$ be the numerical solution of (3.87) obtained by Newton's method for $q = 1, 2, \ldots, N$. Then, there exists a positive constant $\tilde{C}$, independent of both h (as defined in Theorem 3.22) and $\boldsymbol{\alpha}$, such that*

$$\|\boldsymbol{x}(t_q) - \hat{\boldsymbol{x}}^q\|_\infty \le \tilde{C}h^3. \tag{3.108}$$

Proof From Theorem 2.1 in [326], we note that there exists a well-defined number of iterations such that

$$\|\boldsymbol{x}^q - \hat{\boldsymbol{x}}^q\|_\infty = \begin{cases}\mathcal{O}(h_1^2), & \text{if } q = 1,\\ \mathcal{O}(h_q^3), & \text{if } q = 2, 3, \ldots, N,\end{cases} \tag{3.109}$$

where $h_1 \le h^{3/2}$ is as defined in Theorem 3.22. Then, (3.109) can be expressed as

$$\|\boldsymbol{x}^q - \hat{\boldsymbol{x}}^q\|_\infty = \mathcal{O}(h^3), \tag{3.110}$$

where h is as defined in Theorem 3.22.

Combining (3.96) and (3.110), we obtain

$$\|\boldsymbol{x}(t_q) - \hat{\boldsymbol{x}}^q\|_\infty \le \|\boldsymbol{x}(t_q) - \boldsymbol{x}^q\|_\infty + \|\boldsymbol{x}^q - \hat{\boldsymbol{x}}^q\|_\infty = \mathcal{O}(h^3).$$

Therefore, we confirm that (3.108) holds. The proof is complete. □

Remark 3.4 The error estimate (3.108) shows that for $q = 1, 2, \ldots, N$, the numerical solution $\hat{\boldsymbol{x}}^q$ obtained by Newton's method converges to the exact solution $\boldsymbol{x}(t_q)$ with third-order accuracy.

Numerical Examples

The following graded mesh is used to solve the example problems:

$$t_q = \left(\frac{q}{N}\right)^2 t_f, \quad q = 0, 1, \ldots, N,$$

where N is a positive integer. This graded mesh satisfies the condition in Theorem 3.2. The computed maximum error and convergence order are defined as

$$E_{\max} := E_h = \max_{\substack{i\in\{1,\ldots,n\}\\ q\in\{1,\ldots,N\}}} \left\{ \left| x_i(t_q) - \hat{x}_i^q \right| \right\},$$

$$\text{Order} := \log\left(\frac{E_h}{E_{h'}}\right) \Big/ \log\left(\frac{h}{h'}\right),$$

respectively, where h is as defined in Theorem 3.2 under the partition number N and h' is the maximum step size under the partition number N'.

Example 3.3 Consider the following nonlinear FODE [340]:

$$\begin{aligned} {}_0^C\mathrm{D}_t^{\alpha} x(t) &= \frac{\Gamma(4+\alpha)}{6} t^3 + t^{6+2\alpha} - x^2(t), \quad t \in (0, 1], \\ x(0) &= 0. \end{aligned}$$

The exact solution of this example is $x(t) = t^{3+\alpha}$. In contrast to the method proposed in [340] for directly approximating fractional derivatives and the method developed in [19], which employs the three-step Newton polynomial to approximate Volterra integrals, we use our proposed method and the MATLAB code provided in [19] to solve this problem for $\alpha = 0.3$, $\alpha = 0.6$, and $\alpha = 0.9$. The computed numerical results and those reported in [340] are listed in Table 3.3. From Table 3.3, we observe that the maximum error obtained by our method is significantly superior to those obtained in [19] and [340]. Furthermore, we note that our method exhibits a consistent convergence order close to 3 for various fractional orders, indicating its stability and efficiency. We observe that the convergence order obtained in [19] is also close to 3. However, the computational complexity in [19] is higher than that of our method. Specifically, for calculating x^q, the values of x^{q-3}, x^{q-2}, and x^{q-1}, $q = 3, 4, \ldots, N$, are required in [19], while the one-step Euler method and the two-step Adams-Bashforth method are used to obtain the values of x^1 and x^2, respectively. In contrast, our method only requires the values of x^{q-2} and x^{q-1} when calculating x^q for $q = 2, 3, \ldots, N$. The value of x^1 can be calculated by using the second-order Taylor expansion. Thus, it can be seen from Table 3.3 that the computational time in [19] is higher than that in our method. The state trajectories for different fractional orders with $N = 2048$ are depicted in Fig. 3.1. As it can be seen from Fig. 3.1, our method shows satisfactory accuracy.

Example 3.4 Consider the following nonlinear FODEs [247]:

$$\begin{cases} {}_0^C\mathrm{D}_t^{\alpha_1} x_1(t) = \frac{1}{2} x_1(t), \\ {}_0^C\mathrm{D}_t^{\alpha_2} x_2(t) = x_1^2(t) + x_2(t), \quad t \in (0, 1], \\ \boldsymbol{x}(0) = (1, 0)^\top. \end{cases}$$

Table 3.3 Maximum error and convergence order for Example 3.3

$\alpha = 0.3$	Results from [340]		Results from [19]		Our method	
N	$E_{\max}$	Order	$E_{\max}$	Order	$E_{\max}$	Order
128	1.0154e–06	–	1.6389e–06	–	4.1874e–07	–
256	1.4941e–07	2.7647	1.9683e–07	3.0577	5.4184e–08	2.9585
512	2.3628e–08	2.6607	2.3785e–08	3.0488	6.9524e–09	2.9665
1024	3.7049e–09	2.6729	2.8896e–09	3.0411	8.8676e–10	2.9730
2048	5.6279e–10	2.7187	3.5272e–10	3.0343	1.1261e–10	2.9783
$\alpha = 0.6$	Results from [340]		Results from [19]		Our method	
N	$E_{\max}$	Order	$E_{\max}$	Order	$E_{\max}$	Order
128	1.0167e–05	–	1.8931e–06	–	7.2056e–07	–
256	1.9494e–06	2.3828	2.3429e–07	3.0144	9.0807e–08	2.9967
512	3.7176e–07	2.3906	2.9087e–08	3.0098	1.1405e–08	2.9974
1024	7.0707e–08	2.3944	3.6190e–09	3.0067	1.4297e–09	2.9980
2048	1.3497e–08	2.3891	4.5095e–10	3.0046	1.7901e–10	2.9987
$\alpha = 0.9$	Results from [340]		Results from [19]		Our method	
N	$E_{\max}$	Order	$E_{\max}$	Order	$E_{\max}$	Order
128	9.1822e–05	–	2.7547e–06	–	1.0554e–06	–
256	2.1549e–05	2.0911	3.4432e–07	3.0001	1.3237e–07	3.0036
512	5.0416e–06	2.0957	4.3036e–08	3.0001	1.6574e–08	3.0018
1024	1.1777e–06	2.0978	5.3791e–09	3.0001	2.0735e–09	3.0009
2048	2.7492e–07	2.0989	6.7236e–10	3.0001	2.5929e–10	3.0005

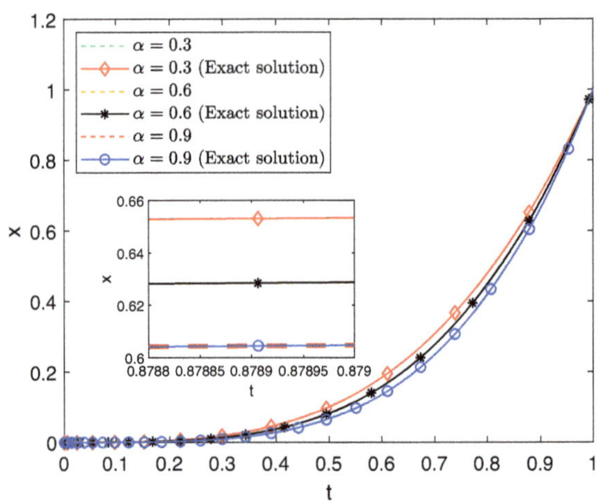

Fig. 3.1 State trajectories when $N = 2048$ for Example 3.3

Table 3.4 Maximum error and convergence order for Example 3.4

α	$(0.7, 0.9)^\top$		$(0.9, 0.7)^\top$		$(1.0, 1.0)^\top$	
N	$E_{\max}$	Order	$E_{\max}$	Order	$E_{\max}$	Order
10	3.6411e–03	–	4.9142e–03	–	1.4588e–03	–
20	5.0444e–04	2.9626	6.8817e–04	2.9465	1.9848e–04	2.9898
40	6.6335e–05	2.9816	9.0991e–05	2.9736	2.5849e–05	2.9959
80	8.5101e–06	2.9897	1.1702e–05	2.9861	3.2967e–06	2.9983
160	1.0751e–06	2.9983	1.4831e–06	2.9936	4.1619e–07	2.9993
320	1.3132e–07	3.0402	1.8506e–07	3.0094	5.2281e–08	2.9997
640	1.5305e–08	3.1045	2.1624e–08	3.1008	6.5512e–09	2.9998

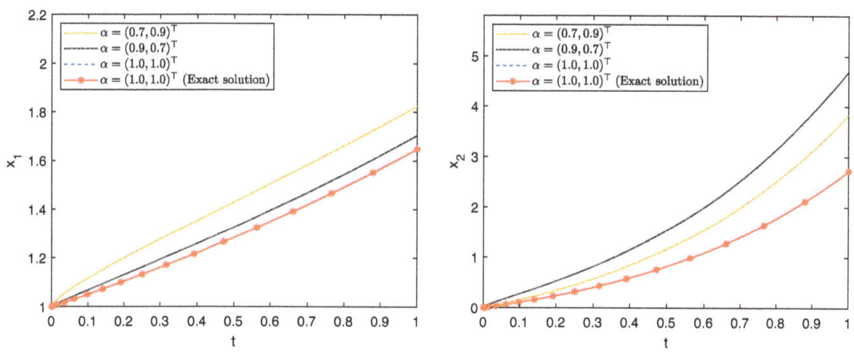

Fig. 3.2 State trajectories when $N = 640$ for Example 3.4

The exact solutions of this example are $x_1(t) = e^{\frac{t}{2}}$ and $x_2(t) = te^t$ for $\alpha = (1.0, 1.0)^\top$. We solve this example for $\alpha = (0.7, 0.9)^\top$, $\alpha = (0.9, 0.7)^\top$, and $\alpha = (1.0, 1.0)^\top$ by our proposed method. Since the exact solutions for the cases $\alpha = (0.7, 0.9)^\top$ and $\alpha = (0.9, 0.7)^\top$ are unknown, we use the numerical solutions for $N = 2560$ as approximations to the exact solutions. The computed maximum error, convergence order, and computational time are presented in Table 3.4. It is observed from Table 3.4 that the value of the maximum error decreases as the value of N increases. For different fractional orders, the computed convergence order of our method is roughly 3. To visualize the numerical results, we plot the state trajectories corresponding to different fractional orders in Fig. 3.2.

3.4 Conclusions

In this chapter, we have reviewed several fractional-order systems described by fractional ordinary differential equations and delay fractional differential equations. Some basic concepts and important properties have been introduced. Then, we have proposed two numerical approaches for solving fractional ordinary differential equations. The convergence rates of these numerical solution approaches have also been analyzed. Finally, numerical examples have been solved to verify the effectiveness and convergence rates of the proposed numerical solution approaches.

Chapter 4
Constrained Mathematical Programming

4.1 Introduction

The fractional optimal control problem can be reduced to a series of finite-dimensional optimization problems by an appropriate discretization scheme, each of which can be viewed as a mathematical programming problem. The solution to the optimal control problem may thus be obtained through solving a sequence of resulting mathematical programming problems, although the computational procedure is much more involved. Hence, understanding the fundamental concepts, theories, and methods of mathematical programming is obviously important.

To begin, we note that the notation used in this chapter is applicable only to this chapter. For example, subsequent appearance of $\boldsymbol{x}$ is not to be confused with the state vector in other chapters. There are already many excellent books on mathematical programming; see, for example, [18, 31, 34, 89, 191, 212, 236, 277, 293]. In this chapter, we summarize some essential concepts and results in constrained mathematical programming. Specifically, in next section, we introduce some basic concepts on constrained mathematical programming problems. Optimality conditions for constrained mathematical programming problems provide adequate theoretical tools to justify stopping criteria for solving a programming problem. In Sect. 4.3, we derive the first-order and second-order optimality conditions to characterize the solutions of the programming problem. This theory is useful for motivating algorithms, but to be truly efficient, optimization techniques, or algorithms, must be developed to find the solution to the programming problem. An important class of methods for constrained programming problem seeks the solution by replacing the original constrained problem by a sequence of unconstrained subproblems or a single unconstrained subproblem. In Sect. 4.4, we describe three important methods in this class, which are the quadratic penalty method, the log-barrier method, and the exact penalty approach. One of the most effective methods for nonlinearly constrained optimization generates steps by solving quadratic subproblems. In Sect. 4.5, we give

C. Liu et al., *Nonlinear Fractional Optimal Control*, Springer Optimization and Its Applications 237, https://doi.org/10.1007/978-981-92-0274-4_4

this sequential quadratic programming (SQP) approach, which is appropriate for small- and medium-sized general nonlinear constrained optimization problems.

4.2 Basic Concepts

As opposed to optimal control problems, mathematical programming problems are static in nature. The general constrained mathematical programming problem is described by

$$\min_{x \in \mathbb{R}^n} \; f(\boldsymbol{x}) \tag{4.1a}$$

$$\text{s.t.} \;\; c_i(\boldsymbol{x}) = 0, \quad i \in \mathcal{E}, \tag{4.1b}$$

$$c_i(\boldsymbol{x}) \geq 0, \quad i \in \mathcal{I}, \tag{4.1c}$$

where functions f and $c_i, i \in \mathcal{E} \cup \mathcal{I}$, are all smooth, real-valued functions on a subset of $\mathbb{R}^n$, and $\mathcal{E}$ and $\mathcal{I}$ are two finite sets of indices. f is the *objective function*, while c_i, $i \in \mathcal{E}$, are the *equality constraints* and c_i, $i \in \mathcal{I}$, are the *inequality constraints*. Note that, if $\mathcal{E} = \mathcal{I} = \emptyset$, the problem (4.1) is called an *unconstrained optimization problem*; if f and c_i are all linear functions, problem (4.1) is called a *linear programming problem*; if c_i, $i \in \mathcal{E} \cup \mathcal{I}$, are all linear functions and f is a quadratic function, problem (4.1) is called a *quadratic programming problem*; otherwise, we call problem (4.1) a *nonlinear programming problem*. We define the *feasible set Ω* to be the set of points $\boldsymbol{x}$ that satisfy the constraints, that is:

$$\Omega = \{\boldsymbol{x} \mid c_i(\boldsymbol{x}) = 0, i \in \mathcal{E}; c_i(\boldsymbol{x}) \geq 0, i \in \mathcal{I}\}, \tag{4.2}$$

so that we can write (4.1) more compactly as

$$\min_{x \in \Omega} \; f(\boldsymbol{x}), \tag{4.3}$$

which means that solution of problem (4.1) is just to find a point $\boldsymbol{x}$ on the feasible set Ω that minimizes the cost function f.

For problem (4.1), we give the following definitions of solutions.

Definition 4.1 A vector $\boldsymbol{x}^*$ is a local solution of problem (4.1) if $\boldsymbol{x}^* \in \Omega$ and there is a neighborhood $\mathcal{N}$ of $\boldsymbol{x}^*$ such that $f(\boldsymbol{x}) \geq f(\boldsymbol{x}^*)$ for $\boldsymbol{x} \in \mathcal{N} \cap \Omega$.

Definition 4.2 A vector $\boldsymbol{x}^*$ is a strict local solution of problem (4.1) if $\boldsymbol{x}^* \in \Omega$ and there is a neighborhood $\mathcal{N}$ of $\boldsymbol{x}^*$ such that $f(\boldsymbol{x}) > f(\boldsymbol{x}^*)$ for $\boldsymbol{x} \in \mathcal{N} \cap \Omega$ with $\boldsymbol{x} \neq \boldsymbol{x}^*$.

Definition 4.3 A vector $\boldsymbol{x}^*$ is a global solution of problem (4.1) if $\boldsymbol{x}^* \in \Omega$ and $f(\boldsymbol{x}) \geq f(\boldsymbol{x}^*)$ for all $\boldsymbol{x} \in \Omega$.

The following definition is an important terminology in the constrained mathematical programming.

Definition 4.4 The active set $\mathcal{A}(\boldsymbol{x})$ at any feasible $\boldsymbol{x}$ consists of the equality constraint indices from $\mathcal{E}$ together with the indices of the inequality constraints i for which $c_i(\boldsymbol{x}) = 0$, that is:

$$\mathcal{A}(\boldsymbol{x}) = \mathcal{E} \cup \{i \in \mathcal{I} |\ c_i(\boldsymbol{x}) = 0\}. \tag{4.4}$$

At a feasible point $\boldsymbol{x}$, the inequality constraint $i \in \mathcal{I}$ is said to be *active* if $c_i(\boldsymbol{x}) = 0$ and *inactive* if the strict inequality $c_i(\boldsymbol{x}) > 0$ is satisfied. Assume that $\mathcal{A}(\boldsymbol{x}^*)$ is the active set at $\boldsymbol{x}^*$. Then, it is enough to solve the following constrained problem:

$$\min_{\boldsymbol{x} \in \mathbb{R}^n} \quad f(\boldsymbol{x}) \tag{4.5}$$

$$\text{s.t.} \quad c_i(\boldsymbol{x}) = 0, \quad i \in \mathcal{A}(\boldsymbol{x}^*). \tag{4.6}$$

In general, it is easier to solve equality-constrained problem (4.5) and (4.6) than to solve the original problem (4.1).

4.3 Optimality Conditions

The principal tool used in analyzing the constrained mathematical programming problem is the Karush-Kuhn-Tucker theory. For completeness, we need the following preliminaries.

Definition 4.5 The point $\boldsymbol{x}^*$ is said to be a regular point of the constraints (4.1b) and (4.1c) if $\boldsymbol{x}^*$ satisfies all the constraints and the gradients of the equality and active inequality constraints

$$\{\nabla c_i(\boldsymbol{x}^*) |\ i \in \mathcal{A}(\boldsymbol{x}^*)\} \tag{4.7}$$

are linearly independent, where

$$\nabla c_i(\boldsymbol{x}^*) := \left. \frac{\partial c_i(\boldsymbol{x})}{\partial \boldsymbol{x}} \right|_{\boldsymbol{x}=\boldsymbol{x}^*}. \tag{4.8}$$

Note that condition (4.7) is known as a *linear independence constraint qualification*.

As a preliminary to stating the necessary conditions for $\boldsymbol{x}^*$ to be a local minimizer, we define the following *Lagrangian function* for problem (4.1):

$$L(\boldsymbol{x}, \lambda) = f(\boldsymbol{x}) - \sum_{i \in \mathcal{E} \cup \mathcal{I}} \lambda_i c_i(\boldsymbol{x}), \tag{4.9}$$

where $\lambda_i \in \mathbb{R}, i \in \mathcal{E} \cup \mathcal{I}$. Note that $\nabla_x L(\boldsymbol{x}, \boldsymbol{\lambda}) = \nabla f(\boldsymbol{x}) - \sum_{i \in \mathcal{E} \cup \mathcal{I}} \lambda_i \nabla c_i(\boldsymbol{x})$.

The conditions defined in the following theorem are called *first-order necessary optimality conditions*. These conditions are the foundation of many algorithms.

Theorem 4.1 (First-Order Necessary Conditions) *Suppose that* $\boldsymbol{x}^*$ *is a local solution of* (4.1) *and also a regular point for the constraints. Then, there is a Lagrange multiplier vector* $\boldsymbol{\lambda}^*$, *with components* $\lambda_i^*, i \in \mathcal{E} \cup \mathcal{I}$, *such that the following conditions are satisfied at* $(\boldsymbol{x}^*, \boldsymbol{\lambda}^*)$:

$$\nabla_x L(\boldsymbol{x}^*, \boldsymbol{\lambda}^*) = \boldsymbol{0}, \tag{4.10a}$$
$$c_i(\boldsymbol{x}^*) = 0, \quad \textit{for all } i \in \mathcal{E}, \tag{4.10b}$$
$$c_i(\boldsymbol{x}^*) \geq 0, \quad \lambda_i^* \geq 0, \quad \textit{for all } i \in \mathcal{I}, \tag{4.10c}$$
$$\lambda_i^* c_i(\boldsymbol{x}^*) = 0, \quad \textit{for all } i \in \mathcal{E} \cup \mathcal{I}. \tag{4.10d}$$

The conditions (4.10) are often known as the Karush-Kuhn-Tucker *conditions* or KKT *conditions* for short. In particular, condition (4.10a) is called a *stationary point condition*, conditions (4.10b) and (4.10c) are called the *feasible conditions*, and (4.10d) is called *complementarity condition* which implies that if the ith inequality constraint is inactive, then $\lambda_i = 0$, and conversely, if $\lambda_i > 0$, then the ith inequality constraint must be active. A special case of this complementarity, called *strict complementarity*, is defined as follows.

Definition 4.6 Given a local solution $\boldsymbol{x}^*$ of (4.1) and a vector $\boldsymbol{\lambda}^*$ satisfying (4.10), we say that the strict complementarity condition holds if exactly one of λ_i^* and $c_i(\boldsymbol{x}^*)$ is zero for each index $i \in \mathcal{I}$. In other words, we have that $\lambda_i^* > 0$ for each $i \in \mathcal{I} \cap \mathcal{A}(\boldsymbol{x}^*)$.

We now in a position to define some sets of feasible directions as follows.

Definition 4.7 Given a feasible point $\boldsymbol{x}$ and $\boldsymbol{0} \neq \boldsymbol{d} \in \mathbb{R}^n$, the set of feasible directions $\mathcal{F}(\boldsymbol{x})$ is

$$\mathcal{F}(\boldsymbol{x}) = \{\boldsymbol{d} \mid \boldsymbol{x} + t\boldsymbol{d} \in \Omega, \text{ for all } t \in [0, \delta]\}, \tag{4.11}$$

where $\delta > 0$ is a given real number.

Definition 4.8 Given a feasible point $\boldsymbol{x}$ and the active constraint set $\mathcal{A}(\boldsymbol{x})$, the set of linearized feasible directions $\mathcal{LF}(\boldsymbol{x})$ is

$$\mathcal{LF}(\boldsymbol{x}) = \{\boldsymbol{d} \mid \boldsymbol{d}^\top \nabla c_i(\boldsymbol{x}) = 0, \text{ for all } i \in \mathcal{E}, \boldsymbol{d}^\top \nabla c_i(\boldsymbol{x}) \geq 0, \text{ for all } i \in \mathcal{A}(\boldsymbol{x}) \cap \mathcal{I}\}. \tag{4.12}$$

Definition 4.9 Given a feasible point $\boldsymbol{x}$ and $\boldsymbol{0} \neq \boldsymbol{d} \in \mathbb{R}^n$, the set of sequential feasible directions $\mathcal{SF}(\boldsymbol{x})$ is

$$\mathcal{SF}(\boldsymbol{x}) = \{\boldsymbol{d} \mid \boldsymbol{x} + \delta_k \boldsymbol{d}_k \in \Omega, k = 1, 2, \ldots, \text{ with } \lim_{k \to \infty} \boldsymbol{d}_k = \boldsymbol{d} \text{ and } \lim_{k \to \infty} \delta_k = \delta\}. \tag{4.13}$$

With the above feasible directions, we present the following first-order sufficient conditions.

Theorem 4.2 (First-Order Sufficient Conditions) *Suppose that* $f(\boldsymbol{x})$ *and* $c_i(\boldsymbol{x})$, $i \in \mathcal{E} \cup \mathcal{I}$, *are differentiable at* $\boldsymbol{x}^* \in \Omega$. *If*

$$\boldsymbol{d}^\top \nabla f(\boldsymbol{x}^*) > 0, \quad \textit{for all } \boldsymbol{0} \neq \boldsymbol{d} \in \mathcal{SF}(\boldsymbol{x}^*), \tag{4.14}$$

then $\boldsymbol{x}^*$ *is a strict local solution of the problem (4.1).*

The next theorem gives necessary conditions involving the second derivatives.

Theorem 4.3 (Second-Order Necessary Conditions) *Suppose that* $\boldsymbol{x}^*$ *is a local solution of* (4.1) *and also a regular point for the constraints. Let* $\boldsymbol{\lambda}^*$ *be the Lagrange multiplier vector for which the* KKT *conditions* (4.10) *are satisfied. Then,*

$$\boldsymbol{d}^\top \nabla^2_{xx} L(\boldsymbol{x}^*, \boldsymbol{\lambda}^*)\boldsymbol{d} \geq 0, \quad \textit{for all } \boldsymbol{d} \in \mathcal{C}(\boldsymbol{x}^*, \boldsymbol{\lambda}^*), \tag{4.15}$$

where $\nabla^2_{xx} L(\boldsymbol{x}^*, \boldsymbol{\lambda}^*) = \nabla^2 f(\boldsymbol{x}^*) - \sum_{i \in \mathcal{E} \cup \mathcal{I}} \lambda_i^* \nabla^2 c_i(\boldsymbol{x}^*)$, *and*

$$\mathcal{C}(\boldsymbol{x}^*, \boldsymbol{\lambda}^*) := \{\boldsymbol{d} \in \mathcal{LF}(\boldsymbol{x}^*) |\ \nabla c_i(\boldsymbol{x}^*)^\top \boldsymbol{d} = 0, \textit{for all } i \in \mathcal{A}(\boldsymbol{x}^*) \cap \mathcal{I}, \lambda_i^* > 0\}. \tag{4.16}$$

The second-order sufficient conditions stated in the next theorem look very much like the necessary conditions just discussed, but it differs in that the constraint qualification is not required, and the inequality in (4.15) is replaced by a strict inequality.

Theorem 4.4 (Second-Order Sufficient Conditions) *Suppose that for some feasible point* $\boldsymbol{x}^* \in \mathbb{R}^n$, *there is a Lagrange multiplier vector* $\boldsymbol{\lambda}^*$ *such that the* KKT *conditions* (4.10) *are satisfied. Suppose also that*

$$\boldsymbol{d}^\top \nabla^2_{xx} L(\boldsymbol{x}^*, \boldsymbol{\lambda}^*)\boldsymbol{d} > 0, \quad \textit{for all } \boldsymbol{0} \neq \boldsymbol{d} \in \mathcal{C}(\boldsymbol{x}^*, \boldsymbol{\lambda}^*). \tag{4.17}$$

Then, $\boldsymbol{x}^*$ *is a strict local solution for* (4.1).

4.4 Penalty Methods

The penalty methods are an important class of methods for constrained mathematical programming problem (4.1). In this class of methods, the original constrained problem is replaced with a sequence of unconstrained subproblems or a single unconstrained problem. The constraints are appended to the objective function via a penalty parameter penalizing any violation of the constraints. By increasing the value of this penalty parameter, the method penalizes constraint violations more severely, and hence forcing the minimizer of the penalty function to move closer to the feasible region of the constrained problem.

The simplest and intuitive penalty method is the *quadratic penalty method*. We describe this approach first in the context of the equality-constrained problem:

$$\min_{\boldsymbol{x}\in\mathbb{R}^n} \; f(\boldsymbol{x}) \tag{4.18a}$$

$$\text{s.t.} \;\; c_i(\boldsymbol{x}) = 0, \quad i \in \mathcal{E}, \tag{4.18b}$$

which is a special case of (4.1). The quadratic penalty function $Q(\boldsymbol{x};\mu)$ for this formulation is

$$Q(\boldsymbol{x};\mu) = f(\boldsymbol{x}) + \frac{\mu}{2}\sum_{i\in\mathcal{E}} c_i^2(\boldsymbol{x}), \tag{4.19}$$

where $\mu > 0$ is the *penalty parameter*. By driving μ to ∞, we penalize the constraint violations with increasing severity. It makes good intuitive sense to consider a sequence of values $\{\mu_k\}$ with $\mu_k \to \infty$ as $k \to \infty$, and to seek the approximate minimizer $\boldsymbol{x}_k$ of $Q(\boldsymbol{x};\mu_k)$ for each k. For the general constrained problem (4.1), we can define the quadratic penalty function as

$$Q(\boldsymbol{x};\mu) = f(\boldsymbol{x}) + \frac{\mu}{2}\sum_{i\in\mathcal{E}} c_i^2(\boldsymbol{x}) + \frac{\mu}{2}\sum_{i\in\mathcal{I}} \Big([c_i(\boldsymbol{x})]_-\Big)^2, \tag{4.20}$$

where $[y]_-$ denotes $\max\{-y, 0\}$. In this case, Q may be less smooth than the objective and constraint functions. For instance, if one of the inequality constraints is $x_1 \geq 0$, then the function $\min^2\{0, x_1\}$ has a discontinuous second derivative, so that Q is no longer twice continuously differentiable.

We describe some convergence properties of the quadratic penalty method in the following two theorems. We restrict our attention to the equality-constrained problem (4.18), for which the quadratic penalty function is defined by (4.19). For the first result, we assume that the penalty function $Q(\boldsymbol{x};\mu_k)$ has a finite minimizer for each value of μ_k.

Theorem 4.5 *Suppose that each* $\boldsymbol{x}_k$ *is the exact global minimizer of* $Q(\boldsymbol{x};\mu_k)$ *defined by* (4.19), *and that* $\mu_k \to \infty$. *Then, every limit point* $\boldsymbol{x}^*$ *of the sequence* $\{\boldsymbol{x}_k\}$ *is a global solution of the problem* (4.18).

Since this result requires us to find the global minimizer for each subproblem, this desirable property of convergence to the global solution of (4.18) cannot be attained in general. The next result concerns convergence properties of the sequence $\{\boldsymbol{x}_k\}$ when we allow inexact (but increasingly accurate) minimizations of $Q(\cdot;\mu_k)$. We make the assumption that the stop test $\|\nabla_{\boldsymbol{x}} Q(\boldsymbol{x};\mu_k)\| \leq \tau_k$ is satisfied for all k.

Theorem 4.6 *Suppose that the tolerances and penalty parameters satisfy* $\tau_k \to 0$ *and* $\mu_k \to \infty$. *Then, if a limit point* $\boldsymbol{x}^*$ *of the sequence* $\{\boldsymbol{x}_k\}$ *is infeasible, then it is a stationary point of the function* $\|\boldsymbol{c}(\boldsymbol{x})\|^2$. *On the other hand, if a limit point* $\boldsymbol{x}^*$ *is* *feasible and also a regular point for the constraints, then* x^* *is a* KKT *point for the problem (4.18). For such points, we have for any infinite subsequence* $\mathcal{K}$ *such that* $\lim_{k\in\mathcal{K}} x_k = x^*$ *that*

$$\lim_{k\in\mathcal{K}} -\mu_k c_i(x_k) = \lambda_i^*, \quad \textit{for all } i \in \mathcal{E}, \tag{4.21}$$

where $\boldsymbol{\lambda}^*$ *is the multiplier vector that satisfies the* KKT *conditions* (4.10) *for the equality-constrained problem* (4.18).

If the penalty function takes values approaching $+\infty$ as x approaches the boundary of the feasible set Ω, it is called the *interior point penalty method* or *barrier function method*. We start by describing the concept of interior point penalty method in terms of inequality-constrained optimization problem:

$$\min_{x\in\mathbb{R}^n} \; f(x) \tag{4.22a}$$

$$\text{s.t.} \quad c_i(x) \geq 0, \quad i \in \mathcal{I}. \tag{4.22b}$$

The *strict feasible set* for problem (4.22) is defined by

$$\mathcal{F}^0 = \{x \in \mathbb{R}^n \mid c_i(x) > 0 \text{ for all } i \in \mathcal{I}\}, \tag{4.23}$$

and we assume that $\mathcal{F}^0$ is nonempty. The most important barrier function is the *logarithmic barrier function*. For the inequality-constrained problem (4.22), the combined objective/barrier function is given by

$$P(x;\mu) = f(x) - \frac{1}{\mu}\sum_{i\in\mathcal{I}} \log(c_i(x)), \tag{4.24}$$

where $1/\mu > 0$ is referred to as the *barrier parameter* and $\log(\cdot)$ denotes the natural logarithm. For the general constrained problem (4.1), one approach for dealing with equality constraints is to include quadratic penalty terms for these constraints in the objective. If we assume for simplicity that the coefficient of the quadratic penalty term is μ, where $1/\mu$ is the barrier parameter, then the log-barrier/quadratic penalty function to be minimized for each value of μ is

$$B(x;\mu) = f(x) - \frac{1}{\mu}\sum_{i\in\mathcal{I}} \log(c_i(x)) + \frac{\mu}{2}\sum_{i\in\mathcal{E}} c_i^2(x). \tag{4.25}$$

As for the logarithmic barrier function (4.24), we can show that the minimizer of $P(x;\mu)$, which we denote by $x(\mu)$, approaches a solution of (4.22) as $\mu \to \infty$, under certain conditions.

Theorem 4.7 *Suppose that* $\mathcal{F}^0$ *is nonempty and that* x^* *is a local solution of (4.22) at which the* KKT *conditions are satisfied for some* $\boldsymbol{\lambda}^*$. *Furthermore, suppose that the linear independence constraint qualification, the strict complementarity condition,*

and that the second-order sufficient conditions hold at $(\boldsymbol{x}^*, \boldsymbol{\lambda}^*)$. *Then, the following statements are true:*
(i) *There is a unique continuously differentiable vector function* $\boldsymbol{x}(\mu)$, *defined for all sufficiently large* μ *by the statement that* $\boldsymbol{x}(\mu)$ *is a local minimizer of* $P(\boldsymbol{x};\mu)$ *in some neighborhood of* $\boldsymbol{x}^*$, *such that* $\lim_{\mu\to\infty} \boldsymbol{x}(\mu) = \boldsymbol{x}^*$.
(ii) *For the function* $\boldsymbol{x}(\mu)$ *in (i), the Lagrange multiplier estimates* $\boldsymbol{\lambda}(\mu)$ *with* $\lambda_i(\mu) = \dfrac{1}{\mu c_i(\boldsymbol{x}(\mu))}$, $i \in \mathcal{I}$, *converge to* $\boldsymbol{\lambda}^*$ *as* $\mu \to \infty$.
(iii) *The Hessian* $\nabla^2_{\boldsymbol{xx}} P(\boldsymbol{x};\mu)$ *is positive definite for all sufficiently large* μ.

If we use an *exact penalty function*, it may be possible to solve general constrained problem (4.1) with one single unconstrained minimization. We now give a more precise definition of exactness for a penalty function. Since a penalty function always makes use of a scalar penalty parameter $\mu > 0$, we will write a general penalty function as $\phi(\boldsymbol{x};\mu)$, and assume that constraints are penalized by increasing μ.

Definition 4.10 A penalty function $\phi(\boldsymbol{x};\mu)$ is exact if there is a positive scalar μ^* such that for any $\mu > \mu^*$, any local solution of the constrained programming problem (4.1) is a local solution of $\phi(\boldsymbol{x};\mu)$.

A typical exact penalty function is defined by

$$\phi_1(\boldsymbol{x};\mu) := f(\boldsymbol{x}) + \mu \sum_{i\in\mathcal{E}} |c_i(\boldsymbol{x})| + \mu \sum_{i\in\mathcal{I}} [c_i(\boldsymbol{x})]_-. \tag{4.26}$$

For all sufficiently large, positive values of μ, one minimization of this unconstrained function will yield the solution of constrained problem (4.1). Nevertheless, it is difficult to determine μ a priori in most practical applications, and rules for adjusting this parameter during the course of the computation must be devised. In addition, minimization of $\phi_1(\boldsymbol{x};\mu)$ is made difficult by the fact that it is non-smooth. Thus, gradient-based optimization techniques are not applicable. The following result establishes the exactness of the penalty function (4.26).

Theorem 4.8 *Suppose that* $\boldsymbol{x}^*$ *is a strict local solution of the nonlinear programming problem (4.1) at which the first-order necessary conditions are satisfied, with Lagrange multipliers* $\boldsymbol{\lambda}^*$, $i \in \mathcal{E} \cup \mathcal{I}$. *Then,* $\boldsymbol{x}^*$ *is a local minimizer of* $\phi_1(\boldsymbol{x};\mu)$ *for all* $\mu > \mu^*$, *where*

$$\mu^* = \|\boldsymbol{\lambda}^*\|_\infty = \max_{i\in\mathcal{E}\cup\mathcal{I}} |\lambda_i^*|. \tag{4.27}$$

If, in addition, the second-order sufficient conditions hold and $\mu > \mu^*$, *then* $\boldsymbol{x}^*$ *is a strict local minimizer of* $\phi_1(\boldsymbol{x};\mu)$.

4.5 Sequential Quadratic Programming Algorithm

SQP is one of the most successful methods for the numerical solution of constrained nonlinear optimization problems. It relies on a profound theoretical foundation and provides powerful algorithmic tools for the solution of small- and medium-sized constrained optimization problems. The theory was initiated by Wilson [322] and was further developed by Han [113, 114], Powell [248].

We begin by considering the equality-constrained problem (4.18). The idea behind the SQP approach is to model (4.18) at the current iterate $\boldsymbol{x}_k$ by a quadratic programming subproblem, then uses the minimizer of this subproblem to define a new iterate $\boldsymbol{x}_{k+1}$. The challenge is to design the quadratic subproblem so that it yields a good step for the nonlinear optimization problem.

From (4.9), we know that the Lagrangian function for this problem is $L(\boldsymbol{x}, \boldsymbol{\lambda}) = f(\boldsymbol{x}) - \sum_{i \in \mathcal{E}} \lambda_i c_i(\boldsymbol{x})$. We use $\boldsymbol{A}(\boldsymbol{x})$ to denote the Jacobian matrix of the constraints, that is:

$$\boldsymbol{A}(\boldsymbol{x}) = (\nabla c_1(\boldsymbol{x}), \nabla c_2(\boldsymbol{x}), \ldots, \nabla c_{|\mathcal{E}|}(\boldsymbol{x}))^\top, \tag{4.28}$$

where $|\mathcal{E}|$ denotes the number of elements in $\mathcal{E}$. The first-order KKT conditions (4.10) of the equality-constrained problem (4.18) can be written as a system of $n + |\mathcal{E}|$ equations in the $n + |\mathcal{E}|$ unknowns $\boldsymbol{x}$ and $\boldsymbol{\lambda}$

$$\boldsymbol{F}(\boldsymbol{x}, \boldsymbol{\lambda}) = \begin{pmatrix} \nabla f(\boldsymbol{x}) - \boldsymbol{A}(\boldsymbol{x})^\top \boldsymbol{\lambda} \\ \boldsymbol{c}(\boldsymbol{x}) \end{pmatrix} = \boldsymbol{0}. \tag{4.29}$$

Any solution $(\boldsymbol{x}^*, \boldsymbol{\lambda}^*)$ of the equality-constrained problem (4.18) for which $\boldsymbol{A}(\boldsymbol{x}^*)$ has full rank satisfies (4.29). One approach that suggests itself is to solve the nonlinear equations (4.29) by using Newton's method.

The Jacobian of (4.29) with respect to $\boldsymbol{x}$ and $\boldsymbol{\lambda}$ is given by

$$\boldsymbol{F}'(\boldsymbol{x}, \boldsymbol{\lambda}) = \begin{pmatrix} \nabla_{\boldsymbol{xx}}^2 L(\boldsymbol{x}, \boldsymbol{\lambda}) & -\boldsymbol{A}(\boldsymbol{x})^\top \\ \boldsymbol{A}(\boldsymbol{x}) & \boldsymbol{0} \end{pmatrix}. \tag{4.30}$$

The Newton step from the iterate $(\boldsymbol{x}_k, \boldsymbol{\lambda}_k)$ is thus given by

$$\begin{pmatrix} \boldsymbol{x}_{k+1} \\ \boldsymbol{\lambda}_{k+1} \end{pmatrix} = \begin{pmatrix} \boldsymbol{x}_k \\ \boldsymbol{\lambda}_k \end{pmatrix} + \begin{pmatrix} \boldsymbol{p}_k \\ \boldsymbol{p}_{\lambda} \end{pmatrix}, \tag{4.31}$$

where $\boldsymbol{p}_k$ and $\boldsymbol{p}_{\lambda}$ solve the Newton–KKT system

$$\begin{pmatrix} \nabla_{\boldsymbol{xx}}^2 L_k & -\boldsymbol{A}_k^\top \\ \boldsymbol{A}_k & \boldsymbol{0} \end{pmatrix} \begin{pmatrix} \boldsymbol{p}_k \\ \boldsymbol{p}_{\lambda} \end{pmatrix} = \begin{pmatrix} -\nabla f_k + \boldsymbol{A}_k^\top \boldsymbol{\lambda}_k \\ -\boldsymbol{c}_k \end{pmatrix}. \tag{4.32}$$

This Newton iteration is well defined when the KKT matrix in (4.32) is non-singular. In fact, this matrix is non-singular if the following assumption holds at $(\boldsymbol{x}, \boldsymbol{\lambda}) = (\boldsymbol{x}_k, \boldsymbol{\lambda}_k)$.

Assumption 4.5.1 (a) The constraint Jacobian $\boldsymbol{A}(\boldsymbol{x})$ has full row rank and (b) the matrix $\nabla^2_{xx} L(\boldsymbol{x}, \boldsymbol{\lambda})$ satisfies that $\boldsymbol{d}^\top \nabla^2_{xx} L(\boldsymbol{x}, \boldsymbol{\lambda})\boldsymbol{d} > 0$ for all $\boldsymbol{d} \neq \boldsymbol{0}$ such that $\boldsymbol{A}^\top(\boldsymbol{x})\boldsymbol{d} = \boldsymbol{0}$.

The first assumption is the linear independence constraint qualification. The second condition holds whenever $(\boldsymbol{x}, \boldsymbol{\lambda})$ is close to the optimum $(\boldsymbol{x}^*, \boldsymbol{\lambda}^*)$ and the second-order sufficient condition is satisfied at the solution.

There is an alternative way to view the iterations (4.31) and (4.32). Suppose that at the iterate $(\boldsymbol{x}_k, \boldsymbol{\lambda}_k)$ we model problem (4.18) using the quadratic program

$$\min_{\boldsymbol{p}} \; f_k + \nabla f_k^\top \boldsymbol{p} + \frac{1}{2}\boldsymbol{p}^\top \nabla^2_{xx} L_k \boldsymbol{p} \tag{4.33a}$$

$$\text{s.t.} \quad \boldsymbol{A}_k \boldsymbol{p} + \boldsymbol{c}_k = \boldsymbol{0}. \tag{4.33b}$$

If Assumption 4.5.1 holds, this problem has a unique solution $(\boldsymbol{p}_k, \boldsymbol{l}_k)$ that satisfies

$$\nabla^2_{xx} L_k \boldsymbol{p}_k + \nabla f_k - \boldsymbol{A}_k^\top \boldsymbol{l}_k = \boldsymbol{0}, \tag{4.34a}$$

$$\boldsymbol{A}_k \boldsymbol{p}_k + \boldsymbol{c}_k = \boldsymbol{0}. \tag{4.34b}$$

The vectors $\boldsymbol{p}_k$ and $\boldsymbol{l}_k$ can be identified with the solution of the Newton equations (4.32). If we subtract $\boldsymbol{A}_k^\top \boldsymbol{\lambda}_k$ from both sides of the first equation in (4.32), we obtain

$$\begin{pmatrix} \nabla^2_{xx} L_k & -\boldsymbol{A}_k^\top \\ \boldsymbol{A}_k & \boldsymbol{0} \end{pmatrix} \begin{pmatrix} \boldsymbol{p}_k \\ \boldsymbol{\lambda}_{k+1} \end{pmatrix} = \begin{pmatrix} -\nabla f_k \\ -\boldsymbol{c}_k \end{pmatrix}. \tag{4.35}$$

Hence, by non-singularity of the coefficient matrix, we have that $\boldsymbol{\lambda}_{k+1} = \boldsymbol{l}_k$ and that $\boldsymbol{p}_k$ solves (4.33) and (4.32). The new iterate $(\boldsymbol{x}_{k+1}, \boldsymbol{\lambda}_{k+1})$ can therefore be defined either as the solution of the quadratic programming problem (4.33) or as the iterate generated by Newton's method (4.31), (4.32) applied to the optimality conditions of the problem.

The above method can be extended to the general constrained mathematical programming problem (4.1). To model this problem, we now linearize both the inequality and equality constraints to obtain

$$\min_{\boldsymbol{p}} \; f_k + \nabla f_k^\top \boldsymbol{p} + \frac{1}{2}\boldsymbol{p}^\top \nabla^2_{xx} L_k \boldsymbol{p} \tag{4.36a}$$

$$\text{s.t.} \quad \nabla c_i(\boldsymbol{x}_k)^\top \boldsymbol{p} + c_i(\boldsymbol{x}_k) = 0, \quad i \in \mathcal{E}, \tag{4.36b}$$

$$\nabla c_i(\boldsymbol{x}_k)^\top \boldsymbol{p} + c_i(\boldsymbol{x}_k) \geq 0, \quad i \in \mathcal{I}. \tag{4.36c}$$

We can use one of the algorithms for quadratic programming to solve this problem. The new iterate is given by $(\boldsymbol{x}_k + \boldsymbol{p}_k, \boldsymbol{\lambda}_{k+1})$ where $\boldsymbol{p}_k$ and $\boldsymbol{\lambda}_{k+1}$ are the solution and the corresponding Lagrange multiplier of (4.36). An SQP method for (4.1) is thus given with the modification that the step is computed from (4.36).

In this inequality-constrained quadratic program approach, the set of active constraints $\mathcal{A}_k$ at the solution of (4.36) constitutes our guess of the active set at the solution of the nonlinear program. If the SQP method is able to correctly identify this optimal active set (and not change its guess at a subsequent iteration), then it will act like a Newton method for equality-constrained optimization and will converge rapidly. The following result gives conditions under which this desirable behavior takes place.

Theorem 4.9 *Suppose that $\boldsymbol{x}^*$ is a local solution of (4.1) at which the* KKT *conditions are satisfied for some $\boldsymbol{\lambda}^*$. Furthermore, suppose that the linear independence constraint qualification, the strict complementarity condition, and the second-order sufficient conditions hold at $(\boldsymbol{x}^*, \boldsymbol{\lambda}^*)$. Then, if $(\boldsymbol{x}_k, \boldsymbol{\lambda}_k)$ is sufficiently close to $(\boldsymbol{x}^*, \boldsymbol{\lambda}^*)$, there is a local solution of the subproblem (4.36) whose active set $\mathcal{A}_k$ is the same as the active set $\mathcal{A}(\boldsymbol{x}^*)$ of the nonlinear program (4.1) at $\boldsymbol{x}^*$.*

There are other important details to develop a realistic and efficient SQP algorithm. For example, the QP approximation involves the Lagrangian Hessian matrix, which involves second derivatives. However, we may not evaluate the Hessian directly but approximate it using a quasi-Newton update, such as the following Broyden-Fletcher-Goldfarb-Shanno (BFGS) update [43, 88, 98, 282]:

$$H_{k+1} = H_k + \frac{\gamma_k(\gamma_k)^\top}{(\gamma_k)^\top \boldsymbol{p}_k} - \frac{H_k \boldsymbol{p}_k(\boldsymbol{p}_k)^\top H_k}{(\boldsymbol{p}_k)^\top H_k \boldsymbol{p}_k}, \tag{4.37}$$

where $\boldsymbol{p}_k = \boldsymbol{x}_{k+1} - \boldsymbol{x}_k$; $L(\boldsymbol{x}, \boldsymbol{\lambda}) = f(\boldsymbol{x}) - \sum_{i \in \mathcal{E} \cup \mathcal{I}} \lambda_i c_i(\boldsymbol{x})$; and $\gamma_k = \nabla_{\boldsymbol{x}} L_{k+1} - \nabla_{\boldsymbol{x}} L_k$. An implementable SQP algorithm is shown in Algorithm 4.1. Furthermore, to illustrate Algorithm 4.1, we give the following example:

Algorithm 4.1 An implementable SQP algorithm.

Step 1. Make a QP approximation to the original problem. For the first iteration, use a Lagrangian Hessian equal to the identity matrix.

Step 2. Solve for the optimum $\boldsymbol{p}$ to the QP problem. As part of this solution, values for the Lagrange multipliers $\boldsymbol{\lambda}$ are obtained.

Step 3. Execute a simple line search by first stepping to the optimum of the QP problem, i.e., $\boldsymbol{x} := \boldsymbol{x} + \boldsymbol{p}$. Let $\mathcal{V}$ be the set of violated constraint indices. Compute $\psi(\boldsymbol{x}; \boldsymbol{\lambda}) = f(\boldsymbol{x}) + \sum_{i \in \mathcal{V}} \lambda_i |c_i(\boldsymbol{x})|$. If $\psi(\boldsymbol{x}; \boldsymbol{\lambda})$ is reduced, then go to Step 4. Otherwise, cut back the step size until the value of $\psi(\boldsymbol{x}; \boldsymbol{\lambda})$ is reduced.

Step 4. Evaluate the Lagrangian gradient at the new point. Calculate the difference in $\boldsymbol{x}$ and in the Lagrangian gradient γ. Update the Lagrangian Hessian using the BFGS update (4.37).

Step 5. Return to Step 1 until $\boldsymbol{p}$ is sufficiently small.

Example 4.1 Find an $\boldsymbol{x} \in \mathbb{R}^2$ such that the objective function

$$f(\boldsymbol{x}) = x_1^4 - 2x_1^2x_2 + x_1^2 + x_2^2 - 2x_1 + 5$$

is minimized subject to

$$c(\boldsymbol{x}) = -(x_1 + 0.25)^2 + 0.75x_2 \geq 0.$$

The optimal solution of this example is $\boldsymbol{x}^* = (0.5, 0.75)^\top$. Suppose we start at the feasible point $\boldsymbol{x}_0 = (-1, 4)^\top$. The relevant values of functions and gradients are

$$f_0 = 17,\ c(\boldsymbol{x}_0) = 2.4375,\ \nabla f_0 = (8, 6)^\top,$$
$$\nabla c(\boldsymbol{x}_0) = (1.5, 0.75)^\top,\ \nabla_{\boldsymbol{xx}}^2 L_0 = \begin{pmatrix} 1 & 0 \\ 0 & 1 \end{pmatrix}.$$

Thus, the first QP problem is

$$\begin{aligned} \min_{\boldsymbol{p}} \ & 17 + 8p_1 + 6p_2 + \frac{1}{2}(p_1^2 + p_2^2) \\ \text{s.t.} \ & 1.5p_1 + 0.75p_2 + 2.4375 \geq 0. \end{aligned}$$

Solving the KKT conditions for the optimum yields

$$\boldsymbol{p} = (-0.5, -2.25)^\top, \text{ and } \lambda = 5.00.$$

The proposed step is $\boldsymbol{x}_1 = \boldsymbol{x}_0 + \boldsymbol{p} = (-1.5, 1.75)^\top$. Since $\psi(\boldsymbol{x}_1; 5) = 11.75 < \psi(\boldsymbol{x}_0; \lambda) = 17$, we accept this full step. Then, the relevant values of functions and gradients are

$$f_1 = 10.5,\ c(\boldsymbol{x}_1) = -0.25,\ \nabla f_1 = (-8, -1)^\top,$$
$$\nabla c(\boldsymbol{x}_1) = (2.5, 0.75)^\top,\ \nabla_{\boldsymbol{xx}}^2 L_1 = \begin{pmatrix} 17.752941 & 5.388235 \\ 5.388235 & 1.913725 \end{pmatrix}.$$

The second QP problem is

$$\begin{aligned} \min_{\boldsymbol{p}} \ & 10.5 - 8p_1 - p_2 + \frac{1}{2}(17.752941p_1^2 + 1.913725p_2^2 + 10.77647p_1p_2) \\ \text{s.t.} \ & 2.5p_1 + 0.75p_2 - 0.25 \geq 0. \end{aligned}$$

Solving the KKT conditions for the optimum yields

$$\boldsymbol{p} = (1.6145, -5.048)^\top, \text{ and } \lambda = -2.615.$$

Table 4.1 Computational results for Example 4.1

Iteration	$x^\top$	$f(x)$	$c(x)$
0	(–1.000000, 4.000000)	17.00000	2.437500
1	(–1.500000, 1.750000)	10.50000	–0.250000
2	(–0.496031, –0.815476)	7.367382	–0.672138
3	(–0.356526, –0.010818)	5.859187	–0.019461
4	(0.253197, –0.158073)	4.607078	–0.371763
5	(0.352204, 0.470464)	4.539642	–0.009802
6	(0.473993, 0.679112)	4.483200	–0.014832
7	(0.495464, 0.740341)	4.499439	–0.000460
8	(0.499890, 0.749754)	4.499973	–0.000019
9	(0.499999, 0.749998)	4.499999	0
10	(0.500000, 0.750000)	4.500000	0

However, since $\lambda = -2.615$ is negative, we need to drop the constraint in the second QP. With the constraint dropped, we obtain

$$\boldsymbol{p} = (2.007, -5.131)^\top, \text{ and } \lambda = 0.$$

This gives a new $\boldsymbol{x}$ of

$$\boldsymbol{x}_2 = \boldsymbol{x}_1 + \boldsymbol{p} = (0.507, -3.381)^\top.$$

Nevertheless, $\psi(\boldsymbol{x}_2; 0) = 17.478449 > \psi(\boldsymbol{x}_1; 5) = 11.75$. We cut the step back with 0.5 times of the original. Thus, the new value of $\boldsymbol{x}$ becomes

$$\boldsymbol{x}_2 = \boldsymbol{x}_1 + 0.5\boldsymbol{p} = (-0.4965, -0.8155)^\top.$$

At this point, $\psi(\boldsymbol{x}_2; 0) = 7.367382 < \psi(\boldsymbol{x}_1; 5) = 11.75$. So we accept this step.

Continue in this fashion until $\boldsymbol{p}$ goes to zero. The whole iteration steps and computed results are summarized in Table 4.1.

Chapter 5
Fractional Optimal Control Problems

5.1 Introduction

In this chapter, we consider necessary optimality conditions, exact penalization, and numerical solution for a general optimal control problem with Caputo fractional derivatives. As stated before, a fractional optimal control problem (FOCP) is an optimal control problem in which the dynamic constraints or the performance index or both of them include at least one fractional derivative term. Many phenomena in engineering, physics, and mathematical finance can be formulated as FOCPs [46, 64, 201–203, 278].

Most of FOCPs of practical importance are not solvable analytically and thus numerical solutions to such problems are usually sought in practice. Recently, development of numerical methods for FOCPs has attracted much attention; see, for example, [12, 13, 36, 225, 252, 253, 274, 298, 299]. However, unlike for conventional optimal control problems, computationally efficient and accurate numerical methods for FOCPs are scarce. Moreover, penalty methods have also been widely used for solving constrained optimization and conventional optimal control problems [72, 167, 169, 313, 315, 323, 324]. In particular, exact penalty methods are very effective approaches for solving constrained optimization and optimal control problems. In such an approach, a constrained optimization problem is approximated by an unconstrained one containing a penalty term with a penalty constant and when the penalty parameter is sufficiently large, both the constrained and penalty problems have the same set of solutions. Although exact penalty methods have been used extensively in solving constrained optimization and conventional optimal control problems, e.g., [72, 167, 323, 324], whether there is any exact penalty method for constrained FOCPs still remains an open problem. In next section, the problem statement for the general constrained FOCP is given. In Sect. 5.3, we discuss the necessary optimality conditions and exact penalty method. In Sect. 5.3.1, we derive the corresponding first-order necessary optimality conditions by using calculus of variation. In Sect. 5.3.2, based on the necessary optimality conditions, we prove that a

C. Liu et al., *Nonlinear Fractional Optimal Control*, Springer Optimization and Its Applications 237, https://doi.org/10.1007/978-981-92-0274-4_5

low-order power penalty method is an exact penalty method for the constrained FOCP by showing that an optimal solution to the penalized problem satisfies the optimality conditions of the original one when the penalty parameter is sufficiently large. In Sect. 5.3.3, an example is provided to verify the necessary optimality conditions. In Sect. 5.4, we develop numerical solution method for the FOCP. In Sect. 5.4.1, based on the second-order numerical integration scheme and the trapezoidal rule, we propose a discretization approach for the FOCP. In Sect. 5.4.2, we develop a gradient-based optimization method to solve the discretized problem. In Sect. 5.4.3, two example problems are solved to illustrate the effectiveness of the proposed optimization method. Finally, we provide some conclusions in Sect. 5.5.

The main references of this chapter are [170, 314].

5.2 Problem Statement

Consider the following constrained FOCP:

$$\min_{\boldsymbol{u}\in\mathcal{U}} \; g_0(\boldsymbol{u}) = \Phi(\boldsymbol{x}(T)) + \int_0^T L(t, \boldsymbol{x}(t), \boldsymbol{u}(t))\, dt, \tag{5.1}$$

$$\text{s.t.} \quad \begin{cases} {}_0^C\mathrm{D}_t^{\boldsymbol{\alpha}}\, \boldsymbol{x}(t) = \boldsymbol{f}(t, \boldsymbol{x}(t), \boldsymbol{u}(t)), & t \in (0, T], \\ \boldsymbol{x}(0) = \boldsymbol{x}^0, \end{cases} \tag{5.2}$$

$$\boldsymbol{g}(t, \boldsymbol{x}(t), \boldsymbol{u}(t)) \le \boldsymbol{0}, \; t \in [0, T] \text{ almost everywhere}, \tag{5.3}$$

where $\boldsymbol{x}(t) = (x_1(t), x_2(t), \ldots, x_n(t))^\top \in \mathbb{R}^n$ is the state vector; $\boldsymbol{u}(t) = (u_1(t), u_2(t), \ldots, u_m(t))^\top \in \mathbb{R}^m$ is the control vector; $T > 0$ is a fixed constant representing the terminal time; $\boldsymbol{f} = (f_1, f_2, \ldots, f_n)^\top$, L, Φ and $g = (g_1, g_2, \ldots, g_p)^\top$ are given functions; $\boldsymbol{x}^0 \in \mathbb{R}^n$ is a given vector; $\mathcal{U}$ is the set of all bounded piecewise continuous functions on $[0, T]$; and

$${}_0^C\mathrm{D}_t^{\boldsymbol{\alpha}}\, \boldsymbol{x}(t)s = \left({}_0^C\mathrm{D}_t^{\alpha_1}\, x_1(t),\; {}_0^C\mathrm{D}_t^{\alpha_2}\, x_2(t), \ldots,\; {}_0^C\mathrm{D}_t^{\alpha_n}\, x_n(t)\right)^\top$$

with ${}_0^C\mathrm{D}_t^{\alpha_i}\, x_i(t)$ denoting the Caputo derivative of $x_i(t)$.

Before further discussion, we assume throughout this chapter that the following conditions are satisfied.

Assumption 5.2.1 For any $i = 1, 2, \ldots, n$, f_i is twice continuously differentiable with respect to all its arguments.

Assumption 5.2.2 The functions L and Φ are twice continuously differentiable with respect to all its arguments.

Assumption 5.2.3 The constraint function $\boldsymbol{g}$ is continuously differentiable with respect to all its arguments.

5.3 Necessary Optimality Conditions and Exact Penalization

5.3.1 *Necessary Optimality Conditions*

We establish the necessary conditions of optimality for the constrained FOCP (5.1)–(5.3) as given in the following theorem.

Theorem 5.1 *Let* $\boldsymbol{u}^*$ *be a minimizer and* $\boldsymbol{x}^*$ *be the corresponding optimal state of problem (5.1)–(5.3). Then, there exist optimal co-state* $\boldsymbol{\lambda}^*(t) = (\lambda_1^*(t), \lambda_2^*(t), \ldots, \lambda_n^*(t))^\top$ *and multiplier* $\boldsymbol{\mu}^*(t) = (\mu_1^*(t), \mu_2^*(t), \ldots, \mu_p^*(t))^\top$ *such that* $\boldsymbol{x}^*$, $\boldsymbol{u}^*$, $\boldsymbol{\lambda}^*$, *and* $\boldsymbol{\mu}^*$ *satisfy the following conditions:*

$$\begin{cases} {}_t^C\mathrm{D}_T^{\boldsymbol{\alpha}}\, \boldsymbol{\lambda}^*(t) = \left.\dfrac{\partial H}{\partial \boldsymbol{x}}\right|_{(t,\boldsymbol{x}^*(t),\boldsymbol{u}^*(t),\boldsymbol{\lambda}^*(t),\boldsymbol{\mu}^*(t))}, \\ {}_0^R\mathrm{D}_t^{\boldsymbol{\alpha}-1}\, \boldsymbol{\lambda}^*(t)\Big|_{t=T} = \left.\dfrac{\partial \Phi}{\partial \boldsymbol{x}}\right|_{(\boldsymbol{x}^*(T))}, \\ \left.\dfrac{\partial H}{\partial \boldsymbol{u}}\right|_{(t,\boldsymbol{x}^*(t),\boldsymbol{u}^*(t),\boldsymbol{\lambda}^*(t),\boldsymbol{\mu}^*(t))} = \boldsymbol{0}, \end{cases} \tag{5.4}$$

$$\begin{cases} {}_0^C\mathrm{D}_t^{\boldsymbol{\alpha}}\, \boldsymbol{x}^*(t) = \boldsymbol{f}(t, \boldsymbol{x}^*(t), \boldsymbol{u}^*(t)), \\ \boldsymbol{x}^*(0) = \boldsymbol{x}^0, \\ \boldsymbol{g}(t, \boldsymbol{x}^*(t), \boldsymbol{u}^*(t)) \le \boldsymbol{0}, \\ \boldsymbol{\mu}^*(t) \ge \boldsymbol{0}, \\ \boldsymbol{\mu}^{*\top}(t)\boldsymbol{g}(t, \boldsymbol{x}^*(t), \boldsymbol{u}^*(t)) = 0 \end{cases} \tag{5.5}$$

for $t \in [0, T]$ *almost everywhere, where*

$$\begin{aligned} &H(t, \boldsymbol{x}(t), \boldsymbol{u}(t), \boldsymbol{\lambda}(t), \boldsymbol{\mu}(t)) \\ &= L(t, \boldsymbol{x}(t), \boldsymbol{u}(t)) + \boldsymbol{\lambda}^\top(t)\boldsymbol{f}(t, \boldsymbol{x}(t), \boldsymbol{u}(t)) + \boldsymbol{\mu}^\top(t)\boldsymbol{g}(t, \boldsymbol{x}(t), \boldsymbol{u}(t)) \end{aligned} \tag{5.6}$$

is the Hamiltonian function of the problem and $\boldsymbol{\alpha} - 1 = (\alpha_1 - 1, \ldots, \alpha_n - 1)^\top$.

Proof Let us consider the following Lagrangian function for FOCP (5.1)–(5.3):

$$\begin{aligned} \bar{g}_0(t, \boldsymbol{x}, \boldsymbol{x}(T), \boldsymbol{u}, \boldsymbol{\lambda}, \boldsymbol{\mu}) = \int_0^T \Big\{ &L(t, \boldsymbol{x}(t), \boldsymbol{u}(t)) + \boldsymbol{\lambda}^\top(t)[\boldsymbol{f}(t, \boldsymbol{x}(t), \boldsymbol{u}(t)) \\ &- {}_0^C\mathrm{D}_t^{\boldsymbol{\alpha}}\, \boldsymbol{x}(t)] + \boldsymbol{\mu}^\top(t)\boldsymbol{g}(t, \boldsymbol{x}(t), \boldsymbol{u}(t)) \Big\} dt + \Phi(\boldsymbol{x}(T)) \end{aligned}$$

with $\boldsymbol{x}(0) = \boldsymbol{x}^0$, where $\boldsymbol{\lambda}(t) = (\lambda_1(t), \lambda_2(t), \ldots, \lambda_n(t))^\top$ and $\boldsymbol{\mu}(t) = (\mu_1(t), \mu_2(t), \ldots, \mu_p(t))^\top$ are multipliers or dual variables of problem (5.1)–(5.3). The multiplier $\boldsymbol{\lambda}$

is also called the co-state variable in the optimal control context. The KKT conditions for an optimal point $(\boldsymbol{x}^*, \boldsymbol{x}^*(T), \boldsymbol{u}^*, \boldsymbol{\lambda}^*, \boldsymbol{\mu}^*)$ of $\bar{g}_0(t, \boldsymbol{x}, \boldsymbol{x}(T), \boldsymbol{u}, \boldsymbol{\lambda}, \boldsymbol{\mu})$ are

$$\delta\bar{g}_0(t, \boldsymbol{x}^*(t), \boldsymbol{x}^*(T), \boldsymbol{u}^*(t), \boldsymbol{\lambda}^*(t), \boldsymbol{\mu}^*(t)) = 0, \tag{5.7}$$

$$\boldsymbol{g}(t, \boldsymbol{x}^*(t), \boldsymbol{u}^*(t)) \leq \boldsymbol{0}, \tag{5.8}$$

$$\boldsymbol{\mu}^*(t) \geq \boldsymbol{0}, \tag{5.9}$$

$$\boldsymbol{\mu}^{*\top}(t)\boldsymbol{g}(t, \boldsymbol{x}^*(t), \boldsymbol{u}^*(t)) = 0 \tag{5.10}$$

for $t \in [0, T]$ almost everywhere, where $\delta\bar{g}_0$ denotes the variation of $\bar{g}_0$ with respect to $\boldsymbol{x}, \boldsymbol{x}(T), \boldsymbol{u}$, and $\boldsymbol{\lambda}$.

We now determine the variation of $\bar{g}_0(t, \boldsymbol{x}, \boldsymbol{x}(T), \boldsymbol{u}, \boldsymbol{\lambda}, \boldsymbol{\mu})$ with respect to $\boldsymbol{x}, \boldsymbol{u}, \boldsymbol{\lambda}$ and $\boldsymbol{x}(T)$. To achieve this, we let

$$\bar{g}_0(t, \boldsymbol{x}, \boldsymbol{x}(T), \boldsymbol{u}, \boldsymbol{\lambda}, \boldsymbol{\mu}) = \bar{g}_0^1(t, \boldsymbol{x}, \boldsymbol{x}(T), \boldsymbol{u}, \boldsymbol{\lambda}, \boldsymbol{\mu}) - \bar{g}_0^2(t, \boldsymbol{x}, \boldsymbol{x}(T), \boldsymbol{u}, \boldsymbol{\lambda}, \boldsymbol{\mu}), \tag{5.11}$$

where

$$\begin{aligned}\bar{g}_0^1(t, \boldsymbol{x}, \boldsymbol{x}(T), \boldsymbol{u}, \boldsymbol{\lambda}, \boldsymbol{\mu}) = \int_0^T \Big\{ & L(t, \boldsymbol{x}(t), \boldsymbol{u}(t)) + \boldsymbol{\lambda}^\top(t)\boldsymbol{f}(t, \boldsymbol{x}(t), \boldsymbol{u}(t)) \\ & + \boldsymbol{\mu}^\top(t)\boldsymbol{g}(t, \boldsymbol{x}(t), \boldsymbol{u}(t))\Big\} dt + \Phi(\boldsymbol{x}(T)), \\ \bar{g}_0^2(t, \boldsymbol{x}, \boldsymbol{x}(T), \boldsymbol{u}, \boldsymbol{\lambda}, \boldsymbol{\mu}) = \int_0^T & \boldsymbol{\lambda}^\top(t)\big[{}_0^C\mathrm{D}_t^\alpha \boldsymbol{x}(t)\big] dt.\end{aligned}$$

The variation of $\bar{g}_0^1(t, \boldsymbol{x}, \boldsymbol{x}(T), \boldsymbol{u}, \boldsymbol{\lambda}, \boldsymbol{\mu})$ with respect to $\boldsymbol{x}, \boldsymbol{x}(T), \boldsymbol{u}$, and $\boldsymbol{\lambda}$ is given as follows:

$$\begin{aligned}&\delta\bar{g}_0^1(t, \boldsymbol{x}, \boldsymbol{x}(T), \boldsymbol{u}, \boldsymbol{\lambda}, \boldsymbol{\mu}) \\ &= \int_0^T \Big[\Big(\frac{\partial L}{\partial \boldsymbol{x}}\Big)^\top + \boldsymbol{\lambda}^\top(t)\frac{\partial \boldsymbol{f}}{\partial \boldsymbol{x}} + \boldsymbol{\mu}^\top(t)\frac{\partial \boldsymbol{g}}{\partial \boldsymbol{x}}\Big]\Big|_{(t,\boldsymbol{x}(t),\boldsymbol{u}(t))} \delta\boldsymbol{x}(t)\, dt \\ &+ \int_0^T \Big\{\Big[\Big(\frac{\partial L}{\partial \boldsymbol{u}}\Big)^\top + \boldsymbol{\lambda}^\top(t)\frac{\partial \boldsymbol{f}}{\partial \boldsymbol{u}} + \boldsymbol{\mu}^\top(t)\frac{\partial \boldsymbol{g}}{\partial \boldsymbol{u}}\Big]\Big|_{(t,\boldsymbol{x}(t),\boldsymbol{u}(t))} \delta\boldsymbol{u}(t) \\ &+ \boldsymbol{f}^\top(t, \boldsymbol{x}(t), \boldsymbol{u}(t))\delta\boldsymbol{\lambda}(t)\Big\} dt + \Big(\frac{\partial \Phi}{\partial \boldsymbol{x}}\Big|_{(\boldsymbol{x}(T))}\Big)^\top \delta\boldsymbol{x}(T).\end{aligned} \tag{5.12}$$

We now calculate the variation of $\bar{g}_0^2(t, \boldsymbol{x}, \boldsymbol{x}(T), \boldsymbol{u}, \boldsymbol{\lambda}, \boldsymbol{\mu})$. It is easy to see that the variation of $\bar{g}_0^2(t, \boldsymbol{x}, \boldsymbol{x}(T), \boldsymbol{u}, \boldsymbol{\lambda}, \boldsymbol{\mu})$ with respect to $\boldsymbol{\lambda}$ is

$$\delta_{\boldsymbol{\lambda}}\bar{g}_0^2(t, \boldsymbol{x}, \boldsymbol{x}(T), \boldsymbol{u}, \boldsymbol{\lambda}, \boldsymbol{\mu}) = \int_0^T \big[{}_0^C\mathrm{D}_t^\alpha \boldsymbol{x}(t)\big]^\top \delta\boldsymbol{\lambda}(t) dt. \tag{5.13}$$

Since $\bar{g}_0^2$ does not contain $\boldsymbol{u}$ explicitly, $\delta_{\boldsymbol{u}}\bar{g}_0^2(t, \boldsymbol{x}, \boldsymbol{x}(T), \boldsymbol{u}, \boldsymbol{\lambda}, \boldsymbol{\mu}) = 0$. To determine the variation of $\bar{g}_0^2(t, \boldsymbol{x}, \boldsymbol{x}(T), \boldsymbol{u}, \boldsymbol{\lambda}, \boldsymbol{\mu})$ with respect to $\boldsymbol{x}$ and $\boldsymbol{x}(T)$, we rewrite it, using (2.25) and (2.32), as the following equivalent form:

$$\begin{aligned}
&\bar{g}_0^2(t, \boldsymbol{x}, \boldsymbol{x}(T), \boldsymbol{u}, \boldsymbol{\lambda}, \boldsymbol{\mu}) \\
&= \int_0^T \left\{\boldsymbol{x}^\top(t)\,{}_t^C\mathrm{D}_T^\alpha\boldsymbol{\lambda}(t) - \boldsymbol{\lambda}^\top(t)\,{}_0^R\mathrm{D}_t^\alpha\boldsymbol{x}^0\right\} dt + \left[\boldsymbol{x}^\top(t)\,{}_0^R\mathrm{D}_t^{\alpha-1}\boldsymbol{\lambda}(t)\right]_0^T \\
&= \int_0^T \left\{\boldsymbol{x}^\top(t)\,{}_t^C\mathrm{D}_T^\alpha\boldsymbol{\lambda}(t) - \boldsymbol{\lambda}^\top(t)\,{}_0^R\mathrm{D}_t^\alpha\boldsymbol{x}^0\right\} dt + \boldsymbol{x}^\top(T)\,{}_0^R\mathrm{D}_T^{\alpha-1}\boldsymbol{\lambda}(T) \\
&\quad - \boldsymbol{x}^{0\top}\,{}_0^R\mathrm{D}_0^{\alpha-1}\boldsymbol{\lambda}(0).
\end{aligned} \tag{5.14}$$

We note that $\boldsymbol{x}^{0\top}\,{}_0^R\mathrm{D}_0^{\alpha-1}\boldsymbol{\lambda}(0)$ is independent of $\boldsymbol{x}(t)$ and $\boldsymbol{x}(T)$. In fact, from Definition 2.4, we have, for $i = 1, 2, \ldots, n$,

$${}_0^R\mathrm{D}_0^{\alpha_i-1}\lambda_i(0) = \lim_{\varepsilon\to 0+} \frac{1}{\Gamma(1-\alpha_i)} \int_0^\varepsilon \frac{\lambda_i(t)}{(\varepsilon - t)^{\alpha_i}} dt.$$

Since the co-state λ_i should be bounded on $[0, T]$ and α_i is a fixed constant in $(0, 1)$ for any feasible i, the limit of the above equality can be estimated as follows.

$$\begin{aligned}
\left|\frac{1}{\Gamma(1-\alpha_i)} \int_0^\varepsilon \frac{\lambda_i(t)dt}{(\varepsilon - t)^{\alpha_i}}\right| &\le \frac{M}{\Gamma(1-\alpha_i)} \int_0^\varepsilon \frac{dt}{(\varepsilon - t)^{\alpha_i}} \\
&= \frac{M\varepsilon^{1-\alpha_i}}{\Gamma(2-\alpha_i)} \to 0
\end{aligned} \tag{5.15}$$

as $\varepsilon \to 0^+$, where M denotes a sufficiently large positive constant. Thus,

$${}_0^R\mathrm{D}_0^{\alpha_i-1}\lambda_i(0) = 0, \quad i = 1, 2, \ldots, n, \tag{5.16}$$

when $\alpha_i \in (0, 1)$. Clearly, if $\alpha_i = 1$, ${}_0^R\mathrm{D}_0^{\alpha_i-1}\lambda_i(0) = \lambda_i(0)$ for $i = 1, \ldots, n$. Using the above result, we see that (5.14) reduces to

$$\begin{aligned}
\bar{g}_0^2(t, \boldsymbol{x}, \boldsymbol{x}(T), \boldsymbol{u}, \boldsymbol{\lambda}, \boldsymbol{\mu}) = &\int_0^T \left\{\boldsymbol{x}^\top(t)\,{}_t^C D_T^\alpha\boldsymbol{\lambda}(t) - \boldsymbol{\lambda}^\top(t)\,{}_0^R\mathrm{D}_t^\alpha\boldsymbol{x}^0\right\} dt \\
&+ \boldsymbol{x}^\top(T)\,{}_0^R\mathrm{D}_T^{\alpha-1}\boldsymbol{\lambda}(T).
\end{aligned} \tag{5.17}$$

Using (5.17), the variations of $\bar{g}_0^2(t, \boldsymbol{x}, \boldsymbol{x}(T), \boldsymbol{u}, \boldsymbol{\lambda}, \boldsymbol{\mu})$ with respect to $\boldsymbol{x}$ and $\boldsymbol{x}(T)$ are, respectively, given by

$$\begin{aligned}
\delta_{\boldsymbol{x}}\bar{g}_0^2(t, \boldsymbol{x}, \boldsymbol{x}(T), \boldsymbol{u}, \boldsymbol{\lambda}, \boldsymbol{\mu}) &= \int_0^T \left[{}_t^C\mathrm{D}_T^\alpha\boldsymbol{\lambda}(t)\right]^\top \delta\boldsymbol{x}(t)dt, \\
\delta_{\boldsymbol{x}(T)}\bar{g}_0^2(t, \boldsymbol{x}, \boldsymbol{x}(T), \boldsymbol{u}, \boldsymbol{\lambda}, \boldsymbol{\mu}) &= \left[{}_0^R\mathrm{D}_T^{\alpha-1}\boldsymbol{\lambda}(T)\right]^\top \delta\boldsymbol{x}(T),
\end{aligned}$$

since ${}_{0}^{R}\mathrm{D}_{t}^{\alpha}\boldsymbol{x}^{0}$ is independent of both $\boldsymbol{x}$ and $\boldsymbol{x}(T)$. Combining the above results with (5.13), we have

$$\delta\bar{g}_{0}^{2}(t,\boldsymbol{x},\boldsymbol{x}(T),\boldsymbol{u},\boldsymbol{\lambda},\boldsymbol{\mu}) = \int_{0}^{T}\Big\{\big[{}_{0}^{C}\mathrm{D}_{t}^{\alpha}\boldsymbol{x}(t)\big]^{\top}\delta\boldsymbol{\lambda}(t) + \big[{}_{t}^{C}\mathrm{D}_{T}^{\alpha}\boldsymbol{\lambda}(t)\big]^{\top}\delta\boldsymbol{x}(t)\Big\}dt \\ + \big[{}_{0}^{R}\mathrm{D}_{T}^{\alpha-1}\boldsymbol{\lambda}(T)\big]^{\top}\delta\boldsymbol{x}(T). \tag{5.18}$$

Finally, from (5.11), (5.12), and (5.18), we obtain

$$\begin{aligned}
&\delta\bar{g}_{0}(t,\boldsymbol{x},\boldsymbol{x}(T),\boldsymbol{u},\boldsymbol{\lambda},\boldsymbol{\mu})\\
&= \int_{0}^{T}\Big[\Big(\frac{\partial L}{\partial \boldsymbol{x}}\Big)^{\top} + \boldsymbol{\lambda}^{\top}(t)\frac{\partial \boldsymbol{f}}{\partial \boldsymbol{x}} - {}_{t}^{C}\mathrm{D}_{T}^{\alpha}\boldsymbol{\lambda}^{\top}(t) + \boldsymbol{\mu}^{\top}(t)\frac{\partial \boldsymbol{g}}{\partial \boldsymbol{x}}\Big]\Big|_{(t,\boldsymbol{x}(t),\boldsymbol{u}(t))}\delta\boldsymbol{x}(t)dt\\
&+ \Big[\frac{\partial \Phi}{\partial \boldsymbol{x}}\Big|_{(\boldsymbol{x}(T))} - {}_{0}^{R}\mathrm{D}_{T}^{\alpha-1}\boldsymbol{\lambda}(T)\Big]^{\top}\delta\boldsymbol{x}(T)\\
&+ \int_{0}^{T}\Big\{\Big[\Big(\frac{\partial L}{\partial \boldsymbol{u}}\Big)^{\top} + \boldsymbol{\lambda}^{\top}(t)\frac{\partial \boldsymbol{f}}{\partial \boldsymbol{u}} + \boldsymbol{\mu}^{\top}(t)\frac{\partial \boldsymbol{g}}{\partial \boldsymbol{u}}\Big]\Big|_{(t,\boldsymbol{x}(t),\boldsymbol{u}(t))}\delta\boldsymbol{u}(t)\\
&+ \big[\boldsymbol{f}(t,\boldsymbol{x}(t),\boldsymbol{u}(t)) - {}_{0}^{C}\mathrm{D}_{t}^{\alpha}\boldsymbol{x}(t)\big]^{\top}\delta\boldsymbol{\lambda}(t)\Big\}dt.
\end{aligned}$$

Using (5.6), we further rewrite the above equality as

$$\begin{aligned}
&\delta\bar{g}_{0}(t,\boldsymbol{x},\boldsymbol{x}(T),\boldsymbol{u},\boldsymbol{\lambda},\boldsymbol{\mu})\\
&= \int_{0}^{T}\Big[\frac{\partial H(t,\boldsymbol{x}(t),\boldsymbol{u}(t),\boldsymbol{\lambda}(t),\boldsymbol{\mu}(t))}{\partial \boldsymbol{x}} - {}_{t}^{C}\mathrm{D}_{T}^{\alpha}\boldsymbol{\lambda}(t)\Big]^{\top}\delta\boldsymbol{x}(t)dt\\
&+ \int_{0}^{T}\Big[\frac{\partial H(t,\boldsymbol{x}(t),\boldsymbol{u}(t),\boldsymbol{\lambda}(t),\boldsymbol{\mu}(t))}{\partial \boldsymbol{u}}\Big]^{\top}\delta\boldsymbol{u}(t)dt\\
&+ \int_{0}^{T}\big[\boldsymbol{f}(t,\boldsymbol{x}(t),\boldsymbol{u}(t)) - {}_{0}^{C}\mathrm{D}_{t}^{\alpha}\boldsymbol{x}(t)\big]^{\top}\delta\boldsymbol{\lambda}(t)dt\\
&+ \Big[\frac{\partial \Phi}{\partial \boldsymbol{x}}\Big|_{(\boldsymbol{x}(T))} - {}_{0}^{R}\mathrm{D}_{T}^{\alpha-1}\boldsymbol{\lambda}(T)\Big]^{\top}\delta\boldsymbol{x}(T).
\end{aligned} \tag{5.19}$$

If $(\boldsymbol{x}^{*},\boldsymbol{x}^{*}(T),\boldsymbol{u}^{*},\boldsymbol{\lambda}^{*},\boldsymbol{\mu}^{*})$ is a minimizer for $\bar{g}_{0}(t,\boldsymbol{x},\boldsymbol{x}(T),\boldsymbol{u},\boldsymbol{\lambda},\boldsymbol{\mu})$, then the coefficients of $\delta\boldsymbol{x}$, $\delta\boldsymbol{u}$, $\delta\boldsymbol{\lambda}$ and $\delta\boldsymbol{x}(T)$ in (5.19) should be zeros at this point, i.e.:

$$\begin{cases} {}_t^C\mathrm{D}_T^\alpha \boldsymbol{\lambda}^*(t) = \left.\dfrac{\partial H}{\partial \boldsymbol{x}}\right|_{(t,\boldsymbol{x}^*(t),\boldsymbol{u}^*(t),\boldsymbol{\lambda}^*(t),\boldsymbol{\mu}^*(t))}, \\ {}_0^R\mathrm{D}_T^{\alpha-1} \boldsymbol{\lambda}^*(T) = \left.\dfrac{\partial \Phi}{\partial \boldsymbol{x}}\right|_{(\boldsymbol{x}^*(T))}, \\ {}_0^C\mathrm{D}_t^\alpha \boldsymbol{x}^*(t) = \boldsymbol{f}(t,\boldsymbol{x}^*(t),\boldsymbol{u}^*(t)), \\ \left.\dfrac{\partial H}{\partial \boldsymbol{u}}\right|_{(t,\boldsymbol{x}^*(t),\boldsymbol{u}^*(t),\boldsymbol{\lambda}^*(t),\boldsymbol{\mu}^*(t))} = \boldsymbol{0}, \quad t \in [0,T], \end{cases} \tag{5.20}$$

$$\text{and } \boldsymbol{x}^*(0) = \boldsymbol{x}^0. \tag{5.21}$$

Therefore, combining (5.20)–(5.21) and (5.7)–(5.10), we have (5.4)–(5.5). This completes the proof. □

Remark 5.1 We comment that a set of first-order optimality conditions is established in [7] for the FOCP when $\Phi = 0$ in (5.1), and (5.3) is not present. In [109], the first-order optimality conditions in [7] are extended to the unconstrained problem (5.1)–(5.2) (i.e., $\Phi \neq 0$). However, some of the conditions (in particular, $\Phi_{xx}(\boldsymbol{x}^*(T)) = \boldsymbol{0}$) established in [109] are unrealistic as, when $\boldsymbol{\alpha} = \boldsymbol{1}$, the transversality condition $\boldsymbol{\lambda}^*(t) = \Phi_x(\boldsymbol{x}^*(T))$ in the conventional optimal control theory cannot be retrieved, which this condition can be recovered from (5.4).

A control $\boldsymbol{u}$ which satisfies the necessary conditions (5.4)–(5.5) is called a *stationary point* as defined below.

Definition 5.1 A control $\boldsymbol{u}^*$ is called a stationary point of the constrained FOCP (5.1)–(5.3) if there exist optimal state $\boldsymbol{x}^*$, co-state $\boldsymbol{\lambda}^*$, and multiplier $\boldsymbol{\mu}^*$ satisfying (5.4)–(5.5).

5.3.2 Exact Penalty Method

Problem (5.1)–(5.3) is a fractional optimal control problem with state and control constraints. We use a general penalty function approach to define the following unconstrained optimal control problem approximating (5.1)–(5.3):

$$\min_{\boldsymbol{u}\in\mathcal{U}} \; \hat{g}_0(\boldsymbol{u},\boldsymbol{r}) := g_0(\boldsymbol{u}) + \int_0^T \boldsymbol{r}^\top(t)\varphi\big([\boldsymbol{g}(t,\boldsymbol{x}(t),\boldsymbol{u}(t))]_+\big)dt \tag{5.22}$$

$$\text{s.t.} \quad \begin{cases} {}_0^C\mathrm{D}_t^\alpha \boldsymbol{x}(t) = \boldsymbol{f}(t,\boldsymbol{x}(t),\boldsymbol{u}(t)), \; t \in (0,T], \\ \boldsymbol{x}(0) = \boldsymbol{x}^0, \end{cases} \tag{5.23}$$

where $[z]_+ = (\max(0, z_1), \max(0, z_2), \ldots, \max(0, z_p))^\top$ for any $z \in \mathbb{R}^p$; $\varphi(z) = (\varphi_1(z), \varphi_2(z), \ldots, \varphi_p(z))^\top$ is a continuous and piecewise smooth vector-valued function of $z \in \mathbb{R}^p$, $z \geq \mathbf{0}$ such that

$$\varphi_i(\mathbf{0}) = 0, \quad \varphi_i(z) \geq z_i \quad i = 1, 2, \ldots, p,$$

and $\boldsymbol{r}(t) = (r_1(t), r_2(t), \ldots, r_p(t))^\top$ with $r_i(t) > 0$ for all feasible i being a chosen penalty function which is usually a constant vector with each component sufficiently large so as to penalize the corresponding components of g that violate (5.3). The most popular choice of φ is $\varphi(z) = z$, which is called a *linear penalty function.* Other examples are $\varphi_i(z_i) = \exp(z_i) - 1$ and

$$\varphi_i(z) = \begin{cases} z_i^{\gamma_1}, & z_i \geq 1, \\ z_i^{\gamma_2}, & z_i \in [0, 1) \end{cases} \tag{5.24}$$

for all feasible i, where $\gamma_1 \geq 1$ and $\gamma_2 \in (0, 1]$ are two constants. The penalty function defined in (5.24) is in fact the low-order power penalty method. We now show that (5.22)–(5.23) is equivalent to (5.1)–(5.3) when $\boldsymbol{r}$ is sufficiently large.

Before stating and proving our main result, we first give the following definitions.

Definition 5.2 Let $\Omega = \{\boldsymbol{u} | \, \boldsymbol{u} \in \mathcal{U}, \boldsymbol{g}(t, \boldsymbol{x}(t), \boldsymbol{u}(t)) \leq \mathbf{0}, t \in [0, T]\}$, where $\boldsymbol{x}$ is the response corresponding to $\boldsymbol{u}$ according to (5.2). Then, $\boldsymbol{u}^*$ is called a local solution to FOCP (5.1)–(5.3) if there exists a positive number ρ such that

$$g_0(\boldsymbol{u}^*) \leq g_0(\boldsymbol{u})$$

when $\boldsymbol{u} \in \Omega \cap B(\rho)$, where $B(\rho) = \{\boldsymbol{u} | \, \|\boldsymbol{u} - \boldsymbol{u}^*\| \leq \rho\}$, and $\| \cdot \|$ denotes an appropriate norm on the function space containing $\mathcal{U}$.

It is clear, from the above definition, that if $\boldsymbol{u}^*$ is an optimal solution of (5.1)–(5.3), then $\boldsymbol{u}^*$ is a local solution of (5.1)–(5.3). Similarly, a local solution of (5.22)–(5.23) is defined as

Definition 5.3 A function $\boldsymbol{u}^*$ is called a local solution to (5.22)–(5.23) if there exists a positive number ρ such that

$$\hat{g}_0(\boldsymbol{u}^*, \boldsymbol{r}) \leq \hat{g}_0(\boldsymbol{u}, \boldsymbol{r})$$

for all $\boldsymbol{u} \in B(\rho)$, where $B(\rho) = \{\boldsymbol{u} | \, \|\boldsymbol{u} - \boldsymbol{u}^*\| \leq \rho\}$.

Definition 5.4 Let $\boldsymbol{u}^*$ be a local solution of the constrained FOCP (5.1)–(5.3). If $\boldsymbol{u}^*$ is also a local solution of the unconstrained FOCP (5.22)–(5.23) when $r_i(t) > r_0$, $t \in [0, T]$, for $i = 1, 2, \ldots, p$, where r_0 is a positive constant, then $\hat{g}_0(\boldsymbol{u}, \boldsymbol{r})$ is called an exact penalty function of (5.1)–(5.3).

With the above definitions, we establish the following results:

Theorem 5.2 *Let $\boldsymbol{u}^*$ be a stationary point of FOCP (5.1)–(5.3) and let $\boldsymbol{x}^*$, $\boldsymbol{\lambda}^*$ and $\boldsymbol{\mu}^*$ be the corresponding state, co-state, and multiplier as defined in Theorem 5.1. Suppose that the following conditions*

(C$_1$) $\Phi^*_{\boldsymbol{xx}} := \left.\dfrac{\partial^2 \Phi}{\partial \boldsymbol{x}^2}\right|_{(\boldsymbol{x}^*(T))}$ *is positive semi-definite;*

(C$_2$) *either $H^*_{\boldsymbol{uu}}$ and $H^*_{\boldsymbol{xx}} - H^*_{\boldsymbol{xu}} H^{*-1}_{\boldsymbol{uu}} H^*_{\boldsymbol{ux}}$ are both positive definite on $[0, T]$, or $H^*_{\boldsymbol{xx}}$ and $H^*_{\boldsymbol{uu}} - H^*_{\boldsymbol{ux}} H^{*-1}_{\boldsymbol{xx}} H^*_{\boldsymbol{xu}}$ are both positive definite, where*

$$H^*_{\boldsymbol{xx}} = \left.\frac{\partial^2 H}{\partial \boldsymbol{x}^2}\right|_{(t,\boldsymbol{x}^*(t),\boldsymbol{u}^*(t),\boldsymbol{\lambda}^*(t),\boldsymbol{\mu}^*(t))}, \qquad H^*_{\boldsymbol{xu}} = \left.\frac{\partial^2 H}{\partial \boldsymbol{x}\partial \boldsymbol{u}}\right|_{(t,\boldsymbol{x}^*(t),\boldsymbol{u}^*(t),\boldsymbol{\lambda}^*(t),\boldsymbol{\mu}^*(t))},$$

$$H^*_{\boldsymbol{uu}} = \left.\frac{\partial^2 H}{\partial \boldsymbol{u}^2}\right|_{(t,\boldsymbol{x}^*(t),\boldsymbol{u}^*(t),\boldsymbol{\lambda}^*(t),\boldsymbol{\mu}^*(t))}, \text{ and } H^*_{\boldsymbol{ux}} = H^{*\top}_{\boldsymbol{xu}}$$

are satisfied. If $r_i(t) \geq \mu_i(t)$, $t \in [0, T]$, for $i = 1, 2, \ldots, p$, then $\boldsymbol{u}^*$ is a local solution of (5.22)–(5.23).

Proof From the definition of the Hamiltonian H in (5.6), we see that $\hat{g}_0(\boldsymbol{u}, \boldsymbol{r})$ in (5.22) can be rewritten as the following equivalent form:

$$\begin{aligned}\hat{g}_0(\boldsymbol{u}, \boldsymbol{r}) = \Phi(\boldsymbol{x}(T)) + \int_0^T \Big\{ & H(t, \boldsymbol{x}(t), \boldsymbol{u}(t), \boldsymbol{\lambda}^*(t), \boldsymbol{\mu}^*(t)) - \boldsymbol{\lambda}^{*\top}(t)\boldsymbol{f}(t, \boldsymbol{x}(t), \boldsymbol{u}(t)) \\ & - \boldsymbol{\mu}^{*\top}(t)\boldsymbol{g}(t, \boldsymbol{x}(t), \boldsymbol{u}(t))\Big\}dt + \int_0^T \boldsymbol{r}^\top(t)\varphi\big([\boldsymbol{g}(t, \boldsymbol{x}(t), \boldsymbol{u}(t))]_+\big)dt,\end{aligned}$$

where $\boldsymbol{\lambda}^*$ and $\boldsymbol{\mu}^*$ are, respectively, the optimal multipliers and co-states defined in Theorem 5.1. From (5.23), we see that $\hat{g}_0(\boldsymbol{u}, \boldsymbol{r})$ can further be written as

$$\begin{aligned}\hat{g}_0(\boldsymbol{u}, \boldsymbol{r}) = & \Phi(\boldsymbol{x}(T)) + \int_0^T \Big\{ H(t, \boldsymbol{x}(t), \boldsymbol{u}(t), \boldsymbol{\lambda}^*(t), \boldsymbol{\mu}^*(t)) - \boldsymbol{\lambda}^{*\top}(t)\, {}_0^C\mathrm{D}_t^\alpha \boldsymbol{x}(t) \\ & - \boldsymbol{\mu}^{*\top}(t)\boldsymbol{g}(t, \boldsymbol{x}(t), \boldsymbol{u}(t))\Big\}dt + \int_0^T \boldsymbol{r}^\top(t)\varphi\big([\boldsymbol{g}(t, \boldsymbol{x}(t), \boldsymbol{u}(t))]_+\big)dt.\end{aligned}$$

Using (2.25), (2.32), and (3.85), we have

$$\begin{aligned}\hat{g}_0(\boldsymbol{u}, \boldsymbol{r}) = \Phi(\boldsymbol{x}(T)) + \int_0^T & \Big\{ H(t, \boldsymbol{x}(t), \boldsymbol{u}(t), \boldsymbol{\lambda}^*(t), \boldsymbol{\mu}^*(t)) - \boldsymbol{x}^\top(t)\, {}_t^C\mathrm{D}_T^\alpha \boldsymbol{\lambda}^*(t) \\ & + \boldsymbol{\lambda}^{*\top}(t)\, {}_0^R\mathrm{D}_t^\alpha \boldsymbol{x}^0 - \boldsymbol{\mu}^{*\top}(t)\boldsymbol{g}(t, \boldsymbol{x}(t), \boldsymbol{u}(t))\Big\}dt - \boldsymbol{x}^\top(T)\, {}_t^R\mathrm{D}_T^{\alpha-1}\boldsymbol{\lambda}^*(T) \\ & + \int_0^T \boldsymbol{r}^\top(t)\varphi\big([\boldsymbol{g}(t, \boldsymbol{x}(t), \boldsymbol{u}(t))]_+\big)dt. \qquad (5.25)\end{aligned}$$

If $\boldsymbol{u}$ is in a neighborhood of $\boldsymbol{u}^*$, we may denote $\boldsymbol{u}$ as $\boldsymbol{u} = \boldsymbol{u}^* + \delta\boldsymbol{u}$, where $\delta\boldsymbol{u} = (\delta u_1, \delta u_2, \ldots, \delta u_m)^\top$ is a small perturbation. Let $\boldsymbol{x} = \boldsymbol{x}^* + \delta\boldsymbol{x}$ with $\delta\boldsymbol{x} = (\delta x_1, \delta x_2, \ldots, \delta x_n)^\top$ be the response corresponding to $\delta\boldsymbol{u}$. We now show that

$$\Delta\hat{g}_0(\boldsymbol{u}^*, \boldsymbol{r}) := \hat{g}_0(\boldsymbol{u}, \boldsymbol{r}) - \hat{g}_0(\boldsymbol{u}^*, \boldsymbol{r}) = \hat{g}_0(\boldsymbol{u}^* + \delta\boldsymbol{u}, \boldsymbol{r}) - \hat{g}_0(\boldsymbol{u}^*, \boldsymbol{r}) \geq 0$$

under the conditions given in Theorem 5.2.
From (5.25), we have

$$\begin{aligned}
&\Delta\hat{g}_0(\boldsymbol{u}^*, \boldsymbol{r}) \\
&= \int_0^T \Big\{ H(t, \boldsymbol{x}^*(t) + \delta\boldsymbol{x}(t), \boldsymbol{u}^*(t) + \delta\boldsymbol{u}(t), \boldsymbol{\lambda}^*(t), \boldsymbol{\mu}^*(t)) - [\boldsymbol{x}^*(t) + \delta\boldsymbol{x}(t)]^\top \\
&\times {}_t^C\mathrm{D}_T^\alpha \boldsymbol{\lambda}^*(t) + \boldsymbol{\lambda}^{*\top}(t) {}_0^R\mathrm{D}_t^\alpha \boldsymbol{x}^0 - \boldsymbol{\mu}^{*\top}(t)\boldsymbol{g}(t, \boldsymbol{x}^*(t) + \delta\boldsymbol{x}(t), \boldsymbol{u}^*(t) + \delta\boldsymbol{u}(t)) \Big\} dt \\
&- [\boldsymbol{x}^*(T) + \delta\boldsymbol{x}(T)]^\top {}_t^R\mathrm{D}_T^{\alpha-1} \boldsymbol{\lambda}^*(T) + \Phi(\boldsymbol{x}^*(T) + \delta\boldsymbol{x}(T)) \\
&- \int_0^T \Big\{ H(t, \boldsymbol{x}^*(t), \boldsymbol{u}^*(t), \boldsymbol{\lambda}^*(t), \boldsymbol{\mu}^*(t)) - \boldsymbol{x}^{*\top}(t) {}_t^C\mathrm{D}_T^\alpha \boldsymbol{\lambda}^*(t) \\
&+ \boldsymbol{\lambda}^{*\top}(t) {}_0^R\mathrm{D}_t^\alpha \boldsymbol{x}^0 - \boldsymbol{\mu}^{*\top}(t)\boldsymbol{g}(t, \boldsymbol{x}^*(t), \boldsymbol{u}^*(t)) \Big\} dt - \boldsymbol{x}^{*\top}(T) {}_t^R\mathrm{D}_T^{\alpha-1} \boldsymbol{\lambda}^*(T) \\
&+ \Phi(\boldsymbol{x}^*(T)) + \int_0^T \boldsymbol{r}^\top(t) \{ \varphi\left([\boldsymbol{g}(t, \boldsymbol{x}^*(t) + \delta\boldsymbol{x}(t), \boldsymbol{u}^*(t) + \delta\boldsymbol{u}(t))]_+ \right) \\
&- \varphi\left([\boldsymbol{g}(t, \boldsymbol{x}^*(t), \boldsymbol{u}^*(t))]_+ \right) \} dt.
\end{aligned}$$

Since $\boldsymbol{u}^*$ is a stationary point of (5.1)–(5.3), it satisfies (5.4)–(5.5), and thus $\boldsymbol{\mu}^{*\top}(t)\boldsymbol{g}(t, \boldsymbol{x}^*(t), \boldsymbol{u}^*(t)) = 0$ and $[\boldsymbol{g}(t, \boldsymbol{x}^*(t), \boldsymbol{u}^*(t))]_+ = \boldsymbol{0}$. Then, we have

$$\begin{aligned}
\Delta\hat{g}_0(\boldsymbol{u}^*, \boldsymbol{r}) = &\int_0^T \Big[H(t, \boldsymbol{x}^*(t) + \delta\boldsymbol{x}(t), \boldsymbol{u}^*(t) + \delta\boldsymbol{u}(t), \boldsymbol{\lambda}^*(t), \boldsymbol{\mu}^*(t)) \\
&- H(t, \boldsymbol{x}^*(t), \boldsymbol{u}^*(t), \boldsymbol{\lambda}^*(t), \boldsymbol{\mu}^*(t)) - \delta\boldsymbol{x}^\top(t) {}_t^C\mathrm{D}_T^\alpha \boldsymbol{\lambda}^*(t) \Big] dt \\
&- \delta\boldsymbol{x}^\top(T) {}_t^R\mathrm{D}_T^{\alpha-1} \boldsymbol{\lambda}^*(T) + [\Phi(\boldsymbol{x}^*(T) + \delta\boldsymbol{x}(T)) - \Phi(\boldsymbol{x}^*(T))] \\
&+ \int_0^T \Big\{ \boldsymbol{r}^\top(t)\varphi\big([\boldsymbol{g}(t, \boldsymbol{x}^*(t) + \delta\boldsymbol{x}(t), \boldsymbol{u}^*(t) + \delta\boldsymbol{u}(t))]_+ \big) dt \\
&- \boldsymbol{\mu}^{*\top}\boldsymbol{g}(t, \boldsymbol{x}^*(t) + \delta\boldsymbol{x}(t), \boldsymbol{u}^*(t) + \delta\boldsymbol{u}(t)) \Big\} dt.
\end{aligned}$$

Note that

$$\begin{aligned}
&\boldsymbol{g}(t, \boldsymbol{x}^*(t) + \delta\boldsymbol{x}(t), \boldsymbol{u}^*(t) + \delta\boldsymbol{u}(t)) \\
&= [\boldsymbol{g}(t, \boldsymbol{x}^*(t) + \delta\boldsymbol{x}(t), \boldsymbol{u}^* + \delta\boldsymbol{u}(t))]_+ - [\boldsymbol{g}(t, \boldsymbol{x}^*(t) + \delta\boldsymbol{x}(t), \boldsymbol{u}^*(t) + \delta\boldsymbol{u}(t))]_-,
\end{aligned}$$

where $[z]_- = \max\{\mathbf{0}, -z\}$ denotes the negative part of a function z. Since the multiplier $\boldsymbol{\mu}^* \geq \mathbf{0}$ by (5.5), we see that

$$-\boldsymbol{\mu}^{*\top}\boldsymbol{g} = -\boldsymbol{\mu}^{*\top}[\boldsymbol{g}]_+ + \boldsymbol{\mu}^{*\top}[\boldsymbol{g}]_- \geq -\boldsymbol{\mu}^{*\top}[\boldsymbol{g}]_+$$

for almost all $t \in [0, T]$. Therefore, using this inequality and Taylor's expansion, we obtain

$$\begin{aligned}
\Delta\hat{g}_0(\boldsymbol{u}^*, \boldsymbol{r}) \geq & \int_0^T \left\{ \left[\left.\frac{\partial H}{\partial \boldsymbol{x}}\right|_{(t,\boldsymbol{x}^*(t),\boldsymbol{u}^*(t),\boldsymbol{\lambda}^*(t),\boldsymbol{\mu}^*(t))} \right]^\top \delta\boldsymbol{x}(t) \right. \\
& + \left[\left.\frac{\partial H}{\partial \boldsymbol{u}}\right|_{(t,\boldsymbol{x}^*(t),\boldsymbol{u}^*(t),\boldsymbol{\lambda}^*(t),\boldsymbol{\mu}^*(t))} \right]^\top \delta\boldsymbol{u}(t) + \frac{1}{2}\big(\delta\boldsymbol{x}^\top(t) H^*_{\boldsymbol{xx}} \delta\boldsymbol{x}(t) \\
& \left. + 2\delta\boldsymbol{x}^\top(t) H^*_{\boldsymbol{xu}} \delta\boldsymbol{u}(t) + \delta\boldsymbol{u}^\top(t) H^*_{\boldsymbol{uu}} \delta\boldsymbol{u}(t)\big) - \delta\boldsymbol{x}^\top(t) {}_t^R\mathrm{D}_T^\alpha \boldsymbol{\lambda}^*(t) \right\} dt \\
& - \delta\boldsymbol{x}^\top(T) {}_t^R\mathrm{D}_T^{\alpha-1} \boldsymbol{\lambda}^*(T) + \left[\delta\boldsymbol{x}^\top(T) \left.\frac{\partial \Phi}{\partial \boldsymbol{x}}\right|_{(\boldsymbol{x}^*(T))} \right. \\
& \left. + \frac{1}{2}\delta\boldsymbol{x}^\top(T) \left.\frac{\partial^2 \Phi}{\partial \boldsymbol{x}^2}\right|_{(\boldsymbol{x}^*(T))} \delta\boldsymbol{x}(T) \right] + \int_0^T (\boldsymbol{r}^\top(t) - \boldsymbol{\mu}^{*\top}(t)) \big\{ \varphi\big([\boldsymbol{g}(t, \boldsymbol{x}^*(t) \\
& + \delta\boldsymbol{x}(t), \boldsymbol{u}^*(t) + \delta\boldsymbol{u}(t))]_+\big) - [\boldsymbol{g}(t, \boldsymbol{x}^*(t) + \delta\boldsymbol{x}(t), \boldsymbol{u}^*(t) + \delta\boldsymbol{u}(t))]_+ \big\} dt \\
& + R(\delta\boldsymbol{x}, \delta\boldsymbol{u}),
\end{aligned} \tag{5.26}$$

where $R(\delta\boldsymbol{x}, \delta\boldsymbol{u})$ is the remainder satisfying

$$|R(\delta\boldsymbol{x}, \delta\boldsymbol{u})| \leq \varepsilon'(\delta\boldsymbol{x}, \delta\boldsymbol{u}) \int_0^T (\delta\boldsymbol{x}^\top(t)\delta\boldsymbol{x}(t) + \delta\boldsymbol{u}^\top(t)\delta\boldsymbol{u}(t))dt \tag{5.27}$$

with $\varepsilon'(\delta\boldsymbol{x}, \delta\boldsymbol{u})$ denoting a function of $\delta\boldsymbol{x}$ and $\delta\boldsymbol{u}$. By the continuous dependence of solution of system (5.2) on parameters, we have both $\|\delta\boldsymbol{x}\| \to 0$ and $\varepsilon'(\delta\boldsymbol{x}, \delta\boldsymbol{u}) \to 0$ as $\|\delta\boldsymbol{u}\| \to 0$.

Since $\boldsymbol{u}^*$ is a stationary point, it satisfies (5.4)–(5.5). Thus, (5.26) reduces to the following form:

$$\begin{aligned}
\Delta\hat{g}_0(\boldsymbol{u}^*, \boldsymbol{r}) \geq & \int_0^T \frac{1}{2}\Big(\delta\boldsymbol{x}^\top(t) H^*_{\boldsymbol{xx}} \delta\boldsymbol{x}(t) + 2\delta\boldsymbol{x}^\top(t) H^*_{\boldsymbol{xu}} \delta\boldsymbol{u}(t) \\
& + \delta\boldsymbol{u}^\top(t) H^*_{\boldsymbol{uu}} \delta\boldsymbol{u}(t)\Big) dt + R(\delta\boldsymbol{x}, \delta\boldsymbol{u}) =: \delta^2\hat{g}_0 + R(\delta\boldsymbol{x}, \delta\boldsymbol{u}),
\end{aligned} \tag{5.28}$$

since $\left.\dfrac{\partial^2 \Phi}{\partial \boldsymbol{x}^2}\right|_{(\boldsymbol{x}^*(T))}$ is positive semi-definite by condition (C_1) and $\varphi(z) \geq z$ element-by-element, where $\delta^2\hat{g}_0$ is the following second-order variation of $\hat{g}_0$ with respect to $\boldsymbol{x}$ and $\boldsymbol{u}$:

$$\delta^2\hat{g}_0 = \frac{1}{2}\int_0^T \left(\delta\boldsymbol{u}^\top(t)\ \delta\boldsymbol{x}^\top(t)\right) \begin{pmatrix} H^*_{\boldsymbol{uu}} & H^{*\top}_{\boldsymbol{xu}} \\ H^*_{\boldsymbol{xu}} & H^*_{\boldsymbol{xx}} \end{pmatrix} \begin{pmatrix} \delta\boldsymbol{u}(t) \\ \delta\boldsymbol{x}(t) \end{pmatrix} dt. \tag{5.29}$$

Next, we show that

$$\delta^2 \hat{g}_0 + R(\delta \boldsymbol{x}, \delta \boldsymbol{u}) \geq 0.$$

Since $H^*_{\boldsymbol{uu}}$ and $H^*_{\boldsymbol{xx}} - H^*_{\boldsymbol{xu}} {H^*_{\boldsymbol{uu}}}^{-1} H^*_{\boldsymbol{ux}}$ are positive definite, there exists a sufficiently small number $w > 0$ such that $H^*_{\boldsymbol{uu}} - w\boldsymbol{I}_p$ and

$$\begin{aligned}
&(H^*_{\boldsymbol{xx}} - w\boldsymbol{I}_n) - H^*_{\boldsymbol{xu}}(H^*_{\boldsymbol{uu}} - w\boldsymbol{I}_p)^{-1} H^*_{\boldsymbol{ux}} \\
&= H^*_{\boldsymbol{xx}} - H^*_{\boldsymbol{xu}} {H^*_{\boldsymbol{uu}}}^{-1} H^*_{\boldsymbol{ux}} - (w\boldsymbol{I}_n + H^*_{\boldsymbol{xu}}(H^*_{\boldsymbol{uu}} - w\boldsymbol{I}_p)^{-1} H^*_{\boldsymbol{ux}} \\
&\quad - H^*_{\boldsymbol{xu}} {H^*_{\boldsymbol{uu}}}^{-1} H^*_{\boldsymbol{ux}})
\end{aligned}$$

are both positive, where $\boldsymbol{I}_p$ and $\boldsymbol{I}_n$ denote the $p \times p$ and $n \times n$ identity matrices, respectively. Thus, since these matrices are symmetric, there exists a matrix $\boldsymbol{Q}$ such that

$$\boldsymbol{Q}^\top \boldsymbol{Q} = (H^*_{\boldsymbol{xx}} - w\boldsymbol{I}) - H^*_{\boldsymbol{xu}}(H^*_{\boldsymbol{uu}} - w\boldsymbol{I})^{-1} H^*_{\boldsymbol{ux}}.$$

Using the above equations, we have

$$\begin{aligned}
&\delta^2 \hat{g}_0 - \frac{w}{2} \int_0^T \left[\delta \boldsymbol{x}^\top(t)\delta \boldsymbol{x}(t) + \delta \boldsymbol{u}^\top(t)\delta \boldsymbol{u}(t)\right] dt \\
&= \frac{1}{2} \int_0^T \begin{bmatrix} \delta \boldsymbol{u}(t) + (H^*_{\boldsymbol{uu}} - w\boldsymbol{I})^{-1} H^*_{\boldsymbol{ux}} \delta \boldsymbol{x}(t) \\ \boldsymbol{Q}\delta \boldsymbol{x}(t) \end{bmatrix}^\top \begin{bmatrix} H^*_{\boldsymbol{uu}} - w\boldsymbol{I} & \boldsymbol{0} \\ \boldsymbol{0} & \boldsymbol{I} \end{bmatrix} \\
&\qquad \times \begin{bmatrix} \delta \boldsymbol{u}(t) + (H^*_{\boldsymbol{uu}} - w\boldsymbol{I})^{-1} H^*_{\boldsymbol{ux}} \delta \boldsymbol{x}(t) \\ \boldsymbol{Q}\delta \boldsymbol{x}(t) \end{bmatrix} dt > 0,
\end{aligned}$$

since $\begin{bmatrix} H^*_{\boldsymbol{uu}} - w\boldsymbol{I} & \boldsymbol{0} \\ \boldsymbol{0} & \boldsymbol{I} \end{bmatrix}$ is positive definite. Therefore, from (5.28), we have

$$\Delta \hat{g}_0(\boldsymbol{u}^*, \boldsymbol{r}) > \frac{w}{2} \int_0^T \left[\delta \boldsymbol{x}^\top(t)\delta \boldsymbol{x}(t) + \delta \boldsymbol{u}^\top(t)\delta \boldsymbol{u}(t))\right] dt + R(\delta \boldsymbol{x}, \delta \boldsymbol{u}).$$

From (5.27) and the above inequality, we have

$$\Delta \hat{g}_0(\boldsymbol{u}^*, \boldsymbol{r}) > \left(\frac{w}{2} - \varepsilon'(\delta \boldsymbol{x}, \delta \boldsymbol{u})\right) \int_0^T \left[\delta \boldsymbol{x}^\top(t)\delta \boldsymbol{x}(t) + \delta \boldsymbol{u}^\top(t)\delta \boldsymbol{u}(t)\right] dt, \qquad (5.30)$$

where $\varepsilon'(\delta \boldsymbol{x}, \delta \boldsymbol{u}) \to 0$ as $\|\delta \boldsymbol{u}\| \to 0$.

By symmetry, if we swap $\delta \boldsymbol{x}$ and $\delta \boldsymbol{u}$, $H^*_{\boldsymbol{xx}}$ and $H^*_{\boldsymbol{uu}}$, and $H^*_{\boldsymbol{ux}}$ and $H^*_{\boldsymbol{xu}}$, the above analysis from (5.28) to (5.29) also holds under the condition that both $H^*_{\boldsymbol{xx}}$ and $H^*_{\boldsymbol{uu}} - H^*_{\boldsymbol{ux}} {H^*_{\boldsymbol{xx}}}^{-1} H^*_{\boldsymbol{xu}}$ are positive definite. Thus, in this case (5.30) also holds true. As a result, $\Delta \hat{g}_0(\boldsymbol{u}^*, \boldsymbol{y}) \geq 0$ when $\|\delta \boldsymbol{u}\|$ is small enough, which implies $\boldsymbol{u}^*$ is a local solution of (5.22)–(5.23). The proof is complete. □

Using Theorem 5.2, we have the following result:

Theorem 5.3 *Under the assumptions in Theorem 5.2, $\boldsymbol{u}^*$ is also a local solution of (5.1)–(5.3).*

Proof By Theorem 5.2, $\boldsymbol{u}^*$ is the local solution of (5.22)–(5.23), which means that there exists a positive number ρ such that $\hat{g}_0(\boldsymbol{u}^*, \boldsymbol{r}) \leq \hat{g}_0(\boldsymbol{u}, \boldsymbol{r})$ for all $\boldsymbol{u} \in B(\rho)$. For $\boldsymbol{u}^*$ and any $\boldsymbol{u} \in B(\rho)$ which satisfies (5.3), we have

$$\hat{g}_0(\boldsymbol{u}^*, \boldsymbol{r}) = g_0(\boldsymbol{u}^*), \quad \hat{g}_0(\boldsymbol{u}, \boldsymbol{r}) = g_0(\boldsymbol{u}).$$

Therefore, there exists a positive number ρ such that $g_0(\boldsymbol{u}^*) \leq g_0(\boldsymbol{u})$ for all $\boldsymbol{u} \in B(\rho)$, which implies $\boldsymbol{u}^*$ is a local solution of (5.1)–(5.3). □

Remark 5.2 From Theorems 5.2 and 5.3, it is clear that under the conditions in Theorem 5.2, $\hat{g}_0(\boldsymbol{u}, \boldsymbol{r})$ is an exact penalty function of (5.1)–(5.3). Therefore, the constrained FOCP (5.1)–(5.3) can be solved via the unconstrained FOCP (5.22)–(5.23).

5.3.3 An Illustrative Example

Consider the following FOCP [274]:

$$\begin{aligned}
\min\ g_0(u) = &\int_0^{20} \left[1 - (x(t) - 0.01t^2 - 1)^2 + u(t) - 2\sqrt{\pi}J_0(4\sqrt{t})\right]^2 dt \\
&+ \left(x(20) - 5 - \sin(8\sqrt{5})\right)^2,
\end{aligned}$$

$$\text{s.t.} \begin{cases} {}_0^C\mathrm{D}_t^\alpha\, x(t) = -(x(t) - 0.01t^2 - 1)^2 + u(t) + 1 + \dfrac{2t^{1.5}}{75\sqrt{\pi}}, & t \in (0, 20], \\ x(0) = 1, \end{cases}$$

where J_0 is the first kind Bessel function of order zero [4]. For $\alpha = 0.5$, the exact solution and the optimal cost of this problem are

$$u_{ex}(t) = -\cos^2(4\sqrt{t}) + 2\sqrt{\pi}J_0(4\sqrt{t}), \tag{5.31}$$

$$x_{ex}(t) = \sin(4\sqrt{t}) + 0.01t^2 + 1, \quad t \in [0, 20], \tag{5.32}$$

$$J(u_{ex}) = 0, \tag{5.33}$$

respectively.

For this problem, define

$$H(x(t), u(t), \lambda(t)) = \left[1 - (x(t) - 0.01t^2 - 1)^2 + u(t) - 2\sqrt{\pi}J_0(4\sqrt{t})\right]^2 + \lambda(t)\left[- (x(t) - 0.01t^2 - 1)^2 + u(t) + 1 + \frac{2t^{1.5}}{75\sqrt{\pi}}\right].$$

Then, by (5.20) and (5.32), we have

$$u(t) = -0.5\lambda(t) - 1 + \sin^2(4\sqrt{t}) + 2\sqrt{\pi}J_0(4\sqrt{t}), \tag{5.34}$$

$${}_t^C\mathrm{D}_{20}^{0.5}\lambda(t) = 0, \quad t \in [0, 20], \tag{5.35}$$

and

$${}_0^R\mathrm{D}_{20}^{-0.5}\lambda(20) = 0. \tag{5.36}$$

Note that Eqs. (5.35) and (5.36) imply that

$$\lambda(t) = 0, \quad t \in [0, 20]. \tag{5.37}$$

Substituting (5.37) into (5.34) yields

$$u(t) = u_{ex}(t), \quad t \in [0, 20],$$

which shows that the control satisfying the optimality conditions (5.20) is just the optimal control. The optimal control and the corresponding optimal state are shown in Fig. 5.1.

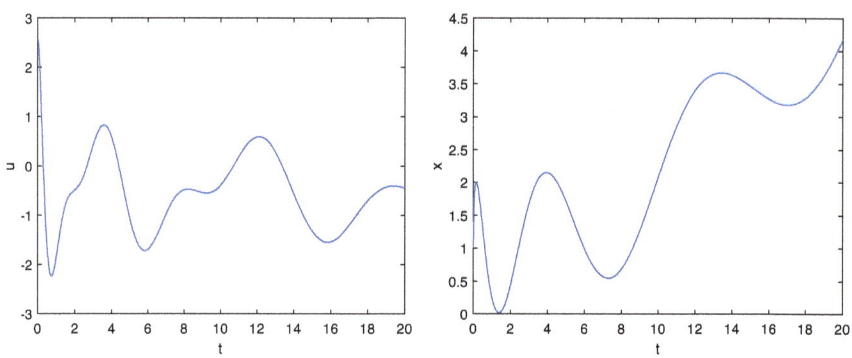

Fig. 5.1 Optimal control and optimal state for example problem

5.4 Numerical Solution Method

From Remark 5.2, we see that constraint (5.3) can be appended to the original objective function (5.1) by the exact penalty method. Since the integrand in the penalty term of $\hat{g}_0$ is smooth, it can be combined with the original objective integrand L to form a new integrand which is still continuously differentiable in $\boldsymbol{x}$ and $\boldsymbol{u}$. Thus, we may develop the numerical solution method to solving Problems (5.1) and (5.2).

5.4.1 *Discretization Scheme*

Let N be a positive integer. We form a partition of $[0, T]$ with the nodes $t_j = jh$ for $j = 0, 1, \ldots, N$, where $h = \dfrac{T}{N}$. We then approximate g_0 in (5.1) by the conventional trapezoidal rule, yielding the following discrete objective:

$$g_0(\boldsymbol{u}) \approx \frac{1}{2}L(t_0, \boldsymbol{x}^0, \boldsymbol{u}^0)h + \sum_{j=1}^{N-1} L(t_j, \boldsymbol{x}^j, \boldsymbol{u}^j)h + \frac{1}{2}L(t_N, \boldsymbol{x}^N, \boldsymbol{u}^N)h + \Phi(\boldsymbol{x}^N), \tag{5.38}$$

where L is the integrand in (5.1), and $(\boldsymbol{x}^j, \boldsymbol{u}^j) := (\boldsymbol{x}(t_j), \boldsymbol{u}(t_j))$ for all feasible j. Since $t_j = jh$, it has been shown in Theorem 3.7 that the system (5.2) is equivalent to

$$x_i^j = x_i^0 + \frac{1}{\Gamma(\alpha_i)} \sum_{k=1}^{j} \int_{(k-1)h}^{kh} (jh - \tau)^{\alpha_i - 1} f_i(\tau, \boldsymbol{x}(\tau), \boldsymbol{u}(\tau))d\tau \tag{5.39}$$

for all feasible i and j. We now consider an approximation for the integral in (3.41). By Assumption 5.2.1, $f_i(t, \boldsymbol{x}(t), \boldsymbol{u}(t))$ is twice continuously differentiable with respect to t, $\boldsymbol{x}$, and $\boldsymbol{u}$. Thus, for $k = 1, 2, \ldots, j$, we apply Taylor's expansion to $f_i(\tau, \boldsymbol{x}(\tau), \boldsymbol{u}(\tau))$ at any point $\tau_{jk}^i \in ((k-1)h, kh)$ to yield

$$f_i(\tau, \boldsymbol{x}(\tau), \boldsymbol{u}(\tau)) = f_i(\tau_{jk}^i, \boldsymbol{x}(\tau_{jk}^i), \boldsymbol{u}(\tau_{jk}^i)) + K_{jk}^i(\tau - \tau_{jk}^i) + c_{jk}^i(\tau - \tau_{jk}^i)^2, \tag{5.40}$$

where K_{jk}^i and c_{jk}^i are, respectively, the coefficients of the first-order and second-order terms of the expansion.

Replacing $f_i(\tau, \boldsymbol{x}(\tau), \boldsymbol{u}(\tau))$ in the integrand of the last term in (5.39) with the right-hand side of (3.42), we have, for any $k = 1, \ldots, j$,

$$\begin{aligned}&\frac{1}{\Gamma(\alpha_i)}\int_{(k-1)h}^{kh}(jh-\tau)^{\alpha_i-1}f_i(\tau,\boldsymbol{x}(\tau),\boldsymbol{u}(\tau))d\tau\\&=\bar{h}_i f_i(\tau^i_{jk},\boldsymbol{x}(\tau^i_{jk}),\boldsymbol{u}(\tau^i_{jk}))[(d_{jk}+1)^{\alpha_i}-d_{jk}^{\alpha_i}]\\&+\frac{K^i_{jk}}{\Gamma(\alpha_i)}\int_{(k-1)h}^{kh}(jh-\tau)^{\alpha_i-1}(\tau-\tau^i_{jk})d\tau+R^i_{jk},\end{aligned}\tag{5.41}$$

where $R^i_{jk}=\dfrac{1}{\Gamma(\alpha_i)}\displaystyle\int_{(k-1)h}^{kh}(jh-\tau)^{\alpha_i-1}c^i_{jk}(\tau-\tau^i_{jk})^2d\tau$; $\bar{h}_i=\dfrac{h^{\alpha_i}}{\Gamma(\alpha_i+1)}$, and $d_{jk}=j-k$.

Using (5.41), we see that τ^i_{jk} should be chosen such that the third term is zero so that the truncation error in (5.41) is R^i_{jk}. This choice of τ^i_{jk} is given as follows.

$$\tau^i_{jk}=h\frac{[(d_{jk}+1)^{\alpha_i+1}-d_{jk}^{\alpha_i+1}]+(\alpha_i+1)[(d_{jk}+1)^{\alpha_i}(k-1)-d_{jk}^{\alpha_i}k]}{(\alpha_i+1)[(d_{jk}+1)^{\alpha_i}-d_{jk}^{\alpha_i}]}.\tag{5.42}$$

We have shown in Theorem 3.18 that $(k-1)h<\tau^i_{jk}<kh$.

Substituting the expression for τ^i_{jk} in (5.42) into (5.41) and combining the resulting expression with (5.39), we have, for $j=1,2,\ldots,N$,

$$x_i^j=x_i^0+\bar{h}_i\sum_{k=1}^{j}f_i(\tau^i_{jk},\boldsymbol{x}(\tau^i_{jk}),\boldsymbol{u}(\tau^i_{jk}))[(d_{jk}+1)^{\alpha_i}-d_{jk}^{\alpha_i}]+R^i_j,\tag{5.43}$$

where τ^i_{jk} is given in (5.42), and $R^i_j=\sum_{k=1}^{j}R^i_{jk}$. It has been shown in Theorem 3.19 that $|R^i_j|\le Ch^2$ for a constant $C>0$, independent of h. Omitting R^i_j, (5.43) becomes an equation approximating (5.39) with the truncation error R^i_j.

Using (5.43), we see that to compute $x_i^j=x_i(t_j)$, we need to calculate $f_i(\tau^i_{jk},\boldsymbol{x}(\tau^i_{jk}),\boldsymbol{u}(\tau^i_{jk}))$. However, $\boldsymbol{x}(\tau^i_{jk})$ and $\boldsymbol{u}(\tau^i_{jk})$ are not available directly from the scheme. Thus, approximations for $\boldsymbol{x}(\tau^i_{jk})$ and $\boldsymbol{u}(\tau^i_{jk})$ need to be determined. Using a linear interpolation and a Taylor expansion, we propose approximations to $\boldsymbol{x}(\tau^i_{jk})$ and $\boldsymbol{u}(\tau^i_{jk})$, respectively.

For any indices $j\ge k$, let

$$\rho^i_{jk}:=\frac{\tau^i_{jk}-t_{k-1}}{h}\in(0,1).$$

We approximate $\boldsymbol{x}(\tau^i_{jk})$ and $\boldsymbol{u}(\tau^i_{jk})$ by $\boldsymbol{x}^i_{jk}$ and $\boldsymbol{u}^i_{jk}$ defined as

$$\boldsymbol{x}^i_{jk}:=\boldsymbol{x}^{k-1}+\rho^i_{jk}(\boldsymbol{x}^k-\boldsymbol{x}^{k-1}),\quad \boldsymbol{u}^i_{jk}:=\boldsymbol{u}^{k-1}+\rho^i_{jk}(\boldsymbol{u}^k-\boldsymbol{u}^{k-1}).$$

The truncation error in the above linear interpolation is of order $\mathcal{O}(h^2)$. Using the above interpolations, we approximate $f_i(\tau^i_{jk}, \boldsymbol{x}(\tau^i_{jk}), \boldsymbol{u}(\tau^i_{jk}))$ by $f_i(\tau^i_{jk}, \boldsymbol{x}(\tau^i_{jk}), \boldsymbol{u}(\tau^i_{jk})) \approx f_i(\tau^i_{jk}, \boldsymbol{x}^i_{jk}, \boldsymbol{u}^i_{jk})$. The truncation error for the above approximation is also of order $\mathcal{O}(h^2)$.

Recall that $\boldsymbol{f}$ is twice continuously differentiable by Assumption 5.2.1. Replacing $f_i(\tau^i_{jk}, \boldsymbol{x}(\tau^i_{jk}), \boldsymbol{u}(\tau^i_{jk}))$ in (5.43) with the above approximation, we have the following scheme for (5.39):

$$x^j_i = x^0_i + \bar{h}_i \sum_{k=1}^{j} \left[f_i(\tau^i_{jk}, \boldsymbol{x}^i_{jk}, \boldsymbol{u}^i_{jk})((j-k+1)^{\alpha_i} - (j-k)^{\alpha_i}) \right] \tag{5.44}$$

for $j = 1, 2, \ldots, N$. Clearly, (5.44) defines a time-stepping scheme for (5.39) with the truncation error $\mathcal{O}(h^2)$.

The above scheme is implicit as it is a nonlinear system in $\boldsymbol{x}^j$. We now define an explicit single-step scheme by further approximating the jth term in the sum in (5.44) by the following Taylor expansion:

$$f_i(\tau^i_{jj}, \boldsymbol{x}^i_{jj}, \boldsymbol{u}^i_{jj}) = f_i(\tau^i_{jj}, \boldsymbol{x}^{j-1}, \boldsymbol{u}^i_{jj}) + \sum_{l=1}^{n} \frac{\partial f_i}{\partial x_l}\bigg|_{(\tau^i_{jj}, \boldsymbol{x}^{j-1}, \boldsymbol{u}^i_{jj})} \rho^i_{jj}(\boldsymbol{x}^j_l - \boldsymbol{x}^{j-1}_l)$$
$$+ \mathcal{O}(h^2).$$

By Assumption 5.2.1, $\dfrac{\partial \boldsymbol{f}}{\partial x_l}$ exists for any given values of $(t, \boldsymbol{x}, \boldsymbol{u})$. Combining the above equality and (5.44), we have the following approximation:

$$x^j_i \approx x^0_i + \bar{h}_i \sum_{k=1}^{j-1} \left[f_i(\tau^i_{jk}, \boldsymbol{x}^i_{jk}, \boldsymbol{u}^i_{jk})((j-k+1)^{\alpha_i} - (j-k)^{\alpha_i}) \right]$$
$$+ \bar{h}_i f_i\left(\tau^i_{jj}, \boldsymbol{x}^{j-1}, \boldsymbol{u}^i_{jj}\right) + \bar{h}_i \sum_{l=1}^{n} \left[\frac{\partial f_i}{\partial x_l}|_{(\tau^i_{jj}, \boldsymbol{x}^{j-1}, \boldsymbol{u}^i_{jj})} (\rho^i_{jj}(x^j_l - x^{j-1}_l)) \right] + \mathcal{O}(h^2).$$

Omitting the $\mathcal{O}(h^2)$ term and re-organizing the result, we have

$$\boldsymbol{B}^j(\boldsymbol{x}^{j-1}, \boldsymbol{u}^{j-1}, \boldsymbol{u}^j)\boldsymbol{x}^j = \boldsymbol{C}^j(\boldsymbol{x}^0, \boldsymbol{x}^1, \boldsymbol{x}^2, \ldots, \boldsymbol{x}^{j-1}, \boldsymbol{u}^0, \boldsymbol{u}^1, \ldots, \boldsymbol{u}^j) \tag{5.45}$$

for $j = 1, 2, \ldots, N$, where $\boldsymbol{B}^j = \boldsymbol{I} - [b^j_{il}]_{n\times n}$ is the $n \times n$ matrix with

$$b^j_{il} = \bar{h}_i \rho^i_{jj} \frac{\partial f_i}{\partial x_l}\bigg|_{(\tau^i_{jj}, \boldsymbol{x}^{j-1}, \boldsymbol{u}^{j-1} + \rho^i_{jj}(\boldsymbol{u}^j - \boldsymbol{u}^{j-1}))} \tag{5.46}$$

for $i = 1, 2, \dots, n$, $l = 1, 2, \dots, n$, and $\boldsymbol{C}^j = (c_1^j, c_2^j, \dots, c_n^j)^\top$ with

$$
\begin{aligned}
c_i^j &= x_i^0 + \bar{h}_i \sum_{k=1}^{j-1} \big[f_i(\tau_{jk}^i, \boldsymbol{x}^{k-1} + \rho_{jk}^i(\boldsymbol{x}^k - \boldsymbol{x}^{k-1}), \boldsymbol{u}^{k-1} + \rho_{jk}^i(\boldsymbol{u}^k - \boldsymbol{u}^{k-1})) \\
&\quad \times ((d_{jk}+1)^{\alpha_i} - d_{jk}^{\alpha_i})\big] + \bar{h}_i f_i(\tau_{jj}^i, \boldsymbol{x}^{j-1}, \boldsymbol{u}^{j-1} + \rho_{jj}^i(\boldsymbol{u}^j - \boldsymbol{u}^{j-1})) \\
&\quad - \sum_{l=1}^{n} x_l^{j-1} b_{il}^j. \qquad (5.47)
\end{aligned}
$$

To calculate $\boldsymbol{x}^j$, we need to solve (5.45)–(5.47). We now show that (5.45)–(5.47) is uniquely solvable when h is sufficiently small.

Theorem 5.4 *The system* (5.45) *has a unique solution if h is sufficiently small.*

Proof We first show that $\boldsymbol{B}^j$ is a strictly diagonally dominant, i.e., for $i = 1, 2, \dots, n$, $1 - b_{ii}^j > \sum_{\substack{l=1\\ l\neq i}}^{n} |b_{il}^j|$. Since $1 - b_{ii}^j \geq 1 - |b_{ii}^j|$, we only need to show that $1 - |b_{ii}^j| > \sum_{\substack{l=1\\ l\neq i}}^{n} |b_{il}^j|$, or, equivalently, $\sum_{l=1}^{n} |b_{il}^j| < 1$ for all i.

Since f_i is twice continuously differentiable in $\boldsymbol{x}$ on $[0, T]$, $\dfrac{\partial f_i}{\partial x_l}$ is bounded on $[0, T]$. Let $M = \max\limits_{\substack{1\leq i\leq n\\ 1\leq l\leq n}} \left| \dfrac{\partial f_i}{\partial x_l} \right|$. Using (5.46), we have

$$
|b_{il}^j| < \bar{h}_i M = \frac{h^{\alpha_i}}{\Gamma(\alpha_i + 1)} M.
$$

Choose $\hat{h}_i = \left(\dfrac{\Gamma(\alpha_i + 1)}{nM} \right)^{\frac{1}{\alpha_i}}$ and $\hat{h} = \min\limits_{1\leq i\leq n} \{\hat{h}_i\}$. When $h < \hat{h}$, we have

$$
\sum_{l=1}^{n} |b_{il}^j| < 1
$$

for all feasible i. Thus, $\boldsymbol{B}^j$ is strictly diagonally dominant. By the Levy-Desplanques theorem [309], we have that $\boldsymbol{B}^j$ is non-singular, and thus (5.45) has a unique solution. □

Introduce $\boldsymbol{X}^j = (x_1^j, x_2^j, \dots, x_n^j)^\top$ and $\boldsymbol{U}^j = (u_1^j, u_2^j, \dots, u_m^j)^\top$, and let $\boldsymbol{X} = (\boldsymbol{X}^0, \boldsymbol{X}^1, \dots, \boldsymbol{X}^N) \in \mathbb{R}^{n\times(N+1)}$ and $\boldsymbol{U} = (\boldsymbol{U}^0, \boldsymbol{U}^1, \dots, \boldsymbol{U}^N) \in \mathbb{R}^{n\times(N+1)}$. Using (5.38) and (5.45), we pose the following discretized optimal control problem:

$$\min_{U\in\mathbb{R}^{m\times(N+1)}} g_0^h(X,U) := \frac{1}{2}\left(L(t_0,X^0,U^0)+L(t_N,X^N,U^N)\right)h$$
$$+\sum_{j=1}^{N-1} L(t_j,X^j,U^j)h+\Phi(X^N), \tag{5.48}$$

$$\text{s.t.}\begin{cases} B^j(X^{j-1},U^{j-1},U^j)X^j = C^j(X^0,\ldots,X^{j-1},U^0,\ldots,U^j),\\ j=1,2,\ldots,N,\ \text{with}\ X^0=x^0. \end{cases} \tag{5.49}$$

5.4.2 Solution Strategy

From the definition of U, we see that the first-order optimality conditions for (5.48) and (5.49) are $\frac{\partial g_0^h}{\partial u_r^j}=0$ for $r=1,2,\ldots,m,\ j=0,1,\ldots,N$. We now determine the gradient $\frac{\partial g_0^h}{\partial u_r^j}$ for all feasible r and j. Using (5.48) and (5.45)–(5.47), we have

$$\frac{\partial g_0^h}{\partial u_r^0} =\frac{1}{2}h\frac{\partial L(t_0,X^0,U^0)}{\partial u_r^0}+h\sum_{p=1}^{N-1}\sum_{i=1}^{n}\frac{\partial L(t_p,X^p,U^p)}{\partial x_i^p}\frac{\partial x_i^p}{\partial u_r^0}$$
$$+\frac{1}{2}h\sum_{i=1}^{n}\frac{\partial L(t_N,X^N,U^N)}{\partial x_i^N}\frac{\partial x_i^N}{\partial u_r^0}+\sum_{i=1}^{n}\frac{\partial \Phi(X_N)}{\partial x_i^N}\frac{\partial x_i^N}{\partial u_r^0},$$

$$\frac{\partial g_0^h}{\partial u_r^j} =h\frac{\partial L(t_j,X^j,U^j)}{\partial u_r^j}+h\sum_{p=1}^{N-1}\sum_{i=1}^{n}\frac{\partial L(t_p,X^p,U^p)}{\partial x_i^p}\frac{\partial x_i^p}{\partial u_r^j}$$
$$+\frac{1}{2}h\sum_{i=1}^{n}\frac{\partial L(t_N,X^N,U^N)}{\partial x_i^N}\frac{\partial x_i^N}{\partial u_r^j}+\sum_{i=1}^{n}\frac{\partial \Phi(X^N)}{\partial x_i^N}\frac{\partial x_i^N}{\partial u_r^j}$$

for $j=1,\ldots,N$, and $r=1,2,\ldots,m$.

Using (5.49), we see that $\frac{\partial x_i^p}{\partial u_r^j}=0$ when $p<j$. Thus, the above expression can be rewritten as

$$\frac{\partial g_0^h}{\partial u_r^0}=\frac{1}{2}h\frac{\partial L}{\partial u_r}\Big|_{(t_0,X^0,U^0)}+h\sum_{p=1}^{N-1}\sum_{i=1}^{n}\frac{\partial L}{\partial x_i}\Big|_{(t_p,X^p,U^p)}\frac{\partial x_i^p}{\partial u_r^0}$$
$$+\frac{1}{2}h\sum_{i=1}^{n}\frac{\partial L}{\partial x_i}\Big|_{(t_N,X^N,U^N)}\frac{\partial x_i^N}{\partial u_r^0}+\sum_{i=1}^{n}\frac{\partial \Phi}{\partial x_i}\Big|_{(X^N)}\frac{\partial x_i^N}{\partial u_r^0}, \tag{5.50}$$

$$\frac{\partial g_0^h}{\partial u_r^j} = h\frac{\partial L}{\partial u_r}\bigg|_{(t_j,X^j,U^j)} + h\sum_{p=j}^{N-1}\sum_{i=1}^{n}\frac{\partial L}{\partial x_i}\bigg|_{(t_p,X^p,U^p)}\frac{\partial x_i^p}{\partial u_r^j}$$
$$+\frac{1}{2}h\sum_{i=1}^{n}\frac{\partial L}{\partial x_i}\bigg|_{(t_N,X^N,U^N)}\frac{\partial x_i^N}{\partial u_r^j} + \sum_{i=1}^{n}\frac{\partial \Phi}{\partial x_i}\bigg|_{(X^N)}\frac{\partial x_i^N}{\partial u_r^j} \tag{5.51}$$

for $j = 1, \ldots, N$, and $r = 1, 2, \ldots, m$. We now need to determine $\frac{\partial x_i^p}{\partial u_r^j}$. By (5.45), we have

$$\boldsymbol{B}^p \boldsymbol{X}^p = \boldsymbol{C}^p$$

for $p = 1, 2, \ldots, N$. Taking the derivative with respect to u_r^j on both sides of the above equation gives

$$\frac{\partial \boldsymbol{B}^p}{\partial u_r^j}\boldsymbol{X}^p + \boldsymbol{B}^p\frac{\partial \boldsymbol{X}^p}{\partial u_r^j} = \frac{\partial \boldsymbol{C}^p}{\partial u_r^j}.$$

Rearranging this equation gives

$$\boldsymbol{B}^p\frac{\partial \boldsymbol{X}^p}{\partial u_r^j} = \frac{\partial \boldsymbol{C}^p}{\partial u_r^j} - \frac{\partial \boldsymbol{B}^p}{\partial u_r^j}\boldsymbol{X}^p, \tag{5.52}$$

where $\boldsymbol{B}^p$ is the system matrix in (5.45); $\frac{\partial \boldsymbol{X}^p}{\partial u_r^j} = \left(\frac{\partial x_1^p}{\partial u_r^j}, \ldots, \frac{\partial x_n^p}{\partial u_r^j}\right)^\top$; $\frac{\partial \boldsymbol{C}^p}{\partial u_r^j} = \left(\frac{\partial c_1^p}{\partial u_r^j}, \ldots, \frac{\partial c_n^p}{\partial u_r^j}\right)^\top$; and $\frac{\partial \boldsymbol{B}^p}{\partial u_r^j} = -\left(\frac{\partial b_{lm}^p}{\partial u_r^j}\right)_{n\times n}$. Using (5.46) and (5.47), we see that the entries of the above matrices are given by

$$\frac{\partial b_{il}^p}{\partial u_r^j} = \rho_{pp}^i \bar{h}_i \frac{\partial\left(\frac{\partial f_i}{\partial x_l}\bigg|_{(\tau_{pp}^i,X^{p-1},U^{p-1}+\rho_{pp}^i(U^p-U^{p-1}))}\right)}{\partial u_r^j},$$
$$\frac{\partial c_i^p}{\partial u_r^j} = \sum_{k=1}^{p-1}\bar{h}_i\Bigg(\frac{\partial f_i(\tau_{pk}^i, \boldsymbol{X}^{k-1} + \rho_{pk}^i(\boldsymbol{X}^k - \boldsymbol{X}^{k-1}), \boldsymbol{U}^{k-1} + \rho_{pk}^i(\boldsymbol{U}^k - \boldsymbol{U}^{k-1}))}{\partial u_r^j}$$
$$\times\left[(p-k+1)^{\alpha_i} - (p-k)^{\alpha_i}\right]\Bigg)$$
$$+\bar{h}_i\frac{\partial f_i\left(\tau_{pp}^i, \boldsymbol{X}^{p-1}, \boldsymbol{U}^{p-1} + \rho_{pp}^i(\boldsymbol{U}^p - \boldsymbol{U}^{p-1})\right)}{\partial u_r^j} - \sum_{l=1}^{n}\left(\frac{\partial x_l^{p-1}}{\partial u_r^j}b_{il}^p + x_l^{p-1}\frac{\partial b_{il}^p}{\partial u_r^j}\right)$$

for $i = 1, 2, \ldots, n$, $l = 1, 2, \ldots, n$. Furthermore, the gradient formulas $\frac{\partial b_{il}^p}{\partial u_r^j}$ and $\frac{\partial c_i^p}{\partial u_r^j}$ are given as follows.

1. When $j = 0$, if $p = 1$, then

$$\frac{\partial b_{il}^1}{\partial u_r^0} = \rho_{11}^i \bar{h}_i \frac{\partial^2 f_i}{\partial x_l \partial u_r}\bigg|_{(\tau_{11}^i, \boldsymbol{X}^0, \boldsymbol{U}^0+\rho_{11}^i(\boldsymbol{U}^1-\boldsymbol{U}^0))} (1-\rho_{11}^i), \tag{5.53}$$

$$\frac{\partial c_i^1}{\partial u_r^0} = \bar{h}_i \frac{\partial f_i}{\partial u_r}\bigg|_{(\tau_{11}^i, \boldsymbol{X}^0, \boldsymbol{U}^0+\rho_{11}^i(\boldsymbol{U}^1-\boldsymbol{U}^0))} (1-\rho_{11}^i) - \sum_{l=1}^{n} x_l^0 \frac{\partial b_{il}^1}{\partial u_r^0}. \tag{5.54}$$

If $p > 1$, then

$$\frac{\partial b_{il}^p}{\partial u_r^0} = \rho_{pp}^i \bar{h}_i \sum_{q=1}^{n} \left(\frac{\partial^2 f_i}{\partial x_l \partial x_q}\bigg|_{(\tau_{pp}^i, \boldsymbol{X}^{p-1}, \boldsymbol{U}^{p-1}+\rho_{pp}^i(\boldsymbol{U}^p-\boldsymbol{U}^{p-1}))} \frac{\partial x_q^{p-1}}{\partial u_r^0} \right), \tag{5.55}$$

$$\begin{aligned}
\frac{\partial c_i^p}{\partial u_r^0} &= \bar{h}_i \frac{\partial f_i}{\partial u_r}\bigg|_{(\tau_{p1}^i, \boldsymbol{X}^0+\rho_{p1}^i(\boldsymbol{X}^1-\boldsymbol{X}^0), \boldsymbol{U}^0+\rho_{p1}^i(\boldsymbol{U}^1-\boldsymbol{U}^0))} (1-\rho_{p1}^i)(p^{\alpha_i}-(p-1)^{\alpha_i}) \\
&\quad + \bar{h}_i \Bigg\{ \sum_{k=2}^{p-1} \Bigg[\sum_{q=1}^{n} \Bigg(\frac{\partial f_i}{\partial x_q}\bigg|_{(\tau_{pk}^i, \boldsymbol{X}^{k-1}+\rho_{pk}^i(\boldsymbol{X}^k-\boldsymbol{X}^{k-1}), \boldsymbol{U}^{k-1}+\rho_{pk}^i(\boldsymbol{U}^k-\boldsymbol{U}^{k-1}))} \\
&\quad \times (1-\rho_{pk}^i) \frac{\partial x_q^{k-1}}{\partial u_r^0} + \frac{\partial f_i}{\partial x_q}\bigg|_{(\tau_{pk}^i, \boldsymbol{X}^{k-1}+\rho_{pk}^i(\boldsymbol{X}^k-\boldsymbol{X}^{k-1}), \boldsymbol{U}^{k-1}+\rho_{pk}^i(\boldsymbol{U}^k-\boldsymbol{U}^{k-1}))} \\
&\quad \times \rho_{pk}^i \frac{\partial x_q^k}{\partial u_r^0} \Bigg) \big((p-k+1)^{\alpha_i} - (p-k)^{\alpha_i}\big) \Bigg] \Bigg\} \\
&\quad + \bar{h}_i \Big(\sum_{q=1}^{n} \frac{\partial f_i}{\partial x_q}\bigg|_{(\tau_{pp}^i, \boldsymbol{X}^{p-1}, \boldsymbol{U}^{p-1}+\rho_{pp}^i(\boldsymbol{U}^p-\boldsymbol{U}^{p-1}))} \frac{\partial x_q^{p-1}}{\partial u_r^0} \Big) \\
&\quad - \sum_{l=1}^{n} \left(\frac{\partial x_l^{p-1}}{\partial u_r^0} b_{il}^p + x_l^{p-1} \frac{\partial b_{il}^p}{\partial u_r^0} \right).
\end{aligned} \tag{5.56}$$

2. For $j = 1, 2, \ldots, N$, if $p < j$, then

$$\frac{\partial b_{il}^p}{\partial u_r^j} = 0 \text{ and } \frac{\partial c_i^p}{\partial u_r^j} = 0. \tag{5.57}$$

If $p = j$, then

$$\frac{\partial b_{il}^p}{\partial u_r^j} = \rho_{jj}^i \bar{h}_i \frac{\partial^2 f_i}{\partial x_l \partial u_r}\bigg|_{(\tau_{pp}^i, \boldsymbol{X}^{j-1}, \boldsymbol{U}^{j-1}+\rho_{jj}^i(\boldsymbol{U}^j-\boldsymbol{U}^{j-1}))} \rho_{jj}^i, \tag{5.58}$$

$$\frac{\partial c_i^p}{\partial u_r^l} = \bar{h}_i \frac{\partial f_i}{\partial u_r}\bigg|_{(\tau_{jj}^i, \boldsymbol{X}^{j-1}, \boldsymbol{U}^{j-1}+\rho_{pp}^i(\boldsymbol{U}^j-\boldsymbol{U}^{j-1}))} \rho_{jj}^i - \sum_{l=1}^{n} \left(x_l^{j-1} \frac{\partial b_{il}^j}{\partial u_r^j} \right). \tag{5.59}$$

If $p = j + 1$, then

$$\frac{\partial b_{il}^p}{\partial u_r^j} = \rho_{pp}^i \bar{h}_i \frac{\partial^2 f_i}{\partial x_l \partial u_r}\bigg|_{(\tau_{pp}^i, \boldsymbol{X}^{p-1}, \boldsymbol{U}^{p-1}+\rho_{pp}^i(\boldsymbol{U}^p-\boldsymbol{U}^{p-1}))}(1-\rho_{pp}^i)$$
$$+ \rho_{pp}^i \bar{h}_i \left(\sum_{q=1}^{n} \frac{\partial^2 f_i}{\partial x_l \partial x_q}\bigg|_{(\tau_{pp}^i, \boldsymbol{X}^{p-1}, \boldsymbol{U}^{p-1}+\rho_{pp}^i(\boldsymbol{U}^p-\boldsymbol{U}^{p-1}))} \frac{\partial x_q^{p-1}}{\partial u_r^j} \right), \tag{5.60}$$

$$\frac{\partial c_i^p}{\partial u_r^j} = \bar{h}_i \left[\frac{\partial f_i}{\partial u_r}\bigg|_{(\tau_{pj}^i, \boldsymbol{X}^{j-1}+\rho_{pj}^i(\boldsymbol{X}^j-\boldsymbol{X}^{j-1}), \boldsymbol{U}^{j-1}+\rho_{pj}^i(\boldsymbol{U}^j-\boldsymbol{U}^{j-1}))} \rho_{pj}^i \right.$$
$$+ \sum_{q=1}^{n} \left(\frac{\partial f_i}{\partial x_q}\bigg|_{(\tau_{pj}^i, \boldsymbol{X}^{j-1}+\rho_{pj}^i(\boldsymbol{X}^j-\boldsymbol{X}^{j-1}), \boldsymbol{U}^{j-1}+\rho_{pj}^i(\boldsymbol{U}^j-\boldsymbol{U}^{j-1}))} \rho_{pj}^i \frac{\partial x_q^{j-1}}{\partial u_r^j} \right)$$
$$\left. \times (2^{\alpha_i}-1) \right] + \bar{h}_i \left[\frac{\partial f_i}{\partial u_r}\bigg|_{(\tau_{pp}^i, \boldsymbol{X}^{p-1}, \boldsymbol{U}^{p-1}+\rho_{pp}^i(\boldsymbol{U}^p-\boldsymbol{U}^{p-1}))} (1-\rho_{pp}^i) \right.$$
$$\left. + \sum_{q=1}^{n} \left(\frac{\partial f_i}{\partial x_q}\bigg|_{(\tau_{pp}^i, \boldsymbol{X}^{p-1}, \boldsymbol{U}^{p-1}+\rho_{pp}^i(\boldsymbol{U}^p-\boldsymbol{U}^{p-1}))} \frac{\partial x_q^{p-1}}{\partial u_r^j} \right) \right]$$
$$- \sum_{l=1}^{n} \left(\frac{\partial x_l^{p-1}}{\partial u_r^j} b_{il}^p + x_l^{p-1} \frac{\partial b_{il}^p}{\partial u_r^j} \right). \tag{5.61}$$

If $p > j+1$, then

$$\frac{\partial b_{il}^p}{\partial u_r^j} = \rho_{pp}^i \bar{h}_i \sum_{q=1}^{n} \frac{\partial^2 f_i}{\partial x_l \partial x_q}\bigg|_{(\tau_{pp}^i, \boldsymbol{X}^{p-1}, \boldsymbol{U}^{p-1}+\rho_{pp}^i(\boldsymbol{U}^p-\boldsymbol{U}^{p-1}))} \frac{\partial x_q^{p-1}}{\partial u_r^j}, \tag{5.62}$$

$$\frac{\partial c_i^p}{\partial u_r^j} = \bar{h}_i \left\{ \sum_{k=j}^{p-1} \left[\frac{\partial f_i}{\partial u_r}\bigg|_{(\tau_{pk}^i, \boldsymbol{X}^{k-1}+\rho_{pk}^i(\boldsymbol{X}^k-\boldsymbol{X}^{k-1}), \boldsymbol{U}^{k-1}+\rho_{pk}^i(\boldsymbol{U}^k-\boldsymbol{U}^{k-1}))} \psi(\rho_{pk}^i) \right. \right.$$
$$+ \sum_{q=1}^{n} \left(\frac{\partial f_i}{\partial x_q}\bigg|_{(\tau_{pk}^i, \boldsymbol{X}^{k-1}+\rho_{pk}^i(\boldsymbol{X}^k-\boldsymbol{X}^{k-1}), \boldsymbol{U}^{k-1}+\rho_{pk}^i(\boldsymbol{U}^k-\boldsymbol{U}^{k-1}))} (1-\rho_{pk}^i) \frac{\partial x_q^{k-1}}{\partial u_r^j} \right.$$
$$+ \frac{\partial f_i}{\partial x_q}\bigg|_{(\tau_{pk}^i, \boldsymbol{X}^{k-1}+\rho_{pk}^i(\boldsymbol{X}^k-\boldsymbol{X}^{k-1}), \boldsymbol{U}^{k-1}+\rho_{pk}^i(\boldsymbol{U}^k-\boldsymbol{U}^{k-1}))} \left. \rho_{pk}^i \frac{\partial x_q^k}{\partial u_r^j} \right)$$
$$\left. \left. \times ((p-k+1)^{\alpha_i} - (p-k)^{\alpha_i}) \right] \right\}$$
$$+ \bar{h}_i \left(\sum_{q=1}^{n} \left(\frac{\partial f_i}{\partial x_q}\bigg|_{(\tau_{pp}^i, \boldsymbol{X}^{p-1}, \boldsymbol{U}^{p-1}+\rho_{pp}^i(\boldsymbol{U}^p-\boldsymbol{U}^{p-1}))} \frac{\partial x_q^{p-1}}{\partial u_r^j} \right) \right)$$
$$- \sum_{l=1}^{n} \left(\frac{\partial x_l^{p-1}}{\partial u_r^j} b_{il}^p + x_l^{p-1} \frac{\partial b_{il}^p}{\partial u_r^j} \right), \tag{5.63}$$

where $\psi(\rho_{pj}^i) = \rho_{pj}^i$, $\psi(\rho_{p(j+1)}^i) = 1 - \rho_{p(j+1)}^i$, and $\psi(\rho_{pk}^i) = 0$ when $k > j+1$.

Note that although gradient formulas (5.53)–(5.63) are complex, they involve only simple function evaluations, not matrix inversions. Thus, calculating these formulas is computationally inexpensive when the exact derivatives of $\boldsymbol{f}$ with respect to its variables are provided. On this basis, we propose Algorithm 5.1 for solving FOCP (5.1) and (5.2).

Algorithm 5.1 Gradient-based optimization algorithm for solving FOCP (5.1) and (5.2).

Step 1. For a given N, let $t_j = jh$ for $j = 0, 1, \ldots, N$, where $h = T/N$.

Step 2. Set $k = 0$. Choose a tolerance $\bar{\varepsilon} > 0$ and an initial value $\boldsymbol{U}^{(0)} = (\boldsymbol{U}^0, \boldsymbol{U}^1, \ldots, \boldsymbol{U}^N)^{(0)} \in \mathbb{R}^{m\times(N+1)}$.

Step 3. Calculate $\boldsymbol{X}^{(k)} = (\boldsymbol{X}^0, \boldsymbol{X}^1, \ldots, \boldsymbol{X}^N)^{(k)} \in \mathbb{R}^{n\times(N+1)}$ using (5.49).

Step 4. Use (5.53)-(5.63) and $(\boldsymbol{X}^{(k)}, \boldsymbol{U}^{(k)})$ to calculate $\left(\dfrac{\partial \boldsymbol{B}^p}{\partial u_r^j}\right)^{(k)}$ and $\left(\dfrac{\partial \boldsymbol{C}^p}{\partial u_r^j}\right)^{(k)}$ for $p = 1, 2, \ldots, N$, $j = 0, 1, 2, \ldots, N$, and $r = 1, 2, \ldots, m$.

Step 5. Solve (5.52) for $\left(\dfrac{\partial \boldsymbol{X}^p}{\partial u_r^j}\right)^{(k)}$.

Step 6. Compute $\nabla g_0^h(\boldsymbol{X}^{(k)}, \boldsymbol{U}^{(k)}) = \left(\dfrac{\partial g_0^h}{\partial \boldsymbol{U}^0}, \dfrac{\partial g_0^h}{\partial \boldsymbol{U}^1}, \ldots, \dfrac{\partial g_0^h}{\partial \boldsymbol{U}^N}\right)^{(k)\top}$ using (5.50) and (5.51), where $\dfrac{\partial g_0^h}{\partial \boldsymbol{U}^j} = \left(\dfrac{\partial g_0^h}{\partial u_1^j}, \dfrac{\partial g_0^h}{\partial u_2^j}, \ldots, \dfrac{\partial g_0^h}{\partial u_m^j}\right)^\top$ for $j = 0, 1, \ldots, N$. If $\|\nabla g_0^h(\boldsymbol{X}^{(k)}, \boldsymbol{U}^{(k)})\| < \bar{\varepsilon}$, go to Step 8. Otherwise, continue.

Step 7. Compute $(\boldsymbol{X}^{(k+1)}, \boldsymbol{U}^{(k+1)})$ using the backtracking line search as follows:

- Step 7a. Choose $\sigma^0 > 0$, $\beta \in (0, 1)$ and $\gamma \in (0, 1)$. Let $l = 1$.
- Step 7b. Update $\sigma^l = \beta\sigma^{l-1}$.
- Step 7c. Let $\hat{\boldsymbol{U}} = \boldsymbol{U}^{(k)} - \sigma^l \nabla g_0^h(\boldsymbol{X}^{(k)}, \boldsymbol{U}^{(k)})$ and compute $\hat{\boldsymbol{X}}$ using (5.49) and $\hat{\boldsymbol{U}}$.
- Step 7d. If $g_0^h(\hat{\boldsymbol{X}}, \hat{\boldsymbol{U}}) \le g_0^h(\boldsymbol{X}^{(k)}, \boldsymbol{U}^{(k)}) - \gamma\sigma^l \|\nabla g_0^h(\boldsymbol{X}^{(k)}, \boldsymbol{U}^{(k)})\|^2$, then set $\boldsymbol{U}^{(k+1)} = \hat{\boldsymbol{U}}$, $\boldsymbol{X}^{(k+1)} = \hat{\boldsymbol{X}}$ and $k = k + 1$, and go to Step 4. Otherwise, let $l = l + 1$ and go to Step 7b.

Step 8. Let $(\boldsymbol{X}^*, \boldsymbol{U}^*) = (\boldsymbol{X}^{(k)}, \boldsymbol{U}^{(k)})$ and evaluate $g_0^h(\boldsymbol{X}^*, \boldsymbol{U}^*)$ using (5.48).

5.4.3 Numerical Examples

We use Algorithm 5.1 with $\beta = 0.6$, $\gamma = 0.05$, and $\sigma^0 = 10$ to solve two non-trivial examples. Computations have been performed under MATLAB environment on a PC with a 3.3GHz Intel Xeon CPU and 16 GB RAM.

Table 5.1 Optimal costs and CPU time when $\boldsymbol{\alpha} = (1.0, 1.0)^\top$ for Example 5.4.3.1

N	20	40	80	160	320
g_0	0.432743	0.432180	0.432036	0.431999	0.431990
CPU(s)	2.7456	4.1808	29.4218	3.5609e+02	4.8756e+03

Table 5.2 Optimal costs and CPU time for Example 5.4.3.1 when $N = 200$

α	$(1.0, 1.0)^\top$	$(0.9, 0.5)^\top$	$(0.2, 0.3)^\top$
g_0	0.431995	0.347784	0.267591
CPU(s)	8.3098e+02	3.3269e+02	96.2214

Example 5.4.3.1 Consider FOCP (5.1)–(5.3) from [13] with $L = \frac{1}{2}(x_1^2 + x_2^2 + u^2)$, $\Phi = 0$, $\boldsymbol{g} = \mathbf{0}$, $\boldsymbol{f} = (-x_1 + x_2 + u, -2x_2)^\top$, $T = 1$, and $\boldsymbol{x}(0) = (1, 1)^\top$. The optimal cost for $\boldsymbol{\alpha} = (1.0, 1.0)^\top$ is $g_0^* = 0.431984$.

We solve the problem for different values of N when $\boldsymbol{\alpha} = (1.0, 1.0)^\top$. The optimal costs are listed in Table 5.1, from which we see that the computed optimal cost approaches g_0^* as N increases. They also over-estimate g_0^*, which is necessary as g_0^* is the theoretical lower bound for any approximations. Furthermore, we solve the problem for various values of $\boldsymbol{\alpha}$. The results are listed in Table 5.2 and Fig. 5.2. From Table 5.2, we see that the total cost decreases as the values of $\boldsymbol{\alpha}$ decrease. A comparison of our results and those in [13] is given in Table 5.3. Clearly, the method in [13] under-estimates the optimal cost g_0^* which should be the lower bound for any approximations.

Example 5.4.3.2 Consider FOCP (5.1)–(5.3) with $L = x_1^2 + x_2^2 + u^2$, $\Phi = x_1^2 + x_2^2$, $\boldsymbol{f} = (x_2, 0.2x_2 - x_1 - 0.1x_1^3 + u)^\top$, $\boldsymbol{g} = \mathbf{0}$, $T = 10$, and $\boldsymbol{x}(0) = (2, 0)^\top$.

For $\boldsymbol{\alpha} = (1.0, 1.0)^\top$, the system is the Duffing equation which has chaotic behavior without any controls. We first demonstrate the convergence of our method for $\boldsymbol{\alpha} = (1.0, 1.0)^\top$ by listing the computed optimal costs in Table 5.4, from which it can be seen that the numerical solution improves as N increases. Furthermore, we solve Example 5.4.3.2 when $\boldsymbol{\alpha} = (0.7, 0.8)^\top$ using the uniform mesh with $N = 1200$. The computed optimal g_0 is 6.9390, and the computed optimals x_1, x_2, and u are shown in Fig. 5.3. From this example, we see that the fractional optimal control can achieve a better optimal solution than its integer counterpart.

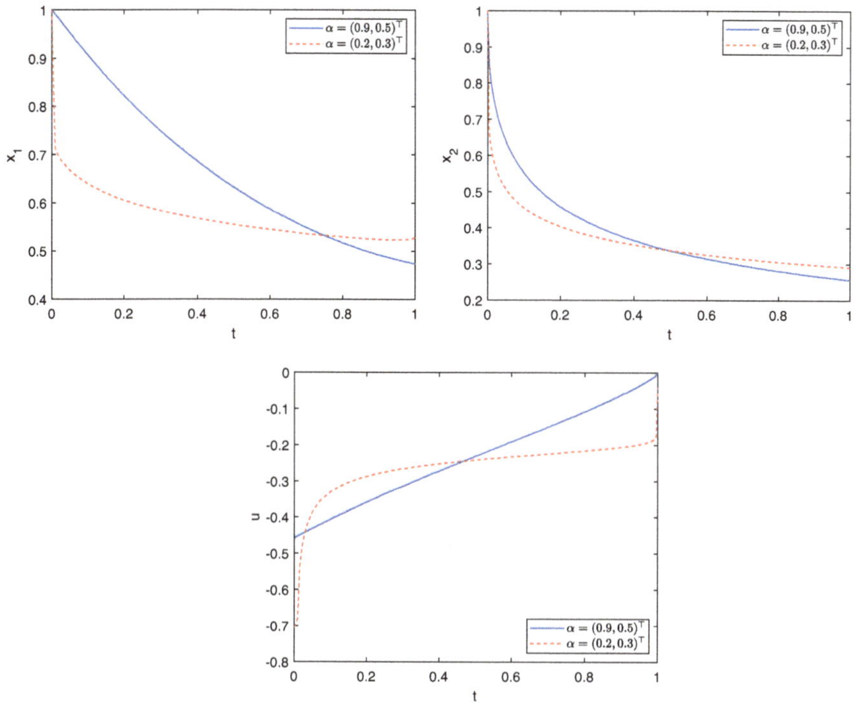

Fig. 5.2 Optimal states and controls for Example 5.4.3.1 when $N = 200$

Table 5.3 Comparison of optimal costs from our method and [13] for Example 5.4.3.1 when $N = 200$

α	g_0^*	F	Error	g_0 from [13]	Error from [13]
$(1.0, 1.0)^\top$	0.431984	0.431995	0.000011	0.431909	–0.000075
$(0.8, 0.8)^\top$	–	0.376305	–	0.376012	–

Table 5.4 Optimal costs of Example 5.4.3.2 when $\alpha = (1.0, 1.0)^\top$

N	300	600	800	1000	1200
g_0	10.5906	9.3284	9.2114	9.1659	9.1464

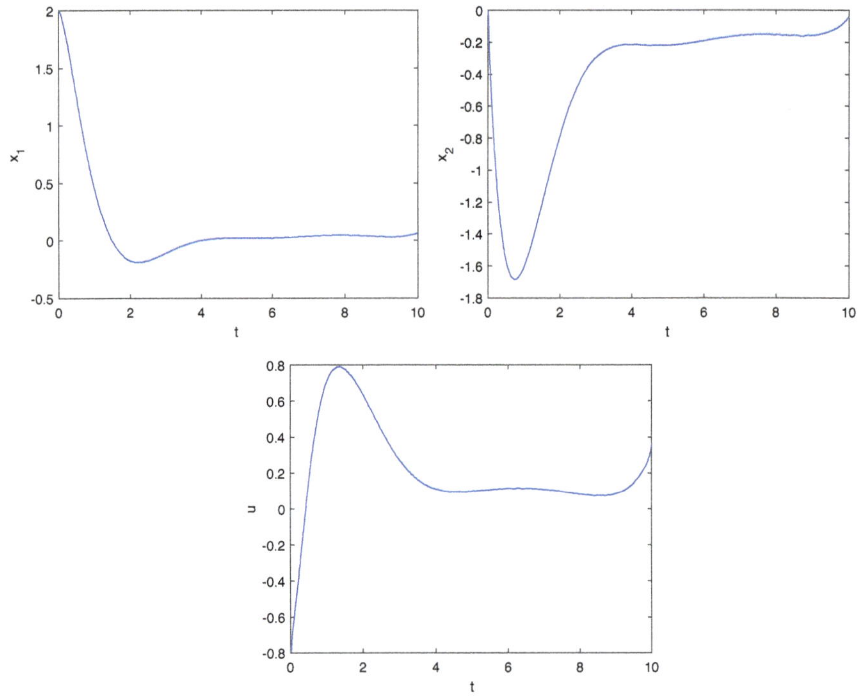

Fig. 5.3 Optimal state and control for Example 5.4.3.2 with $\alpha = (0.7, 0.8)^\top$ when $N = 1200$

5.5 Conclusions

In this chapter, the first-order necessary optimality conditions for path-constrained fractional optimal control problems have been deduced. Then, the exact penalty method for the constrained fractional optimal control problem has been discussed. Based on the numerical integration technique, we have proposed the discretization scheme for the fractional optimal control problem. Furthermore, the gradient of the discretized cost function with respect to the discretized controls has been derived. On this basis, the gradient-based optimization algorithm has been proposed for solving the resulting optimal control problem. Numerical examples have been solved to verify the necessary optimality conditions and to illustrate the effectiveness of the proposed optimization algorithm.

Chapter 6
Delay Fractional Optimal Control Problems

6.1 Introduction

In this chapter, we consider numerical solution of a class of delay fractional optimal control problems (DFOCPs). As a special kind of fractional optimal control problems, DFOCPs involve fractional time-delay systems. DFOCPs arise in many real-world applications such as mechanical engineering [27], signal processing [157], epidemiology [295], and finance [318].

Time-delays occur in almost all real-life systems. The existence of time-delays often makes control and analysis of such systems much more difficult. It should, however, be noted that most of the fractional time-delay systems are highly nonlinear and it is often impossible to obtain analytical solutions of the DFOCPs in general. Thus, finding robust and accurate numerical methods for solving DFOCPs has become an active research topic. Recently, numerical methods have been developed for solving DFOCPs based on basis function approximations of the state and control in [37, 124, 125, 204–206, 254, 272]. To apply these methods, an operational matrix for fractional derivatives is first derived. Then, it is utilized to reduce the DFOCPs to a system of algebra equations which is solved by iteration method, where Legendre orthonormal polynomials [37], block-pulse functions [124, 204–206], Chelyshkov wavelets [223], Bernoulli wavelets [254] and Bernstein polynomials [272] are used. A spectral collocation method based on Müntz-Legendre polynomials is employed in [125]. Nevertheless, it is well known that approximating the solution of a differential equation by high-order basis functions often results in ill-conditioned algebraic systems and numerical instability. In addition, the fractional orders are taken to be the same in the above-mentioned DFOCPs. In next section, we give a class of DFOCPs involving nonlinear time-delay systems with different fractional orders and subject to state inequality constraints. In Sect. 6.3, we develop numerical solution algorithm for the DFOCPs. In Sect. 6.3.1, based on Taylor expansion and linear interpolation, we propose an explicit numerical scheme for nonlinear fractional time-delay system and prove the convergence rate of the numerical solution to the exact solution. This gives

C. Liu et al., *Nonlinear Fractional Optimal Control*, Springer Optimization and Its Applications 237, https://doi.org/10.1007/978-981-92-0274-4_6

rise to a discrete-time optimal control problem. In Sect. 6.3.2, we derive gradient formulas of the cost and constraint functions with respect to decision variables. On this basis, a gradient-based optimization approach is developed to solve the discrete-time optimal control problem. In Sect. 6.4, numerical results of four non-trivial example problems are provided to demonstrate the effectiveness of our proposed solution approach. Finally, the conclusions are provided in Sect. 6.5.

The main reference of this chapter is [179].

6.2 Problem Formulation

Consider the following nonlinear fractional time-delay system:

$$
{}_0^C\mathrm{D}_t^{\boldsymbol{\alpha}} \boldsymbol{x}(t) = \boldsymbol{f}(t, \boldsymbol{x}(t), \boldsymbol{x}(t-\tau), \boldsymbol{u}(t), \boldsymbol{u}(t-\sigma)), \ t \in (0, T], \tag{6.1}
$$

$$
\boldsymbol{x}(t) = \boldsymbol{\phi}(t), \quad t \in [-\tau, 0], \tag{6.2}
$$

$$
\boldsymbol{u}(t) = \boldsymbol{\psi}(t), \quad t \in [-\sigma, 0], \tag{6.3}
$$

where $\boldsymbol{x}(t) = (x_1(t), \ldots, x_n(t))^\top \in \mathbb{R}^n$ and $\boldsymbol{u}(t) = (u_1(t), \ldots, u_r(t))^\top \in \mathbb{R}^r$ are, respectively, the state and control vectors; $\tau > 0$ is a given state-delay; $\sigma > 0$ is a given control-delay; T is a given terminal time; $\boldsymbol{f} : [0, T] \times \mathbb{R}^n \times \mathbb{R}^n \times \mathbb{R}^r \times \mathbb{R}^r \to \mathbb{R}^n$, $\boldsymbol{\phi} : [-\tau, 0] \to \mathbb{R}^n$, and $\boldsymbol{\psi} : [-\sigma, 0] \to \mathbb{R}^r$ are given functions; and ${}_0^C\mathrm{D}_t^{\boldsymbol{\alpha}} \boldsymbol{x}(t) = ({}_0^C\mathrm{D}_t^{\alpha_1} x_1(t), \ldots, {}_0^C\mathrm{D}_t^{\alpha_n} x_n(t))^\top$ with ${}_0^C\mathrm{D}_t^{\alpha_i} x_i(t)$ denoting the Caputo fractional derivative of order $\alpha_i \in (0, 1]$ of $x_i(t)$. For ease of discussion, we only allow one state-delay and one control-delay in (6.1). However, the results in this chapter can be extended in a straightforward manner to the problems involving multiple state-delays and/or multiple control-delays.

Let W be a compact and convex subset of $\mathbb{R}^r$. A piecewise continuous function $\boldsymbol{u} : [-\sigma, T] \to \mathbb{R}^r$ is said to be an *admissible control* iff $\boldsymbol{u}(t) \in W$ for almost all $t \in [0, T]$, and $\boldsymbol{u}(t) = \boldsymbol{\psi}(t)$ for all $t \in [-\sigma, 0]$. Let $\mathcal{U}$ denote the class of all such admissible controls.

We assume throughout this chapter that the following conditions are satisfied.

Assumption 6.2.1 The functions $\boldsymbol{f}$ and $\boldsymbol{\phi}$ are twice continuously differentiable with respect to all their arguments.

Assumption 6.2.2 The function $\boldsymbol{\psi}$ is continuous with respect to t.

Assumptions 6.2.1 and 6.2.2 ensure that system (6.1)–(6.3) admits a unique continuous solution $\boldsymbol{x}(\cdot|\boldsymbol{u})$ satisfying the fractional differential equation (6.1) almost everywhere on $[0, T]$, and the initial condition (6.2) everywhere on $[-\tau, 0]$ corresponding to each $\boldsymbol{u} \in \mathcal{U}$. We now state our optimal control problem as follows:

Problem (P) *Given the fractional time-delay system (6.1)–(6.3), find an admissible control* $\boldsymbol{u} \in \mathcal{U}$ *such that the cost function*

$$g_0(\boldsymbol{u}) = \Phi_0(\boldsymbol{x}(T|\boldsymbol{u})) \tag{6.4}$$

is minimized subject to the state inequality constraints

$$g_j(\boldsymbol{u}) = \Phi_j(\boldsymbol{x}(T|\boldsymbol{u})) \leq 0, \quad j = 1, \ldots, N_i, \tag{6.5}$$

where $\Phi_j : \mathbb{R}^n \to \mathbb{R}$, $j = 0, 1, \ldots, N_i$, *are given continuously differentiable functions.*

Note that, in (6.4) and (6.5), functions Φ_j, $j = 0, 1, \ldots, N_i$, only include terminal state of system (6.1). However, we can easily transform any integral form into (6.4) and (6.5) by introducing additional state variables. For example, consider the following integral term:

$$\int_0^T L(t, \boldsymbol{x}(t), \boldsymbol{x}(t-\tau), \boldsymbol{u}(t), \boldsymbol{u}(t-\sigma))dt,$$

where $L : [0, T] \times \mathbb{R}^n \times \mathbb{R}^n \times \mathbb{R}^r \times \mathbb{R}^r \to \mathbb{R}$ is a given continuously differentiable function. Clearly, this term can be replaced by $x_{n+1}(T)$, where x_{n+1} is a new state variable satisfying the dynamics

$$\begin{aligned} \dot{x}_{n+1}(t) &= L(t, \boldsymbol{x}(t), \boldsymbol{x}(t-\tau), \boldsymbol{u}(t), \boldsymbol{u}(t-\sigma)), \quad t \in (0, T], \\ x_{n+1}(t) &= 0, \quad t \in [-\tau, 0]. \end{aligned}$$

Thus, there is no loss of generality in ignoring the integral terms in (6.4) and (6.5).

6.3 Numerical Solution Algorithm

6.3.1 Numerical Integration

By Theorem 3.13, it follows that, for each $\boldsymbol{u} \in \mathcal{U}$, Eq. (6.1) is equivalent to the following Volterra integral equation:

$$x_i(t) = \phi_i(0) + \frac{1}{\Gamma(\alpha_i)} \int_0^t (t-s)^{\alpha_i - 1} f_i(s, \boldsymbol{x}(s), \boldsymbol{x}(s-\tau), \boldsymbol{u}(s), \boldsymbol{u}(s-\sigma))ds \tag{6.6}$$

for $t \in (0, T]$, $i = 1, \ldots, n$. Now, consider a uniform grid $\{t_k = kh |\ k = 0, 1, \cdots, N\}$ on $[0, T]$, where N is an integer such that $h = \dfrac{T}{N}$. Then, (6.6) can be rewritten as

$$x_i(t_k) = \phi_i(0) + \frac{1}{\Gamma(\alpha_i)}$$
$$\times \sum_{l=1}^{k} \int_{(l-1)h}^{lh} (kh - s)^{\alpha_i - 1} f_i(s, \boldsymbol{x}(s), \boldsymbol{x}(s-\tau), \boldsymbol{u}(s), \boldsymbol{u}(s-\sigma))ds \tag{6.7}$$

for all feasible i and k. We now consider an approximation of the integral in (6.7). By Assumption 6.2.1, $f_i(t, \boldsymbol{x}(t), \boldsymbol{x}(t-\tau), \boldsymbol{u}(t), \boldsymbol{u}(t-\sigma))$ is twice continuously differentiable with respect to all its arguments. Thus, for $l = 1, \ldots, k$, we apply the Taylor expansion to $f_i(s, \boldsymbol{x}(s), \boldsymbol{x}(s-\tau), \boldsymbol{u}(s), \boldsymbol{u}(s-\sigma))$ at any point $s_{kl}^i \in ((l-1)h, lh)$ to yield

$$\begin{aligned} &f_i(s, \boldsymbol{x}(s), \boldsymbol{x}(s-\tau), \boldsymbol{u}(s), \boldsymbol{u}(s-\sigma)) \\ &= f_i(s_{kl}^i, \boldsymbol{x}(s_{kl}^i), \boldsymbol{x}(s_{kl}^i - \tau), \boldsymbol{u}(s_{kl}^i), \boldsymbol{u}(s_{kl}^i - \sigma)) + K_{kl}^i(s - s_{kl}^i) \\ &\quad + c_{kl}^i(s - s_{kl}^i)^2, \end{aligned} \tag{6.8}$$

where K_{kl}^i and c_{kl}^i are the coefficients of first-order and second-order terms of the expansion. Substituting the right hand-side of (6.8) into the integrand of the last term in (6.7), we obtain that, for $l = 1, \ldots, k$,

$$\begin{aligned} &\frac{1}{\Gamma(\alpha_i)} \int_{(l-1)h}^{lh} (kh - s)^{\alpha_i - 1} f_i(s, \boldsymbol{x}(s), \boldsymbol{x}(s-\tau), \boldsymbol{u}(s), \boldsymbol{u}(s-\sigma))ds \\ &= \bar{h}_i f_i(s_{kl}^i, \boldsymbol{x}(s_{kl}^i), \boldsymbol{x}(s_{kl}^i - \tau), \boldsymbol{u}(s_{kl}^i), \boldsymbol{u}(s_{kl}^i - \sigma))[(d_{kl} + 1)^{\alpha_i} - d_{kl}^{\alpha_i}] \\ &\quad + \frac{K_{kl}^i}{\Gamma(\alpha_i)} \int_{(l-1)h}^{lh} (kh - s)^{\alpha_i - 1}(s - s_{kl}^i)ds + R_{kl}^i, \end{aligned} \tag{6.9}$$

where $d_{kl} = (k - l)$;

$$\bar{h}_i = \frac{h^{\alpha_i}}{\Gamma(\alpha_i + 1)}; \tag{6.10}$$

and

$$R_{kl}^i = \frac{c_{kl}^i}{\Gamma(\alpha_i)} \int_{(l-1)h}^{lh} (kh - s)^{\alpha_i - 1}(s - s_{kl}^i)^2 ds. \tag{6.11}$$

To achieve the truncation error in (6.9) being R_{kl}^i, we see that s_{kl}^i should be chosen such that the second term on the right hand-side of (6.9) becomes zero. This choice of s_{kl}^i is given as follows:

$$s_{kl}^i = h \frac{[(d_{kl} + 1)^{\alpha_i + 1} - d_{kl}^{\alpha_i + 1}] + (\alpha_i + 1)[(d_{kl} + 1)^{\alpha_i}(l - 1) - d_{kl}^{\alpha_i} l]}{(\alpha_i + 1)[(d_{kl} + 1)^{\alpha_i} - d_{kl}^{\alpha_i}]}. \tag{6.12}$$

Obviously, $s_{kl}^i \in ((l-1)h, lh)$.

Combining the resulting expression with (6.7), we have, for $k = 1, \ldots, N$,

$$x_i(t_k) = \phi_i(0) + \bar{h}_i \sum_{l=1}^{k} \{f_i(s_{kl}^i, \boldsymbol{x}(s_{kl}^i), \boldsymbol{x}(s_{kl}^i - \tau), \boldsymbol{u}(s_{kl}^i), \boldsymbol{u}(s_{kl}^i - \sigma)) \\ \times [(d_{kl}+1)^{\alpha_i} - d_{kl}^{\alpha_i}]\} + R_k^i, \tag{6.13}$$

where s_{kl}^i is given by (6.12); and $R_k^i = \sum_{l=1}^{k} R_{kl}^i$.

Let $\boldsymbol{x}^k$ and $\boldsymbol{u}^k$ be the numerical counterparts of $\boldsymbol{x}(t_k)$ and $\boldsymbol{u}(t_k)$. For any $k \geq l$, let $\rho_{kl}^i = \dfrac{s_{kl}^i - t_{l-1}}{h} \in (0, 1)$. Then, $\boldsymbol{x}(s_{kl}^i)$ and $\boldsymbol{u}(s_{kl}^i)$ can be approximated by $\boldsymbol{x}_{kl}^i$ and $\boldsymbol{u}_{kl}^i$, respectively, where

$$\boldsymbol{x}_{kl}^i = \boldsymbol{x}^{l-1} + \rho_{kl}^i(\boldsymbol{x}^l - \boldsymbol{x}^{l-1}); \text{ and } \boldsymbol{u}_{kl}^i = \boldsymbol{u}^{l-1} + \rho_{kl}^i(\boldsymbol{u}^l - \boldsymbol{u}^{l-1}). \tag{6.14}$$

In the numerical integration (6.13), it is essential to establish the approximation of the delayed terms. For this, we approximate $\boldsymbol{x}(s_{kl}^i - \tau)$ and $\boldsymbol{u}(s_{kl}^i - \sigma)$ by

$$\tilde{\boldsymbol{x}}_{kl_\tau}^i = \begin{cases} \phi(s_{kl}^i - \tau), & \text{if } s_{kl}^i - \tau \leq 0, \\ \boldsymbol{x}^{l_\tau - 1} + \bar{\rho}_{kl_\tau}^i(\boldsymbol{x}^{l_\tau} - \boldsymbol{x}^{l_\tau - 1}), & \text{if } s_{kl}^i - \tau \in (t_{l_\tau - 1}, t_{l_\tau}] \\ & \text{for some } l_\tau \in \{1, \ldots, l\}, \end{cases} \tag{6.15}$$

and

$$\tilde{\boldsymbol{u}}_{kl_\sigma}^i = \begin{cases} \psi(s_{kl}^i - \sigma), & \text{if } s_{kl}^i - \sigma < 0, \\ \boldsymbol{u}^{l_\sigma - 1} + \tilde{\rho}_{kl_\sigma}^i(\boldsymbol{u}^{l_\sigma} - \boldsymbol{u}^{l_\sigma - 1}), & \text{if } s_{kl}^i - \sigma \in [t_{l_\sigma - 1}, t_{l_\sigma}) \\ & \text{for some } l_\sigma \in \{1, \ldots, l\}, \end{cases} \tag{6.16}$$

respectively, where $\bar{\rho}_{kl_\tau}^i = \dfrac{s_{kl}^i - \tau - t_{l_\tau - 1}}{h} \in (0, 1]$; and $\tilde{\rho}_{kl_\sigma}^i = \dfrac{s_{kl}^i - \sigma - t_{l_\sigma - 1}}{h} \in [0, 1)$. Note that, in (6.15) and (6.16), $l_\tau = k$ and/or $l_\sigma = k$ are possible. They are called overlapping in the numerical computation of delay differential equations [165]. Furthermore, the truncation error in the above linear interpolation is of order $\mathcal{O}(h^2)$. Using the above interpolations, we approximate $f_i(s_{kl}^i, \boldsymbol{x}(s_{kl}^i), \boldsymbol{x}(s_{kl}^i - \tau), \boldsymbol{u}(s_{kl}^i), \boldsymbol{u}(s_{kl}^i - \sigma))$ by $f_i(s_{kl}^i, \boldsymbol{x}_{kl}^i, \tilde{\boldsymbol{x}}_{kl_\tau}^i, \boldsymbol{u}_{kl}^i, \tilde{\boldsymbol{u}}_{kl_\sigma}^i)$. The truncation error for the above approximation is also of order $\mathcal{O}(h^2)$.

Now, we have the following approximate scheme for (6.7):

$$x_i^k = x_i^0 + \bar{h}_i \sum_{l=1}^{k} \left[f_i(s_{kl}^i, \boldsymbol{x}_{kl}^i, \tilde{\boldsymbol{x}}_{kl_\tau}^i, \boldsymbol{u}_{kl}^i, \tilde{\boldsymbol{u}}_{kl_\sigma}^i)((k-l+1)^{\alpha_i} - (k-l)^{\alpha_i}) \right] \tag{6.17}$$

for $k = 1, \ldots, N$.

The above scheme is implicit since it is a nonlinear system in $\boldsymbol{x}^k$. It is shown in [108] that the implicit scheme for solving delay differential equations is not stable numerically. Thus, by introducing indicator representation of overlapping, an explicit single-step scheme is obtained by applying Taylor expansion to approximate the kth term in the summation appeared in (6.17):

$$
\begin{aligned}
&f_i(s^i_{kk}, \boldsymbol{x}^i_{kk}, \tilde{\boldsymbol{x}}^i_{kk_\tau}, \boldsymbol{u}^i_{kk}, \tilde{\boldsymbol{u}}^i_{kk_\sigma})\\
&= f_i(s^i_{kk}, \boldsymbol{x}^{k-1}, \delta^i_{kk_\tau}\boldsymbol{x}^{k-1} + (1-\delta^i_{kk_\tau})\tilde{\boldsymbol{x}}^i_{kk_\tau}, \boldsymbol{u}^i_{kk}, \tilde{\boldsymbol{u}}^i_{kk_\sigma})\\
&\quad + \sum_{p=1}^{n}\left[\frac{\partial f_i}{\partial x_p}\bigg|_{(s^i_{kk},\boldsymbol{x}^{k-1},\delta^i_{kk_\tau}\boldsymbol{x}^{k-1}+(1-\delta^i_{kk_\tau})\tilde{\boldsymbol{x}}^i_{kk_\tau},\boldsymbol{u}^i_{kk},\tilde{\boldsymbol{u}}^i_{kk_\sigma})} \rho^i_{kk}(x^k_p - x^{k-1}_p)\right]\\
&\quad + \sum_{p=1}^{n}\left[\frac{\partial f_i}{\partial \tilde{x}_p}\bigg|_{(s^i_{kk},\boldsymbol{x}^{k-1},\delta^i_{kk_\tau}\boldsymbol{x}^{k-1}+(1-\delta^i_{kk_\tau})\tilde{\boldsymbol{x}}^i_{kk_\tau},\boldsymbol{u}^i_{kk},\tilde{\boldsymbol{u}}^i_{kk_\sigma})} \delta^i_{kk_\tau}\bar{\rho}^i_{kk_\tau}(x^k_p - x^{k-1}_p)\right]\\
&\quad + \mathcal{O}(h^2),
\end{aligned}
\tag{6.18}
$$

where $\partial\tilde{\boldsymbol{x}}$ denotes the partial differentiation with respect to $\boldsymbol{x}(t-\tau)$; and $\delta^i_{kk_\tau} = 1$ in the case that overlapping occurs for the state-delay τ, and $\delta^i_{kk_\tau} = 0$ otherwise.

Combining (6.18) with (6.17), we have the following approximation:

$$
\begin{aligned}
x^k_i &= x^0_i + \bar{h}_i \sum_{l=1}^{k-1}\left[f_i(s^i_{kl}, \boldsymbol{x}^i_{kl}, \tilde{\boldsymbol{x}}^i_{kl_\tau}, \boldsymbol{u}^i_{kl}, \tilde{\boldsymbol{u}}^i_{kl_\sigma})((k-l+1)^{\alpha_i} - (k-l)^{\alpha_i})\right]\\
&\quad + \bar{h}_i f_i(s^i_{kk}, \boldsymbol{x}^{k-1}, \delta^i_{kk_\tau}\boldsymbol{x}^{k-1} + (1-\delta^i_{kk_\tau})\tilde{\boldsymbol{x}}^i_{kk_\tau}, \boldsymbol{u}^i_{kk}, \tilde{\boldsymbol{u}}^i_{kk_\sigma})\\
&\quad + \bar{h}_i \sum_{p=1}^{n}\left[\frac{\partial f_i}{\partial x_p}\bigg|_{(s^i_{kk},\boldsymbol{x}^{k-1},\delta^i_{kk_\tau}\boldsymbol{x}^{k-1}+(1-\delta^i_{kk_\tau})\tilde{\boldsymbol{x}}^i_{kk_\tau},\boldsymbol{u}^i_{kk},\tilde{\boldsymbol{u}}^i_{kk_\sigma})} \rho^i_{kk}(x^k_p - x^{k-1}_p)\right]\\
&\quad + \bar{h}_i \sum_{p=1}^{n}\left[\frac{\partial f_i}{\partial \tilde{x}_p}\bigg|_{(s^i_{kk},\boldsymbol{x}^{k-1},\delta^i_{kk_\tau}\boldsymbol{x}^{k-1}+(1-\delta^i_{kk_\tau})\tilde{\boldsymbol{x}}^i_{kk_\tau},\boldsymbol{u}^i_{kk},\tilde{\boldsymbol{u}}^i_{kk_\sigma})} \delta^i_{kk_\tau}\bar{\rho}^i_{kk_\tau}(x^k_p - x^{k-1}_p)\right]\\
&\quad + \mathcal{O}(h^{2+\alpha_i}).
\end{aligned}
\tag{6.19}
$$

Omitting the $\mathcal{O}(h^{2+\alpha_i})$ in (6.19) and reorganizing the result, we have

$$
\begin{aligned}
&\boldsymbol{B}^k(\{\tilde{\boldsymbol{x}}^i_{kk_\tau}\}^n_{i=1}, \boldsymbol{x}^{k-1}, \{\tilde{\boldsymbol{u}}^i_{kk_\sigma}\}^n_{i=1}, \boldsymbol{u}^{k-1}, \boldsymbol{u}^k)\boldsymbol{x}^k\\
&= \boldsymbol{C}^k(\{\{\tilde{\boldsymbol{x}}^i_{kl_\tau}\}^n_{i=1}\}^{k-1}_{l=1}, \{\boldsymbol{x}^l\}^{k-1}_{l=1}, \{\{\tilde{\boldsymbol{u}}^i_{kl_\sigma}\}^n_{i=1}\}^{k-1}_{l=1}, \{\boldsymbol{u}^\ell\}^k_{\ell=0})
\end{aligned}
\tag{6.20}
$$

for $k = 1, \ldots, N$, where $\boldsymbol{B}^k = \boldsymbol{I} - [b_{ip}^k]_{n\times n}$ is the $n \times n$ matrix with

$$b_{ip}^k = \bar{h}_i \rho_{kk}^i \left.\frac{\partial f_i}{\partial x_p}\right|_{(s_{kk}^i, \boldsymbol{x}^{k-1}, \delta_{kk_\tau}^i \boldsymbol{x}^{k-1} + (1-\delta_{kk_\tau}^i)\tilde{\boldsymbol{x}}_{kk_\tau}^i, \boldsymbol{u}_{kk}^i, \tilde{\boldsymbol{u}}_{kk_\sigma}^i)}$$
$$+ \bar{h}_i \delta_{kk_\tau}^i \bar{\rho}_{kk_\tau}^i \left.\frac{\partial f_i}{\partial \tilde{x}_p}\right|_{(s_{kk}^i, \boldsymbol{x}^{k-1}, \delta_{kk_\tau}^i \boldsymbol{x}^{k-1} + (1-\delta_{kk_\tau}^i)\tilde{\boldsymbol{x}}_{kk_\tau}^i, \boldsymbol{u}_{kk}^i, \tilde{\boldsymbol{u}}_{kk_\sigma}^i)} \tag{6.21}$$

for $i = 1, \ldots, n$, $p = 1, \ldots, n$; and $\boldsymbol{C}^k = (c_1^k, \ldots, c_n^k)^\top$ with

$$c_i^k = x_i^0 + \bar{h}_i \sum_{l=1}^{k-1} \left[f_i(s_{kl}^i, \boldsymbol{x}_{kl}^i, \tilde{\boldsymbol{x}}_{kl_\tau}^i, \boldsymbol{u}_{kl}^i, \tilde{\boldsymbol{u}}_{kl_\sigma}^i)((k-l+1)^{\alpha_i} - (k-l)^{\alpha_i}) \right]$$
$$+ \bar{h}_i f_i(s_{kk}^i, \boldsymbol{x}^{k-1}, \delta_{kk_\tau}^i \boldsymbol{x}^{k-1} + (1-\delta_{kk_\tau}^i)\tilde{\boldsymbol{x}}_{kk_\tau}^i, \boldsymbol{u}_{kk}^i, \tilde{\boldsymbol{u}}_{kk_\sigma}^i)$$
$$- \sum_{p=1}^{n} b_{ip}^k x_p^{k-1}. \tag{6.22}$$

Thus, the approximate solution of (6.7) can be obtained by solving the discrete-time system (6.20) for each $\boldsymbol{u} \in \mathcal{U}$.

Using the linear interpolation theory and Taylor's theorem, we are able to prove that, for any $k = 1, \ldots, N$, approximation solution $\boldsymbol{x}^k$ converges to the exact solution $\boldsymbol{x}(t_k)$ at the rate of $\mathcal{O}(h^2)$. This is given in the following theorem.

Theorem 6.1 *Let $\boldsymbol{x}(t)$ be the exact solution of (6.6). Then, there exists an $\hat{h}$ such that $h < \hat{h}$ implies*

$$\|\boldsymbol{x}(t_k) - \boldsymbol{x}^k\|_\infty \le C h^2, \quad k = 1, \ldots, N, \tag{6.23}$$

where $\|\cdot\|_\infty$ is the infinity norm; $\{\boldsymbol{x}^k\}$ is the sequence generated by (6.20); and $C > 0$ is a constant independent of h and $\boldsymbol{\alpha}$.

Proof We will prove the theorem in the following steps.
Step 1. Preliminaries

For $k = 1, \ldots, N$, define $\tilde{\boldsymbol{C}}^k = (\tilde{c}_1^k, \cdots, \tilde{c}_n^k)^\top$, where

$$\tilde{c}_i^k = x_i^0 + \bar{h}_i \sum_{l=1}^{k-1} \left[f_i(s_{kl}^i, \boldsymbol{x}_{kl}^i, \tilde{\boldsymbol{x}}_{kl_\tau}^i, \boldsymbol{u}_{kl}^i, \tilde{\boldsymbol{u}}_{kl_\sigma}^i) \cdot ((k-l+1)^{\alpha_i} - (k-l)^{\alpha_i}) \right]$$
$$+ \bar{h}_i f_i(s_{kk}^i, \boldsymbol{x}^{k-1}, \delta_{kk_\tau}^i \boldsymbol{x}^{k-1} + (1-\delta_{kk_\tau}^i)\tilde{\boldsymbol{x}}_{kk_\tau}^i, \boldsymbol{u}_{kk}^i, \tilde{\boldsymbol{u}}_{kk_\sigma}^i)$$
$$- \sum_{p=1}^{n} b_{ip}^k x_p(t_{k-1}) + \mathcal{O}(h^2).$$

Note that this vector can be regarded as a disturbance of vector $\boldsymbol{C}^k$ defined in (6.22). Furthermore, by using linear interpolation, $\boldsymbol{x}(s_{kl}^i)$, $\boldsymbol{x}(s_{kl}^i - \tau)$, $\boldsymbol{u}(s_{kl}^i)$ and $\boldsymbol{u}(s_{kl}^i - \sigma)$ in (6.13) can be, respectively, expressed as

$$x_p(s_{kl}^i) = x_p(t_{l-1}) + \rho_{kl}^i(x_p(t_l) - x_p(t_{l-1})) + \mathcal{O}(h^2), \quad p = 1, \ldots, n, \tag{6.24}$$

$$\begin{aligned} & x_p(s_{kl}^i - \tau) \\ & = \begin{cases} \phi_p(s_{kl}^i - \tau), & \text{if } s_{kl}^i - \tau \le 0, \\ x_p(t_{l_\tau - 1}) + \bar{\rho}_{kl_\tau}^i(x_p(t_{l_\tau}) - x_p(t_{l_\tau - 1})) + \mathcal{O}(h^2), & \text{if } s_{kl}^i - \tau \in (t_{l_\tau - 1}, t_{l_\tau}], \end{cases} \\ & \qquad\qquad\qquad\qquad p = 1, \ldots, n, \end{aligned} \tag{6.25}$$

$$\boldsymbol{u}(s_{kl}^i) = \boldsymbol{u}_{kl}^i + \mathbf{1}^r \mathcal{O}(h^2), \tag{6.26}$$

and

$$\boldsymbol{u}(s_{kl}^i - \sigma) = \tilde{\boldsymbol{u}}_{kl_\sigma}^i + \delta_{kl_\sigma}^i \mathbf{1}^r \mathcal{O}(h^2) \tag{6.27}$$

for all feasible i, k and l, where $\boldsymbol{u}_{kl}^i$ and $\tilde{\boldsymbol{u}}_{kl_\sigma}^i$ are as defined in (6.14) and (6.16), respectively; and $\mathbf{1}^r$ is a column vector of all ones in $\mathbb{R}^r$.

Step 2. The truncation error R_k^i is of order $\mathcal{O}(h^2)$

Let $i \in \{1, \ldots, n\}$, $k \in \{1, \ldots, N\}$ and $l \in \{1, \ldots, k\}$. From (6.11), we have

$$\begin{aligned} |R_{kl}^i| &= \left| \frac{c_{kl}^i}{\Gamma(\alpha_i)} \int_{(l-1)h}^{lh} (kh - s)^{\alpha_i - 1}(s - s_{kl}^i) ds \right| \\ &\le \frac{|c_{kl}^i| h^2}{\Gamma(\alpha_i)} \int_{(l-1)h}^{lh} (kh - s)^{\alpha_i - 1} ds \\ &= \frac{|c_{kl}^i| h^2}{\Gamma(\alpha_i + 1)} \left\{ [(k - l + 1)h]^{\alpha_i} - [(k - l)h]^{\alpha_i} \right\}. \end{aligned}$$

By Assumption 6.2.1, $\boldsymbol{f}$ is twice continuously differentiable. Thus, c_{kl}^i is bounded on $[-\tau, T]$. Let $\bar{c} = \max_{\substack{i \in \{1,\ldots,n\} \\ k \in \{1,\ldots,N\}}} \max_{l \in \{1,\ldots,k\}} |c_{kl}^i|$. Then, from the definition of R_k^i, we have

$$\begin{aligned} |R_k^i| &\le \sum_{l=1}^{k} |R_{kl}^i| \\ &\le \frac{\bar{c} h^2}{\Gamma(\alpha_i + 1)} \sum_{l=1}^{k} \left\{ [(k - l + 1)h]^{\alpha_i} - [(k - l)h]^{\alpha_i} \right\} \\ &= \frac{\bar{c} h^2}{\Gamma(\alpha_i + 1)} (kh)^{\alpha_i} \le \frac{\bar{c} h^2}{\Gamma(\alpha_i + 1)} T^{\alpha_i}. \end{aligned}$$

Recall that $\alpha_i \in (0, 1]$ for $i = 1, \dots, n$. Then, $\Gamma(\alpha_i + 1) \in [2^{\alpha_i - 1}, 1]$. Also, it is obvious that $T^{\alpha_i} \leq \max\{1, T\}$. Choosing $\bar{C} = 2\bar{c}\max\{1, T\}$ gives

$$|R_k^i| \leq \bar{C}h^2 = \mathcal{O}(h^2). \tag{6.28}$$

Step 3. $\boldsymbol{B}^k$ is non-singular

By Assumption 6.2.1, $\boldsymbol{f}$ is twice continuously differentiable in $\boldsymbol{x}(t)$ and $\boldsymbol{x}(t-\tau)$. Thus, $\dfrac{\partial f_i}{\partial x_p}$ and $\dfrac{\partial f_i}{\partial \tilde{x}_p}$ are bounded on $[0, T]$ for $i = 1, \dots, n$. Let

$$M = \max_{\substack{i \in \{1,\dots,n\} \\ p \in \{1,\dots,n\}}} \left\{ \left| \frac{\partial f_i}{\partial x_p} \right|, \left| \frac{\partial f_i}{\partial \tilde{x}_p} \right| \right\}.$$

Note that $\rho_{kk}^i \in (0, 1)$ and $\bar{\rho}_{kk_\tau}^i \in (0, 1]$ for all feasible i and k. It follows from (6.21) that

$$|b_{ip}^k| \leq 2\bar{h}_i M = \frac{2h^{\alpha_i} M}{\Gamma(\alpha_i + 1)}.$$

Let $\hat{h}_i = \left(\dfrac{\Gamma(\alpha_i + 1)}{2nM} \right)^{\frac{1}{\alpha_i}}$ and $\hat{h} = \min_{i \in \{1,\dots,n\}} \hat{h}_i$. Then, for all $h < \hat{h}$, $\sum_{p=1}^n |b_{ip}^k| < 1$, $i = 1, \dots, n$. Since $1 - b_{ii}^k \geq 1 - |b_{ii}^k|$, we have, for all $i = 1, \dots, n$,

$$\sum_{\substack{p=1 \\ p \neq i}}^n |b_{ip}^k| < 1 - |b_{ii}^k| \leq 1 - b_{ii}^k,$$

which implies that matrix $\boldsymbol{B}^k$ is strictly diagonally dominant. By the Levy-Desplanques theorem [309], $\boldsymbol{B}^k$ is non-singular.

Step 4. Base case for (6.23)

For $k = 1$, it follows from (6.13) that, for $i = 1, \dots, n$,

$$x_i(t_1) = x_i^0 + \bar{h}_i f_i(s_{11}^i, \boldsymbol{x}(s_{11}^i), \boldsymbol{x}(s_{11}^i - \tau), \boldsymbol{u}(s_{11}^i), \boldsymbol{u}(s_{11}^i - \sigma)) + R_1^i. \tag{6.29}$$

Substituting (6.24)–(6.27) with $k = l = 1$ into (6.29) and applying Taylor expansion yields

$$\begin{aligned} x_i(t_1) &= \phi_i(0) + \bar{h}_i f_i(s_{11}^i, \boldsymbol{x}^0, \delta_{11_\tau}^i \boldsymbol{x}^0 + (1 - \delta_{11_\tau}^i)\boldsymbol{\phi}(s_{11}^i - \tau), \boldsymbol{u}_{11}^i, \tilde{\boldsymbol{u}}_{11_\sigma}^i) \\ &+ \bar{h}_i \sum_{p=1}^n \left\{ \frac{\partial f_i}{\partial x_p} \Big|_{(s_{11}^i, \boldsymbol{x}^0, \delta_{11_\tau}^i \boldsymbol{x}^0 + (1-\delta_{11_\tau}^i)\boldsymbol{\phi}(s_{11}^i - \tau), \boldsymbol{u}_{11}^i, \tilde{\boldsymbol{u}}_{11_\sigma}^i)} \Big[\rho_{11}^i (x_p(t_1) - x_p^0) \right. \\ &\left. + \mathcal{O}(h^2) \Big] \right\} + \bar{h}_i \sum_{p=1}^n \left\{ \frac{\partial f_i}{\partial \tilde{x}_p} \Big|_{(s_{11}^i, \boldsymbol{x}^0, \delta_{11_\tau}^i \boldsymbol{x}^0 + (1-\delta_{11_\tau}^i)\boldsymbol{\phi}(s_{11}^i - \tau), \boldsymbol{u}_{11}^i, \tilde{\boldsymbol{u}}_{11_\sigma}^i)} \right. \end{aligned}$$

$$\times\,\delta^i_{11_\tau}\Big[\bar{\rho}^i_{11_\tau}(x_p(t_1)-x^0_p)+\mathcal{O}(h^2)\Big]\Big\}$$
$$+\bar{h}_i\sum_{m=1}^{r}\frac{\partial f_i}{\partial u_m}\Big|_{(s^i_{11},\boldsymbol{x}^0,\delta^i_{11_\tau}\boldsymbol{x}^0+(1-\delta^i_{11_\tau})\phi(s^i_{11}-\tau),\boldsymbol{u}^i_{11},\tilde{\boldsymbol{u}}^i_{11_\sigma})}\mathcal{O}(h^2)$$
$$+\bar{h}_i\sum_{m=1}^{r}\frac{\partial f_i}{\partial \tilde{u}_m}\Big|_{(s^i_{11},\boldsymbol{x}^0,\delta^i_{11_\tau}\boldsymbol{x}^0+(1-\delta^i_{11_\tau})\phi(s^i_{11}-\tau),\boldsymbol{u}^i_{11},\tilde{\boldsymbol{u}}^i_{11_\sigma})}\delta^i_{11_\sigma}\mathcal{O}(h^2)+R^i_1. \tag{6.30}$$

Recall that W is compact. By Assumption 6.2.1 and (6.28), (6.30) can be simplified as

$$x_i(t_1)=\phi_i(0)+\bar{h}_i f_i(s^i_{11},\boldsymbol{x}^0,\delta^i_{11_\tau}\boldsymbol{x}^0+(1-\delta^i_{11_\tau})\phi(s^i_{11}-\tau),\boldsymbol{u}^i_{11},\tilde{\boldsymbol{u}}^i_{11_\sigma})$$
$$+\sum_{p=1}^{n}\Big[\frac{\partial f_i}{\partial x_p}\Big|_{(s^i_{11},\boldsymbol{x}^0,\delta^i_{11_\tau}\boldsymbol{x}^0+(1-\delta^i_{11_\tau})\phi(s^i_{11}-\tau),\boldsymbol{u}^i_{11},\tilde{\boldsymbol{u}}^i_{11_\sigma})}\bar{h}_i\rho^i_{11}(x_p(t_1)-x^0_p)\Big]$$
$$+\sum_{p=1}^{n}\Big[\frac{\partial f_i}{\partial \tilde{x}_p}\Big|_{(s^i_{11},\boldsymbol{x}^0,\delta^i_{11_\tau}\boldsymbol{x}^0+(1-\delta^i_{11_\tau})\phi(s^i_{11}-\tau),\boldsymbol{u}^i_{11},\tilde{\boldsymbol{u}}^i_{11_\sigma})}\bar{h}_i\delta^i_{11_\tau}\bar{\rho}^i_{11_\tau}(x_p(t_1)-x^0_p)\Big]$$
$$+\mathcal{O}(h^2). \tag{6.31}$$

Thus, $\boldsymbol{x}(t_1)$ is a solution of the following linear equations:

$$\boldsymbol{B}^1\boldsymbol{x}(t_1)=\tilde{\boldsymbol{C}}^1. \tag{6.32}$$

Since matrix $\boldsymbol{B}^1$ is non-singular when $h<\hat{h}$ (recall that $\hat{h}$ is defined in Step 3), (6.20) as well as (6.32) has a unique solution. Thus, combining (6.32) with (6.20) yields

$$\|\boldsymbol{x}(t_1)-\boldsymbol{x}^1\|_\infty\le\|(\boldsymbol{B}^1)^{-1}\|_\infty\|\tilde{\boldsymbol{C}}^1-C^1\|_\infty=\mathcal{O}(h^2). \tag{6.33}$$

This proves that (6.23) holds for $k=1$ when $h<\hat{h}$.

Step 5. Inductive step for (6.23)

We now consider the case of $k\ge 2$. Assume that, for $l=1,\ldots,k-1$,

$$\|\boldsymbol{x}(t_l)-\boldsymbol{x}^l\|_\infty=\mathcal{O}(h^2), \tag{6.34}$$

when $h<\hat{h}$.

From (6.13) and Step 2, we have

$$x_i(t_k)=\phi_i(0)+R^i_k$$
$$+\bar{h}_i\sum_{l=1}^{k-1}f_i(s^i_{kl},\boldsymbol{x}(s^i_{kl}),\boldsymbol{x}(s^i_{kl}-\tau),\boldsymbol{u}(s^i_{kl}),\boldsymbol{u}(s^i_{kl}-\sigma))\big[(d_{kl}+1)^{\alpha_i}-d_{kl}^{\alpha_i}\big]$$
$$+\bar{h}_i f_i(s^i_{kk},\boldsymbol{x}(s^i_{kk}),\boldsymbol{x}(s^i_{kk}-\tau),\boldsymbol{u}(s^i_{kk}),\boldsymbol{u}(s^i_{kk}-\sigma))$$
$$=\phi_i(0)+A^i_{k-1}+D^i_k+\mathcal{O}(h^2), \tag{6.35}$$

where

$$A_{k-1}^i = \bar{h}_i \sum_{l=1}^{k-1} f_i(s_{kl}^i, \boldsymbol{x}(s_{kl}^i), \boldsymbol{x}(s_{kl}^i - \tau), \boldsymbol{u}(s_{kl}^i), \boldsymbol{u}(s_{kl}^i - \sigma))[(d_{kl}+1)^{\alpha_i} - d_{kl}^{\alpha_i}];$$

and

$$D_k^i = \bar{h}_i f_i(s_{kk}^i, \boldsymbol{x}(s_{kk}^i), \boldsymbol{x}(s_{kk}^i - \tau), \boldsymbol{u}(s_{kk}^i), \boldsymbol{u}(s_{kk}^i - \sigma)).$$

Using (6.24)–(6.27) and applying Taylor expansion gives

$$\begin{aligned}
D_k^i &= \bar{h}_i f_i(s_{kk}^i, \boldsymbol{x}(t_{k-1}), \delta_{kk_\tau}^i \boldsymbol{x}(t_{k-1}) + (1-\delta_{kk_\tau}^i)\boldsymbol{x}(s_{kk}^i - \tau), \boldsymbol{u}_{kk}^i, \tilde{\boldsymbol{u}}_{kk_\sigma}^i) \\
&+ \bar{h}_i \sum_{p=1}^{n} \left\{ \frac{\partial f_i}{\partial x_p} \Big|_{(s_{kk}^i, \boldsymbol{x}(t_{k-1}), \delta_{kk_\tau}^i \boldsymbol{x}(t_{k-1}) + (1-\delta_{kk_\tau}^i)\boldsymbol{x}(s_{kk}^i - \tau), \boldsymbol{u}_{kk}^i, \tilde{\boldsymbol{u}}_{kk_\sigma}^i)} \right. \\
&\left. \times \left[\rho_{kk}^i (x_p(t_k) - x_p(t_{k-1})) + \mathcal{O}(h^2) \right] \right\} \\
&+ \bar{h}_i \sum_{p=1}^{n} \left\{ \frac{\partial f_i}{\partial \tilde{x}_p} \Big|_{(s_{kk}^i, \boldsymbol{x}(t_{k-1}), \delta_{kk_\tau}^i \boldsymbol{x}(t_{k-1}) + (1-\delta_{kk_\tau}^i)\boldsymbol{x}(s_{kk}^i - \tau), \boldsymbol{u}_{kk}^i, \tilde{\boldsymbol{u}}_{kk_\sigma}^i)} \right. \\
&\left. \times \left[\delta_{kk_\tau}^i \bar{\rho}_{kk_\tau}^i (x_p(t_k) - x_p(t_{k-1})) + \mathcal{O}(h^2) \right] \right\} \\
&+ \bar{h}_i \sum_{m=1}^{r} \frac{\partial f_i}{\partial u_m} \Big|_{(s_{kk}^i, \boldsymbol{x}(t_{k-1}), \delta_{kk_\tau}^i \boldsymbol{x}(t_{k-1}) + (1-\delta_{kk_\tau}^i)\boldsymbol{x}(s_{kk}^i - \tau), \boldsymbol{u}_{kk}^i, \tilde{\boldsymbol{u}}_{kk_\sigma}^i)} \mathcal{O}(h^2) \\
&+ \bar{h}_i \sum_{m=1}^{r} \frac{\partial f_i}{\partial \tilde{u}_m} \Big|_{(s_{kk}^i, \boldsymbol{x}(t_k), \delta_{kk_\tau}^i \boldsymbol{x}(t_{k-1}) + (1-\delta_{kk_\tau}^i)\boldsymbol{x}(s_{kk}^i - \tau), \boldsymbol{u}_{kk}^i, \tilde{\boldsymbol{u}}_{kk_\sigma}^i)} \mathcal{O}(h^2). \qquad (6.36)
\end{aligned}$$

By Assumption 6.2.1, Step 2, and (6.10), we can simplify (6.36) as

$$\begin{aligned}
D_k^i &= \bar{h}_i f_i(s_{kk}^i, \boldsymbol{x}(t_{k-1}), \delta_{kk_\tau}^i \boldsymbol{x}(t_{k-1}) + (1-\delta_{kk_\tau}^i)\boldsymbol{x}(s_{kk}^i - \tau), \boldsymbol{u}_{kk}^i, \tilde{\boldsymbol{u}}_{kk_\sigma}^i) \\
&\quad + \bar{h}_i \sum_{p=1}^{n} \left[\frac{\partial f_i}{\partial x_p} \Big|_{(s_{kk}^i, \boldsymbol{x}(t_{k-1}), \delta_{kk_\tau}^i \boldsymbol{x}(t_{k-1}) + (1-\delta_{kk_\tau}^i)\boldsymbol{x}(s_{kk}^i - \tau), \boldsymbol{u}_{kk}^i, \tilde{\boldsymbol{u}}_{kk_\sigma}^i)} \right. \\
&\quad \left. \times \rho_{kk}^i (x_p(t_k) - x_p(t_{k-1})) \right] \\
&\quad + \bar{h}_i \sum_{p=1}^{n} \left\{ \frac{\partial f_i}{\partial \tilde{x}_p} \Big|_{(s_{kk}^i, \boldsymbol{x}(t_{k-1}), \delta_{kk_\tau}^i \boldsymbol{x}(t_{k-1}) + (1-\delta_{kk_\tau}^i)\boldsymbol{x}(s_{kk}^i - \tau), \boldsymbol{u}_{kk}^i, \tilde{\boldsymbol{u}}_{kk_\sigma}^i)} \right. \\
&\quad \left. \times \left[\delta_{kk_\tau}^i \bar{\rho}_{kl_\tau}^i (x_p(t_k) - x_p(t_{k-1})) \right] \right\} + \mathcal{O}(h^{2+\alpha_i}). \qquad (6.37)
\end{aligned}$$

Furthermore, using a Taylor expansion, we obtain

$$\begin{aligned}
&f_i(s^i_{kk}, \boldsymbol{x}(t_{k-1}), \delta^i_{kk_\tau}\boldsymbol{x}(t_{k-1}) + (1-\delta^i_{kk_\tau})\boldsymbol{x}(s^i_{kk}-\tau), \boldsymbol{u}^i_{kk}, \tilde{\boldsymbol{u}}^i_{kk_\sigma}) \\
&= f_i(s^i_{kk}, \boldsymbol{x}^{k-1}, \delta^i_{kk_\tau}\boldsymbol{x}^{k-1} + (1-\delta^i_{kk_\tau})\tilde{\boldsymbol{x}}^i_{kk_\tau}, \boldsymbol{u}^i_{kk}, \tilde{\boldsymbol{u}}^i_{kk_\sigma}) + \bar{R}^i_k,
\end{aligned} \tag{6.38}$$

where

$$\begin{aligned}
\bar{R}^i_k &= \frac{\partial f_i}{\partial x}\Big|_{(s^i_{kk}, \boldsymbol{\xi}^{k-1}, \delta^i_{kk_\tau}\zeta^{k-1}+(1-\delta^i_{kk_\tau})\boldsymbol{\nu}^{k-1}, \boldsymbol{u}^i_{kk}, \tilde{\boldsymbol{u}}^i_{kk_\sigma})} (\boldsymbol{x}(t_{k-1}) - \boldsymbol{x}^{k-1}) \\
&+ \frac{\partial f_i}{\partial \tilde{x}}\Big|_{(s^i_{kk}, \boldsymbol{\xi}^{k-1}, \delta^i_{kk_\tau}\zeta^{k-1}+(1-\delta^i_{kk_\tau})\boldsymbol{\nu}^{k-1}, \boldsymbol{u}^i_{kk}, \tilde{\boldsymbol{u}}^i_{kk_\sigma})} \delta^i_{kk_\tau}(\boldsymbol{x}(t_{k-1}) - \boldsymbol{x}^{k-1}) \\
&+ \frac{\partial f_i}{\partial \tilde{x}}\Big|_{(s^i_{kk}, \boldsymbol{\xi}^{k-1}, \delta^i_{kk_\tau}\zeta^{k-1}+(1-\delta^i_{kk_\tau})\boldsymbol{\nu}^{k-1}, \boldsymbol{u}^i_{kk}, \tilde{\boldsymbol{u}}^i_{kk_\sigma})} (1-\delta^i_{kk_\tau})(\boldsymbol{x}(s^i_{kk}-\tau) - \tilde{\boldsymbol{x}}^i_{kk_\tau});
\end{aligned}$$

$\boldsymbol{\xi}^{k-1}$ and ζ^{k-1} are two points between $\boldsymbol{x}(t_{k-1})$ and $\boldsymbol{x}^{k-1}$; and $\boldsymbol{\nu}^{k-1}$ is a point between $\boldsymbol{x}(s^i_{kk}-\tau)$ and $\tilde{\boldsymbol{x}}^i_{kk_\tau}$. By Assumption 6.2.1 and (6.34), $|\bar{R}^i_k| = \mathcal{O}(h^2)$. Similarly, we have, for $p = 1, \ldots, n$,

$$\begin{aligned}
&\frac{\partial f_i}{\partial x_p}\Big|_{(s^i_{kk}, \boldsymbol{x}(t_{k-1}), \delta^i_{kk_\tau}\boldsymbol{x}(t_{k-1})+(1-\delta^i_{kk_\tau})\boldsymbol{x}(s^i_{kk}-\tau), \boldsymbol{u}^i_{kk}, \tilde{\boldsymbol{u}}^i_{kk_\sigma})} \\
&= \frac{\partial f_i}{\partial x_p}\Big|_{(s^i_{kk}, \boldsymbol{x}^{k-1}, \delta^i_{kk_\tau}\boldsymbol{x}^{k-1}+(1-\delta^i_{kk_\tau})\boldsymbol{x}^i_{kk_\tau}, \boldsymbol{u}^i_{kk}, \tilde{\boldsymbol{u}}^i_{kk_\sigma})} + \mathcal{O}(h^2),
\end{aligned} \tag{6.39}$$

and

$$\begin{aligned}
&\frac{\partial f_i}{\partial \tilde{x}_p}\Big|_{(s^i_{kk}, \boldsymbol{x}(t_{k-1}), \delta^i_{kk_\tau}\boldsymbol{x}(t_{k-1})+(1-\delta^i_{kk_\tau})\boldsymbol{x}(s^i_{kk}-\tau), \boldsymbol{u}^i_{kk}, \tilde{\boldsymbol{u}}^i_{kk_\sigma})} \\
&= \frac{\partial f_i}{\partial \tilde{x}_p}\Big|_{(s^i_{kk}, \boldsymbol{x}^{k-1}, \delta^i_{kk_\tau}\boldsymbol{x}^{k-1}+(1-\delta^i_{kk_\tau})\boldsymbol{x}^i_{kk_\tau}, \boldsymbol{u}^i_{kk}, \tilde{\boldsymbol{u}}^i_{kk_\sigma})} + \mathcal{O}(h^2).
\end{aligned} \tag{6.40}$$

Using estimations (6.38)–(6.40), (6.37) can be rewritten as

$$\begin{aligned}
D^i_k &= \bar{h}_i f_i(s^i_{kk}, \boldsymbol{x}^{k-1}, \delta^i_{kk_\tau}\boldsymbol{x}^{k-1} + (1-\delta^i_{kk_\tau})\boldsymbol{x}^i_{kk_\tau}, \boldsymbol{u}^i_{kk}, \tilde{\boldsymbol{u}}^i_{kk_\sigma}) \\
&+ \bar{h}_i \sum_{p=1}^{n} \Bigg[\frac{\partial f_i}{\partial x_p}\Big|_{(s^i_{kk}, \boldsymbol{x}^{k-1}, \delta^i_{kk_\tau}\boldsymbol{x}^{k-1}+(1-\delta^i_{kk_\tau})\boldsymbol{x}^i_{kk_\tau}, \boldsymbol{u}^i_{kk}, \tilde{\boldsymbol{u}}^i_{kk_\sigma})} \\
&\times \rho^i_{kk}(x_p(t_k) - x_p(t_{k-1}))\Bigg] \\
&+ \bar{h}_i \sum_{p=1}^{n} \Bigg\{ \frac{\partial f_i}{\partial \tilde{x}_p}\Big|_{(s^i_{kk}, \boldsymbol{x}^{k-1}, \delta^i_{kk_\tau}\boldsymbol{x}^{k-1}+(1-\delta^i_{kk_\tau})\boldsymbol{x}^i_{kk_\tau}, \boldsymbol{u}^i_{kk}, \tilde{\boldsymbol{u}}^i_{kk_\sigma})} \\
&\times \Big[\delta^i_{kk_\tau}\bar{\rho}^i_{kl_\tau}(x_p(t_k) - x_p(t_{k-1}))\Big]\Bigg\} + \mathcal{O}(h^{2+\alpha_i}).
\end{aligned} \tag{6.41}$$

Now, we consider the estimation for A^i_{k-1}. By Assumption 6.2.1 and (6.24)–(6.26), we have

$$f_i(s^i_{kl}, \boldsymbol{x}(s^i_{kl}), \boldsymbol{x}(s^i_{kl} - \tau), \boldsymbol{u}(s^i_{kl}), \boldsymbol{u}(s^i_{kl} - \sigma)) = f_i(s^i_{kl}, \boldsymbol{x}^i_{kl}, \tilde{\boldsymbol{x}}^i_{kl}, \boldsymbol{u}^i_{kl}, \tilde{\boldsymbol{u}}^i_{kl}) + \tilde{R}^i_{kl},$$

where

$$\begin{aligned}
\tilde{R}^i_l = &\sum_{p=1}^{n} \left\{ \frac{\partial f_i}{\partial x_p} \Big|_{(s^i_{kl}, \boldsymbol{x}^i_{kl}, \tilde{\boldsymbol{x}}^i_{kl}, \boldsymbol{u}^i_{kl}, \tilde{\boldsymbol{u}}^i_{kl})} \Big[\big(x_p(t_{l-1}) \right. \\
&\left. + \rho^i_{kl}(x_p(t_l) - x_p(t_{l-1})) + \mathcal{O}(h^2) \big) - \big(x_p^{l-1} + \rho^i_{kl}(x_p^l - x_p^{l-1}) \big) \Big] \right\} \\
&+ \sum_{p=1}^{n} \left\{ \frac{\partial f_i}{\partial \tilde{x}_p} \Big|_{(s^i_{kl}, \boldsymbol{x}^i_{kl}, \tilde{\boldsymbol{x}}^i_{kl}, \boldsymbol{u}^i_{kl}, \tilde{\boldsymbol{u}}^i_{kl})} \theta^i(l_\tau - 1) \Big[\big(x_p(t_{l_\tau - 1}) + \bar{\rho}^i_{kl}(x_p(t_{l_\tau}) \right. \\
&\left. - x_p(t_{l_\tau - 1})) + \mathcal{O}(h^2) \big) - \big(x_p^{l_\tau - 1} + \bar{\rho}^i_{kl_\tau}(x_p^{l_\tau} - x_p^{l_\tau - 1}) \big) \Big] \right\} \\
&+ \sum_{m=1}^{r} \frac{\partial f_i}{\partial u_m} \Big|_{(s^i_{kl}, \boldsymbol{x}^i_{kl}, \tilde{\boldsymbol{x}}^i_{kl}, \boldsymbol{u}^i_{kl}, \tilde{\boldsymbol{u}}^i_{kl})} \mathcal{O}(h^2) + \sum_{m=1}^{r} \frac{\partial f_i}{\partial \tilde{u}_m} \Big|_{(s^i_{kl}, \boldsymbol{x}^i_{kl}, \tilde{\boldsymbol{x}}^i_{kl}, \boldsymbol{u}^i_{kl}, \tilde{\boldsymbol{u}}^i_{kl})} \mathcal{O}(h^2) \\
&+ \mathcal{O}(h^2);
\end{aligned}$$

and $\theta^i(\cdot)$ is as defined in (6.53). From (6.34) and Assumption 6.2.1, we obtain $|\tilde{R}^i_{kl}| = \mathcal{O}(h^2)$. Furthermore, using a similar proof as given for Step 2, we have

$$\sum_{l=1}^{k-1} \tilde{R}^i_{kl}[(d_{kl} + 1)^{\alpha_i} - d_{kl}^{\alpha_i}] = \mathcal{O}(h^2).$$

Thus,

$$A^i_{k-1} = \bar{h}_i \sum_{l=1}^{k-1} f_i(s^i_{kl}, \boldsymbol{x}^i_{kl}, \tilde{\boldsymbol{x}}^i_{kl}, \boldsymbol{u}^i_{kl}, \tilde{\boldsymbol{u}}^i_{kl})[(d_{kl} + 1)^{\alpha_i} - d_{kl}^{\alpha_i}] + \mathcal{O}(h^2). \tag{6.42}$$

Substituting (6.41) and (6.42) into (6.35) gives

$$\begin{aligned}
x_i(t_k) = &\ \phi_i(0) + \bar{h}_i f_i(s^i_{kk}, \boldsymbol{x}^{k-1}, \delta^i_{kk_\tau} \boldsymbol{x}^{k-1} + (1 - \delta^i_{kk_\tau}) \boldsymbol{x}^i_{kk_\tau}, \boldsymbol{u}^i_{kk}, \tilde{\boldsymbol{u}}^i_{kk_\sigma}) \\
&+ \bar{h}_i \sum_{p=1}^{n} \left[\frac{\partial f_i}{\partial x_p} \Big|_{(s^i_{kk}, \boldsymbol{x}^{k-1}, \delta^i_{kk_\tau} \boldsymbol{x}^{k-1} + (1 - \delta^i_{kk_\tau}) \boldsymbol{x}^i_{kk_\tau}, \boldsymbol{u}^i_{kk}, \tilde{\boldsymbol{u}}^i_{kk_\sigma})} \right. \\
&\left. \times \rho^i_{kk}(x_p(t_k) - x_p(t_{k-1})) \right]
\end{aligned}$$

$$
\begin{aligned}
&+ \bar{h}_i \sum_{p=1}^{n} \left\{ \frac{\partial f_i}{\partial \tilde{x}_p} \Big|_{(s_{kk}^i, \boldsymbol{x}^{k-1}, \delta_{kk_\tau}^i \boldsymbol{x}^{k-1} + (1-\delta_{kk_\tau}^i) \boldsymbol{x}_{kk_\tau}^i, \boldsymbol{u}_{kk}^i, \tilde{\boldsymbol{u}}_{kk_\sigma}^i)} \right. \\
&\times \left. \left[\delta_{kk_\tau}^i \bar{\rho}_{kl_\tau}^i (x_p(t_k) - x_p(t_{k-1})) \right] \right\} + \mathcal{O}(h^{2+\alpha_i}) \\
&+ \bar{h}_i \sum_{l=1}^{k-1} f_i(s_{kl}^i, \boldsymbol{x}_{kl}^i, \tilde{\boldsymbol{x}}_{kl}^i, \boldsymbol{u}_{kl}^i, \tilde{\boldsymbol{u}}_{kl}^i)[(d_{kl}+1)^{\alpha_i} - d_{kl}^{\alpha_i}] + \mathcal{O}(h^2).
\end{aligned}
$$

Combining this equation with (6.20) and using (6.34), we have

$$
\|\boldsymbol{x}(t_k) - \boldsymbol{x}^k\|_\infty \le \|\boldsymbol{B}^k\|_\infty \|\tilde{\boldsymbol{C}}^k - \boldsymbol{C}^k\|_\infty = \mathcal{O}(h^2),
$$

from which we immediately infer that (6.23) holds for $l = k$. □

Define $\boldsymbol{X} = (\boldsymbol{X}^0, \boldsymbol{X}^1, \ldots, \boldsymbol{X}^N) \in \mathbb{R}^{n \times (N+1)}$ and $\boldsymbol{U} = (\boldsymbol{U}^0, \boldsymbol{U}^1, \ldots, \boldsymbol{U}^N) \in \mathbb{R}^{r \times (N+1)}$, where $\boldsymbol{X}^k = (x_1^k, \ldots, x_n^k)^\top$ and $\boldsymbol{U}^k = (u_1^k, \ldots, u_r^k)^\top$. Then, after applying the above numerical integration technique, Problem (P) can be approximated by the following discrete-time optimal control problem.

Problem (P(h)). *Given the discrete-time system (6.20), find $\boldsymbol{U} \in W^{N+1}$ such that the cost function*

$$
g_0^h(\boldsymbol{U}) = \Phi_0(\boldsymbol{X}^N) \tag{6.43}
$$

is minimized subject to the state inequality constraints

$$
g_j^h(\boldsymbol{U}) = \Phi_j(\boldsymbol{X}^N) \le 0, \quad j = 1, \ldots, N_i. \tag{6.44}
$$

6.3.2 Gradient Computation

Note that the cost function (6.43) is in a similar form to the constraint functions (6.44). Hence, the gradients of the cost and constraint functions can be computed in a unified fashion. For $j = 0, 1, \ldots, N_i$, we have

$$
\frac{\partial g_j^h(\boldsymbol{U})}{\partial u_m^q} = \frac{\partial \Phi_j(\boldsymbol{X}^N)}{\partial u_m^q} = \sum_{i=1}^{n} \frac{\partial \Phi_j(\boldsymbol{X}^N)}{\partial x_i^N} \frac{\partial x_i^N}{\partial u_m^q}, \quad m = 1, \ldots, r, \; q = 0, 1, \ldots, N. \tag{6.45}
$$

To determine $\frac{\partial x_i^N}{\partial u_m^q}$ in (6.45), we need to solve

$$
\frac{\partial \boldsymbol{B}^k}{\partial u_m^q} \boldsymbol{X}^k + \boldsymbol{B}^k \frac{\partial \boldsymbol{X}^k}{\partial u_m^q} = \frac{\partial \boldsymbol{C}^k}{\partial u_m^q} \tag{6.46}
$$

for $k = 1, \ldots, N$, where $\boldsymbol{B}^k$ is the system matrix in (6.21). Rearranging equation (6.46) yields

$$\boldsymbol{B}^k \frac{\partial \boldsymbol{X}^k}{\partial u_m^q} = \frac{\partial \boldsymbol{C}^k}{\partial u_m^q} - \frac{\partial \boldsymbol{B}^k}{\partial u_m^q} \boldsymbol{X}^k. \tag{6.47}$$

Using (6.21) and (6.22), the gradient formulas $\frac{\partial b_{ip}^k}{\partial u_m^q}$ and $\frac{\partial c_i^k}{\partial u_m^q}$ for $i = 1, \ldots, n$, $p = 1, \ldots, n$, are given as below:

1. When $q = 0$, if $k = 1$, then

$$\begin{aligned}
\frac{\partial b_{ip}^1}{\partial u_m^0} &= \bar{h}_i \rho_{11}^i \frac{\partial^2 f_i}{\partial x_p \partial u_m} \Big|_{(s_{11}^i, \boldsymbol{X}^0, \delta_{11_\tau}^i \boldsymbol{X}^0 + (1-\delta_{11_\tau}^i)\tilde{\boldsymbol{x}}_{11_\tau}^i, \boldsymbol{u}_{11}^i, \tilde{\boldsymbol{u}}_{11_\sigma}^i)} (1 - \rho_{11}^i) \\
&+ \bar{h}_i \delta_{11_\tau}^i \bar{\rho}_{11_\tau}^i \frac{\partial^2 f_i}{\partial \tilde{x}_p \partial u_m} \Big|_{(s_{11}^i, \boldsymbol{X}^0, \delta_{11_\tau}^i \boldsymbol{X}^0 + (1-\delta_{11_\tau}^i)\tilde{\boldsymbol{x}}_{11_\tau}^i, \boldsymbol{u}_{11}^i, \tilde{\boldsymbol{u}}_{11_\sigma}^i)} (1 - \rho_{11}^i) \\
&+ \bar{h}_i \rho_{11}^i \frac{\partial^2 f_i}{\partial x_p \partial \tilde{u}_m} \Big|_{(s_{11}^i, \boldsymbol{X}^0, \delta_{11_\tau}^i \boldsymbol{X}^0 + (1-\delta_{11_\tau}^i)\tilde{\boldsymbol{x}}_{11_\tau}^i, \boldsymbol{u}_{11}^i, \tilde{\boldsymbol{u}}_{11_\sigma}^i)} \eta_{11_\sigma}^i (1 - \tilde{\rho}_{11_\sigma}^i), \\
&+ \bar{h}_i \delta_{11_\tau}^i \bar{\rho}_{11_\tau}^i \frac{\partial^2 f_i}{\partial \tilde{x}_p \partial \tilde{u}_m} \Big|_{(s_{11}^i, \boldsymbol{X}^0, \delta_{11_\tau}^i \boldsymbol{X}^0 + (1-\delta_{11_\tau}^i)\tilde{\boldsymbol{x}}_{11_\tau}^i, \boldsymbol{u}_{11}^i, \tilde{\boldsymbol{u}}_{11_\sigma}^i)} \\
&\times \eta_{11_\sigma}^i (1 - \tilde{\rho}_{11_\sigma}^i),
\end{aligned} \tag{6.48}$$

$$\begin{aligned}
\frac{\partial c_i^1}{\partial u_m^0} &= \bar{h}_i \frac{\partial f_i}{\partial u_m} \Big|_{(s_{11}^i, \boldsymbol{X}^0, \delta_{11_\tau}^i \boldsymbol{X}^0 + (1-\delta_{11_\tau}^i)\tilde{\boldsymbol{x}}_{11_\tau}^i, \boldsymbol{u}_{11}^i, \tilde{\boldsymbol{u}}_{11_\sigma}^i)} (1 - \rho_{11}^i) \\
&+ \bar{h}_i \frac{\partial f_i}{\partial \tilde{u}_m} \Big|_{(s_{11}^i, \boldsymbol{X}^0, \delta_{11_\tau}^i \boldsymbol{X}^0 + (1-\delta_{11_\tau}^i)\tilde{\boldsymbol{x}}_{11_\tau}^i, \boldsymbol{u}_{11}^i, \tilde{\boldsymbol{u}}_{11_\sigma}^i)} \eta_{11_\sigma}^i (1 - \tilde{\rho}_{11_\sigma}^i) \\
&- \sum_{p=1}^n \frac{\partial b_{ip}^1}{\partial u_m^0} x_p^0,
\end{aligned} \tag{6.49}$$

where, for two given integers $\jmath$ and $\jmath'$,

$$\eta_{\jmath\jmath'}^i = \begin{cases} 1, & \jmath' = \jmath, \\ 0, & \text{otherwise}. \end{cases} \tag{6.50}$$

If $k > 1$, then

$$\begin{aligned}
\frac{\partial b_{ip}^k}{\partial u_m^0} &= \bar{h}_i \sum_{\varsigma=1}^n \Bigg(\rho_{kk}^i \frac{\partial^2 f_i}{\partial x_p \partial x_\varsigma} + \delta_{kk_\tau}^i \bar{\rho}_{kk_\tau}^i \frac{\partial^2 f_i}{\partial \tilde{x}_p \partial x_\varsigma} + \delta_{kk_\tau}^i \rho_{kk}^i \frac{\partial^2 f_i}{\partial x_p \partial \tilde{x}_\varsigma} \\
&+ (\delta_{kk_\tau}^i)^2 \bar{\rho}_{kk_\tau}^i \frac{\partial^2 f_i}{\partial \tilde{x}_p \partial \tilde{x}_\varsigma} \Bigg) \Big|_{(s_{kk}^i, \boldsymbol{X}^{k-1}, \delta_{kk_\tau}^i \boldsymbol{X}^{k-1} + (1-\delta_{kk_\tau}^i)\tilde{\boldsymbol{x}}_{kk_\tau}^i, \boldsymbol{u}_{kk}^i, \tilde{\boldsymbol{u}}_{kk_\sigma}^i)} \frac{\partial x_\varsigma^{k-1}}{\partial u_m^0} \\
&+ \bar{h}_i \sum_{\varsigma=1}^n \rho_{kk}^i \frac{\partial^2 f_i}{\partial x_p \partial \tilde{x}_\varsigma} \Big|_{(s_{kk}^i, \boldsymbol{X}^{k-1}, \delta_{kk_\tau}^i \boldsymbol{X}^{k-1} + (1-\delta_{kk_\tau}^i)\tilde{\boldsymbol{x}}_{kk_\tau}^i, \boldsymbol{u}_{kk}^i, \tilde{\boldsymbol{u}}_{kk_\sigma}^i)}
\end{aligned}$$

$$\times (1-\delta^i_{kk_\tau})\Bigg[\eta^i_{1k_\tau}\bar{\rho}^i_{kk_\tau}\frac{\partial x_\varsigma^{k_\tau}}{\partial u_m^0}+\theta^i(k_\tau-2)\Bigg((1-\bar{\rho}^i_{kk_\tau})\frac{\partial x_\varsigma^{k_\tau-1}}{\partial u_m^0}$$
$$+\bar{\rho}^i_{kk_\tau}\frac{\partial x_\varsigma^{k_\tau}}{\partial u_m^0}\Bigg)\Bigg]+\bar{h}_i\Bigg(\rho^i_{kk}\frac{\partial^2 f_i}{\partial x_p\partial\tilde{u}_m}$$
$$+\delta^i_{kk_\tau}\bar{\rho}^i_{kk_\tau}\frac{\partial^2 f_i}{\partial\tilde{x}_p\partial\tilde{u}_m}\Bigg)\Bigg|_{(s^i_{kk},\boldsymbol{X}^{k-1},\delta^i_{kk_\tau}\boldsymbol{X}^{k-1}+(1-\delta^i_{kk_\tau})\tilde{\boldsymbol{x}}^i_{kk_\tau},\boldsymbol{u}^i_{kk},\tilde{\boldsymbol{u}}^i_{kk_\sigma})}$$
$$\times\eta^i_{1k_\sigma}(1-\tilde{\rho}^i_{kk_\sigma}), \tag{6.51}$$

$$\frac{\partial c_i^k}{\partial u_m^0}=\bar{h}_i\Bigg[\sum_{\varsigma=1}^{n}\Bigg(\rho^i_{k1}\frac{\partial f_i}{\partial x_\varsigma}+\eta^i_{11_\tau}\bar{\rho}^i_{k1_\tau}\frac{\partial f_i}{\partial\tilde{x}_\varsigma}\Bigg)\Bigg|_{(s^i_{k1},\boldsymbol{x}^i_{k1},\tilde{\boldsymbol{x}}^i_{k1_\tau},\boldsymbol{u}^i_{k1},\tilde{\boldsymbol{u}}^i_{k1_\sigma})}\frac{\partial x^1_\varsigma}{\partial u_m^0}$$
$$+\Bigg((1-\rho^i_{k1})\frac{\partial f_i}{\partial u_m}+\eta^i_{11_\sigma}(1-\tilde{\rho}^i_{k1_\sigma})\frac{\partial f_i}{\partial\tilde{u}_m}\Bigg)\Bigg|_{(s^i_{k1},\boldsymbol{x}^i_{k1},\tilde{\boldsymbol{x}}^i_{k1_\tau},\boldsymbol{u}^i_{k1},\tilde{\boldsymbol{u}}^i_{k1_\sigma})}\Bigg]$$
$$\times(k^{\alpha_i}-(k-1)^{\alpha_i})+\bar{h}_i\sum_{l=2}^{k-1}\Bigg\{\sum_{\varsigma=1}^{n}\frac{\partial f_i}{\partial x_\varsigma}\Bigg|_{(s^i_{kl},\boldsymbol{x}^i_{kl},\tilde{\boldsymbol{x}}^i_{kl_\tau},\boldsymbol{u}^i_{kl},\tilde{\boldsymbol{u}}^i_{kl_\sigma})}\Bigg(\rho^i_{kl}\frac{\partial x^l_\varsigma}{\partial u_m^0}$$
$$+(1-\rho^i_{kl})\frac{\partial x^{l-1}_\varsigma}{\partial u_m^0}\Bigg)+\sum_{\varsigma=1}^{n}\frac{\partial f_i}{\partial\tilde{x}_\varsigma}\Bigg|_{(s^i_{kl},\boldsymbol{x}^i_{kl},\tilde{\boldsymbol{x}}^i_{kl_\tau},\boldsymbol{u}^i_{kl},\tilde{\boldsymbol{u}}^i_{kl_\sigma})}\Bigg[\eta^i_{1l_\tau}\bar{\rho}^i_{kl_\tau}\frac{\partial x^{l_\tau}_\varsigma}{\partial u_m^0}$$
$$+\theta^i(l_\tau-2)\Bigg(\bar{\rho}^i_{kl_\tau}\frac{\partial x^{l_\tau}_\varsigma}{\partial u_m^0}+(1-\bar{\rho}^i_{kl_\tau})\frac{\partial x^{l_\tau-1}_\varsigma}{\partial u_m^0}\Bigg)\Bigg]$$
$$+\frac{\partial f_i}{\partial\tilde{u}_m}\Bigg|_{(s^i_{kl},\boldsymbol{x}^i_{kl},\tilde{\boldsymbol{x}}^i_{kl_\tau},\boldsymbol{u}^i_{kl},\tilde{\boldsymbol{u}}^i_{kl_\sigma})}\eta^i_{1l_\sigma}(1-\tilde{\rho}^i_{kl_\sigma})\Bigg\}\big[(k-l+1)^{\alpha_i}-(k-l)^{\alpha_i}\big]$$
$$+\bar{h}_i\sum_{\varsigma=1}^{n}\Bigg(\frac{\partial f_i}{\partial x_\varsigma}+\delta^i_{kk_\tau}\frac{\partial f_i}{\partial\tilde{x}_\varsigma}\Bigg)\Bigg|_{(s^i_{kk},\boldsymbol{X}^{k-1},\delta^i_{kk_\tau}\boldsymbol{X}^{k-1}+(1-\delta^i_{kk_\tau})\tilde{\boldsymbol{x}}^i_{kk_\tau},\boldsymbol{u}^i_{kk},\tilde{\boldsymbol{u}}^i_{kk_\sigma})}$$
$$\times\frac{\partial x^{k-1}_\varsigma}{\partial u_m^0}+\bar{h}_i(1-\delta^i_{kk_\tau})\sum_{\varsigma=1}^{n}\Bigg\{\frac{\partial f_i}{\partial\tilde{x}_\varsigma}\Bigg|_{(s^i_{kk},\boldsymbol{X}^{k-1},\delta^i_{kk_\tau}\boldsymbol{X}^{k-1}+(1-\delta^i_{kk_\tau})\tilde{\boldsymbol{x}}^i_{kk_\tau},\boldsymbol{u}^i_{kk},\tilde{\boldsymbol{u}}^i_{kk_\sigma})}$$
$$\times\Bigg(\eta^i_{1k_\tau}\bar{\rho}^i_{kk_\tau}\frac{\partial x^{k_\tau}_\varsigma}{\partial u_m^0}+\theta^i(k_\tau-2)\Bigg[(1-\bar{\rho}^i_{kk_\tau})\frac{\partial x^{k_\tau-1}_\varsigma}{\partial u_m^0}+\bar{\rho}^i_{kk_\tau}\frac{\partial x^{k_\tau}_\varsigma}{\partial u_m^0}\Bigg]\Bigg)\Bigg\}$$
$$+\bar{h}_i\frac{\partial f_i}{\partial\tilde{u}_m}\Bigg|_{(s^i_{kk},\boldsymbol{X}^{k-1},\delta^i_{kk_\tau}\boldsymbol{X}^{k-1}+(1-\delta^i_{kk_\tau})\tilde{\boldsymbol{x}}^i_{kk_\tau},\boldsymbol{u}^i_{kk},\tilde{\boldsymbol{u}}^i_{kk_\sigma})}\eta^i_{1k_\sigma}(1-\tilde{\rho}^i_{kk_\sigma})$$
$$-\sum_{p=1}^{n}\Bigg[\frac{\partial b^k_{ip}}{\partial u_m^0}x_p^{k-1}+b^k_{ip}\frac{\partial x_p^{k-1}}{\partial u_m^0}\Bigg], \tag{6.52}$$

where, for two given integers $\jmath$ and $\jmath'$, $\eta^i_{\jmath\jmath'}$ is as defined in (6.50); and for a given π,

$$\theta^i(\pi)=\begin{cases}1, & \pi\geq 0,\\ 0, & \text{otherwise.}\end{cases} \tag{6.53}$$

2. For $q = 1, \ldots, N$, if $q > k$, then

$$\frac{\partial b_{ip}^{k}}{\partial u_m^q} = 0 \text{ and } \frac{\partial c_i^k}{\partial u_m^q} = 0. \tag{6.54}$$

If $q = k$, then

$$\begin{aligned}
\frac{\partial b_{ip}^{k}}{\partial u_m^q} &= \bar{h}_i (\rho_{kk}^i)^2 \frac{\partial^2 f_i}{\partial x_p \partial u_m} \Big|_{(s_{kk}^i, X^{k-1}, \delta_{kk_\tau}^i X^{k-1} + (1-\delta_{kk_\tau}^i)\tilde{x}_{kk_\tau}^i, u_{kk}^i, \tilde{u}_{kk_\sigma}^i)} \\
&+ \bar{h}_i \delta_{kk_\tau}^i \bar{\rho}_{kk_\tau}^i \rho_{kk}^i \frac{\partial^2 f_i}{\partial \tilde{x}_p \partial u_m} \Big|_{(s_{kk}^i, X^{k-1}, \delta_{kk_\tau}^i X^{k-1} + (1-\delta_{kk_\tau}^i)\tilde{x}_{kk_\tau}^i, u_{kk}^i, \tilde{u}_{kk_\sigma}^i)} \\
&+ \bar{h}_i \rho_{kk}^i \frac{\partial^2 f_i}{\partial x_p \partial \tilde{u}_m} \Big|_{(s_{kk}^i, X^{k-1}, \delta_{kk_\tau}^i X^{k-1} + (1-\delta_{kk_\tau}^i)\tilde{x}_{kk_\tau}^i, u_{kk}^i, \tilde{u}_{kk_\sigma}^i)} \eta_{qk_\sigma}^i \tilde{\rho}_{kk_\sigma}^i \\
&+ \bar{h}_i \delta_{kk_\tau}^i \bar{\rho}_{kk_\tau}^i \frac{\partial^2 f_i}{\partial \tilde{x}_p \partial \tilde{u}_m} \Big|_{(s_{kk}^i, X^{k-1}, \delta_{kk_\tau}^i X^{k-1} + (1-\delta_{kk_\tau}^i)\tilde{x}_{kk_\tau}^i, u_{kk}^i, \tilde{u}_{kk_\sigma}^i)} \\
&\times \eta_{qk_\sigma}^i \tilde{\rho}_{kk_\sigma}^i,
\end{aligned} \tag{6.55}$$

$$\begin{aligned}
\frac{\partial c_i^k}{\partial u_m^q} &= \bar{h}_i \rho_{kk}^i \frac{\partial f_i}{\partial u_m} \Big|_{(s_{kk}^i, X^{k-1}, \delta_{kk_\tau}^i X^{k-1} + (1-\delta_{kk_\tau}^i)\tilde{x}_{kk_\tau}^i, u_{kk}^i, \tilde{u}_{kk_\sigma}^i)} \\
&+ \bar{h}_i \eta_{kk_\sigma}^i \tilde{\rho}_{kk_\sigma}^i \frac{\partial f_i}{\partial \tilde{u}_m} \Big|_{(s_{kk}^i, X^{k-1}, \delta_{kk_\tau}^i X^{k-1} + (1-\delta_{kk_\tau}^i)\tilde{x}_{kk_\tau}^i, u_{kk}^i, \tilde{u}_{kk_\sigma}^i)} \\
&- \sum_{p=1}^{n} \frac{\partial b_{ip}^k}{\partial u_m^q} x_p^{k-1},
\end{aligned} \tag{6.56}$$

where, for two given integers j and j', $\eta_{jj'}^i$ is as defined in (6.50); and for a given π, $\theta^i(\pi)$ is as defined in (6.53).

If $q = k-1$, then

$$\begin{aligned}
\frac{\partial b_{ip}^{k}}{\partial u_m^q} &= \bar{h}_i \sum_{\varsigma=1}^{n} \Bigg\{ \Bigg(\rho_{kk}^i \frac{\partial^2 f_i}{\partial x_p \partial x_\varsigma} + \rho_{kk}^i \delta_{kk_\tau}^i \frac{\partial^2 f_i}{\partial x_p \partial \tilde{x}_\varsigma} + \delta_{kk_\tau}^i \bar{\rho}_{kk_\tau}^i \frac{\partial^2 f_i}{\partial \tilde{x}_p \partial x_\varsigma} \\
&+ (\delta_{kk_\tau}^i)^2 \bar{\rho}_{kk_\tau}^i \frac{\partial^2 f_i}{\partial \tilde{x}_p \partial \tilde{x}_\varsigma} \Bigg) \Big|_{(s_{kk}^i, X^q, \delta_{kk_\tau}^i X^q + (1-\delta_{kk_\tau}^i)\tilde{x}_{kk_\tau}^i, u_{kk}^i, \tilde{u}_{kk_\sigma}^i)} \frac{\partial x_\varsigma^{k-1}}{\partial u_m^q} \\
&+ \rho_{kk}^i (1 - \delta_{kk_\tau}^i) \frac{\partial^2 f_i}{\partial x_p \partial \tilde{x}_\varsigma} \Big|_{(s_{kk}^i, X^q, \delta_{kk_\tau}^i X^q + (1-\delta_{kk_\tau}^i)\tilde{x}_{kk_\tau}^i, u_{kk}^i, \tilde{u}_{kk_\sigma}^i)} \\
&\times \eta_{qk_\tau}^i \bar{\rho}_{kk_\tau}^i \frac{\partial x_\varsigma^{k_\tau}}{\partial u_m^q} \Bigg\} + \bar{h}_i \Bigg(\rho_{kk}^i \frac{\partial^2 f_i}{\partial x_p \partial u_m} \\
&+ \delta_{kk_\tau}^i \bar{\rho}_{kk_\tau}^i \frac{\partial^2 f_i}{\partial \tilde{x}_p \partial u_m} \Bigg) \Big|_{(s_{kk}^i, X^q, \delta_{kk_\tau}^i X^q + (1-\delta_{kk_\tau}^i)\tilde{x}_{kk_\tau}^i, u_{kk}^i, \tilde{u}_{kk_\sigma}^i)} (1 - \rho_{kk}^i)
\end{aligned}$$

$$+\bar{h}_i\rho^i_{kk}\frac{\partial^2 f_i}{\partial x_p\partial\tilde{u}_m}\Big|_{(s^i_{kk},\boldsymbol{X}^q,\delta^i_{kk_\tau}\boldsymbol{X}^q+(1-\delta^i_{kk_\tau})\tilde{\boldsymbol{x}}^i_{kk_\tau},\boldsymbol{u}^i_{kk},\tilde{\boldsymbol{u}}^i_{kk_\sigma})}\big[\eta^i_{qk_\sigma}\tilde{\rho}^i_{kk_\sigma}$$
$$+\eta^i_{kk_\sigma}(1-\tilde{\rho}^i_{kk_\sigma})\big]$$
$$+\bar{h}_i\delta^i_{kk_\tau}\bar{\rho}^i_{kk_\tau}\frac{\partial^2 f_i}{\partial\tilde{x}_p\partial\tilde{u}_m}\Big|_{(s^i_{kk},\boldsymbol{X}^q,\delta^i_{kk_\tau}\boldsymbol{X}^q+(1-\delta^i_{kk_\tau})\tilde{\boldsymbol{x}}^i_{kk_\tau},\boldsymbol{u}^i_{kk},\tilde{\boldsymbol{u}}^i_{kk_\sigma})}\big[\eta^i_{qk_\sigma}\tilde{\rho}^i_{kk_\sigma}$$
$$+\eta^i_{kk_\sigma}(1-\tilde{\rho}^i_{kk_\sigma})\big], \tag{6.57}$$

$$\frac{\partial c^k_i}{\partial u^q_m}=\bar{h}_i\Bigg[\sum_{\varsigma=1}^{n}\left(\rho^i_{kq}\frac{\partial f_i}{\partial x_\varsigma}+\eta^i_{qq_\tau}\bar{\rho}^i_{kq_\tau}\frac{\partial f_i}{\partial\tilde{x}_\varsigma}\right)\Big|_{(s^i_{kq},\boldsymbol{x}^i_{kq},\tilde{\boldsymbol{x}}^i_{kq_\tau},\boldsymbol{u}^i_{kq},\tilde{\boldsymbol{u}}^i_{kq_\sigma})}\frac{\partial x^{k-1}_\varsigma}{\partial u^q_m}$$
$$+\left(\rho^i_{kq}\frac{\partial f_i}{\partial u_m}+\eta^i_{qq_\sigma}\tilde{\rho}^i_{kq_\sigma}\frac{\partial f_i}{\partial\tilde{u}_m}\right)\Big|_{(s^i_{kq},\boldsymbol{x}^i_{kq},\tilde{\boldsymbol{x}}^i_{kq_\tau},\boldsymbol{u}^i_{kq},\tilde{\boldsymbol{u}}^i_{kq_\sigma})}\Bigg](2^{\alpha_i}-1)$$
$$+\bar{h}_i\Bigg[\sum_{\varsigma=1}^{n}\left(\frac{\partial f_i}{\partial x_\varsigma}+\delta^i_{kk_\tau}\frac{\partial f_i}{\partial\tilde{x}_\varsigma}\right)\Big|_{(s^i_{kk},\boldsymbol{X}^q,\delta^i_{kk_\tau}\boldsymbol{X}^q+(1-\delta^i_{kk_\tau})\tilde{\boldsymbol{x}}^i_{kk_\tau},\boldsymbol{u}^i_{kk},\tilde{\boldsymbol{u}}^i_{kk_\sigma})}$$
$$\times\frac{\partial x^{k-1}_\varsigma}{\partial u^q_m}+(1-\delta^i_{kk_\tau})\sum_{\varsigma=1}^{n}\frac{\partial f_i}{\partial\tilde{x}_\varsigma}\Big|_{(s^i_{kk},\boldsymbol{X}^q,\delta^i_{kk_\tau}\boldsymbol{X}^q+(1-\delta^i_{kk_\tau})\tilde{\boldsymbol{x}}^i_{kk_\tau},\boldsymbol{u}^i_{kk},\tilde{\boldsymbol{u}}^i_{kk_\sigma})}$$
$$\times\eta^i_{qk_\tau}\bar{\rho}^i_{kk_\tau}\frac{\partial x^{k_\tau}_\varsigma}{\partial u^q_m}+\Bigg((1-\rho^i_{kk})\frac{\partial f_i}{\partial u_m}+\big[\eta^i_{kk_\sigma}(1-\tilde{\rho}^i_{kk_\sigma})$$
$$+\eta^i_{qk_\sigma}\tilde{\rho}^i_{kk_\sigma}\big]\frac{\partial f_i}{\partial\tilde{u}_m}\Bigg)\Big|_{(s^i_{kk},\boldsymbol{X}^q,\delta^i_{kk_\tau}\boldsymbol{X}^q+(1-\delta^i_{kk_\tau})\tilde{\boldsymbol{x}}^i_{kk_\tau},\boldsymbol{u}^i_{kk},\tilde{\boldsymbol{u}}^i_{kk_\sigma})}\Bigg]$$
$$-\sum_{p=1}^{n}\left[\frac{\partial b^k_{ip}}{\partial u^q_m}x^q_p+b^k_{ip}\frac{\partial x^q_p}{\partial u^q_m}\right], \tag{6.58}$$

where, for two given integers $\jmath$ and $\jmath'$, $\eta^i_{\jmath\jmath'}$ is as defined in (6.50); and for a given π, $\theta^i(\pi)$ is as defined in (6.53).

If $q<k-1$, then

$$\frac{\partial b^k_{ip}}{\partial u^q_m}=\bar{h}_i\sum_{\varsigma=1}^{n}\Bigg(\rho^i_{kk}\frac{\partial^2 f_i}{\partial x_p\partial x_\varsigma}+\rho^i_{kk}\delta^i_{kk_\tau}\frac{\partial^2 f_i}{\partial x_p\partial\tilde{x}_\varsigma}+\delta^i_{kk_\tau}\bar{\rho}^i_{kk_\tau}\frac{\partial^2 f_i}{\partial\tilde{x}_p\partial x_\varsigma}$$
$$+(\delta^i_{kk_\tau})^2\bar{\rho}^i_{kk_\tau}\frac{\partial^2 f_i}{\partial\tilde{x}_p\partial\tilde{x}_\varsigma}\Bigg)\Big|_{(s^i_{kk},\boldsymbol{X}^{k-1},\delta^i_{kk_\tau}\boldsymbol{X}^{k-1}+(1-\delta^i_{kk_\tau})\tilde{\boldsymbol{x}}^i_{kk_\tau},\boldsymbol{u}^i_{kk},\tilde{\boldsymbol{u}}^i_{kk_\sigma})}\frac{\partial x^{k-1}_\varsigma}{\partial u^q_m}$$
$$+\bar{h}_i\sum_{\varsigma=1}^{n}\Bigg\{\rho^i_{kk}(1-\delta^i_{kk_\tau})\frac{\partial^2 f_i}{\partial x_p\partial\tilde{x}_\varsigma}\Big|_{(s^i_{kk},\boldsymbol{X}^{k-1},\delta^i_{kk_\tau}\boldsymbol{X}^{k-1}+(1-\delta^i_{kk_\tau})\tilde{\boldsymbol{x}}^i_{kk_\tau},\boldsymbol{u}^i_{kk},\tilde{\boldsymbol{u}}^i_{kk_\sigma})}$$
$$\times\left[\eta^i_{qk_\tau}\bar{\rho}^i_{kk_\tau}\frac{\partial x^{k_\tau}_\varsigma}{\partial u^q_m}+\theta^i(k_\tau-q-1)\left(\bar{\rho}^i_{kk_\tau}\frac{\partial x^{k_\tau}_\varsigma}{\partial u^q_m}+(1-\bar{\rho}^i_{kk_\tau})\frac{\partial x^{k_\tau-1}_\varsigma}{\partial u^q_m}\right)\right]\Bigg\}$$

$$
\begin{aligned}
&+ \bar{h}_i \rho^i_{kk} \frac{\partial^2 f_i}{\partial x_p \partial \tilde{u}_m} \Big|_{(s^i_{kk}, \boldsymbol{X}^{k-1}, \delta^i_{kk_\tau} \boldsymbol{X}^{k-1} + (1-\delta^i_{kk_\tau}) \tilde{\boldsymbol{x}}^i_{kk_\tau}, \boldsymbol{u}^i_{kk}, \tilde{\boldsymbol{u}}^i_{kk_\sigma})} \big[\eta^i_{qk_\sigma} \tilde{\rho}^i_{kk_\sigma} \\
&+ \eta^i_{(q+1)k_\sigma} (1 - \tilde{\rho}^i_{kk_\sigma})\big] \\
&+ \bar{h}_i \delta^i_{kk_\tau} \bar{\rho}^i_{kk_\tau} \frac{\partial^2 f_i}{\partial \tilde{x}_p \partial \tilde{u}_m} \Big|_{(s^i_{kk}, \boldsymbol{X}^{k-1}, \delta^i_{kk_\tau} \boldsymbol{X}^{k-1} + (1-\delta^i_{kk_\tau}) \tilde{\boldsymbol{x}}^i_{kk_\tau}, \boldsymbol{u}^i_{kk}, \tilde{\boldsymbol{u}}^i_{kk_\sigma})} \\
&\times \big[\eta^i_{qk_\sigma} \tilde{\rho}^i_{kk_\sigma} + \eta^i_{(q+1)k_\sigma} (1 - \tilde{\rho}^i_{kk_\sigma})\big],
\end{aligned}
\tag{6.59}
$$

$$
\begin{aligned}
\frac{\partial c^k_i}{\partial u^q_m} = \bar{h}_i &\Bigg[\sum_{\varsigma=1}^{n} \left(\rho^i_{kq} \frac{\partial f_i}{\partial x_\varsigma} + \eta^i_{qq_\tau} \bar{\rho}^i_{kq_\tau} \frac{\partial f_i}{\partial \tilde{x}_\varsigma} \right) \Big|_{(s^i_{kq}, \boldsymbol{x}^i_{kq}, \tilde{\boldsymbol{x}}^i_{kq_\tau}, \boldsymbol{u}^i_{kq}, \tilde{\boldsymbol{u}}^i_{kq_\sigma})} \frac{\partial x^q_\varsigma}{\partial u^q_m} \\
&+ \left(\rho^i_{kq} \frac{\partial f_i}{\partial u_m} + \eta^i_{qq_\sigma} \tilde{\rho}^i_{kq_\sigma} \frac{\partial f_i}{\partial \tilde{u}_m} \right) \Big|_{(s^i_{kq}, \boldsymbol{x}^i_{kq}, \tilde{\boldsymbol{x}}^i_{kq_\tau}, \boldsymbol{u}^i_{kq}, \tilde{\boldsymbol{u}}^i_{kq_\sigma})} \Bigg] \big[(k-q+1)^{\alpha_i} \\
&- (k-q)^{\alpha_i}\big] + \bar{h}_i \sum_{l=q+1}^{k-1} \Bigg\{ \sum_{\varsigma=1}^{n} \frac{\partial f_i}{\partial x_\varsigma} \Big|_{(s^i_{kl}, \boldsymbol{x}^i_{kl}, \tilde{\boldsymbol{x}}^i_{kl_\tau}, \boldsymbol{u}^i_{kl}, \tilde{\boldsymbol{u}}^i_{kl_\sigma})} \Bigg(\rho^i_{kl} \frac{\partial x^l_\varsigma}{\partial u^q_m} \\
&+ (1 - \rho^i_{kl}) \frac{\partial x^{l-1}_\varsigma}{\partial u^q_m} \Bigg) + \sum_{\varsigma=1}^{n} \frac{\partial f_i}{\partial \tilde{x}_\varsigma} \Big|_{(s^i_{kl}, \boldsymbol{x}^i_{kl}, \tilde{\boldsymbol{x}}^i_{kl_\tau}, \boldsymbol{u}^i_{kl}, \tilde{\boldsymbol{u}}^i_{kl_\sigma})} \Bigg[\eta^i_{ql_\tau} \bar{\rho}^i_{kl_\tau} \frac{\partial x^{l_\tau}_\varsigma}{\partial u^q_m} \\
&+ \theta^i (l_\tau - q - 1) \left((1 - \bar{\rho}^i_{kl_\tau}) \frac{\partial x^{l_\tau - 1}_\varsigma}{\partial u^q_m} + \bar{\rho}^i_{kk_\tau} \frac{\partial x^{l_\tau}_\varsigma}{\partial u^q_m} \right) \Bigg] \\
&+ \frac{\partial f_i}{\partial u_m} \Big|_{(s^i_{kl}, \boldsymbol{x}^i_{kl}, \tilde{\boldsymbol{x}}^i_{kl_\tau}, \boldsymbol{u}^i_{kl}, \tilde{\boldsymbol{u}}^i_{kl_\sigma})} \big[\eta^i_{(q+1)l} (1 - \rho^i_{kl})\big] \\
&+ \frac{\partial f_i}{\partial \tilde{u}_m} \Big|_{(s^i_{kl}, \boldsymbol{x}^i_{kl}, \tilde{\boldsymbol{x}}^i_{kl_\tau}, \boldsymbol{u}^i_{kl}, \tilde{\boldsymbol{u}}^i_{kl_\sigma})} \big[\eta^i_{ql_\sigma} \tilde{\rho}^i_{kl_\sigma} + \eta^i_{(q+1)l_\sigma} (1 - \tilde{\rho}^i_{kl_\sigma})\big] \Bigg\} \\
&\times \big[(k-l+1)^{\alpha_i} - (k-l)^{\alpha_i} + \bar{h}_i \sum_{\varsigma=1}^{n} \left(\frac{\partial f_i}{\partial x_\varsigma} \right] \\
&+ \delta^i_{kk_\tau} \frac{\partial f_i}{\partial \tilde{x}_\varsigma} \Bigg) \Big|_{(s^i_{kk}, \boldsymbol{X}^{k-1}, \delta^i_{kk_\tau} \boldsymbol{X}^{k-1} + (1-\delta^i_{kk_\tau}) \tilde{\boldsymbol{x}}_{kk_\tau}, \boldsymbol{u}^i_{kk}, \tilde{\boldsymbol{u}}_{kk_\sigma})} \frac{\partial x^{k-1}_\varsigma}{\partial u^q_m} \\
&+ \bar{h}_i \sum_{\varsigma=1}^{n} (1 - \delta^i_{kk_\tau}) \frac{\partial f_i}{\partial \tilde{x}_\varsigma} \Big|_{(s^i_{kk}, \boldsymbol{X}^{k-1}, \delta^i_{kk_\tau} \boldsymbol{X}^{k-1} + (1-\delta^i_{kk_\tau}) \tilde{\boldsymbol{x}}_{kk_\tau}, \boldsymbol{u}^i_{kk}, \tilde{\boldsymbol{u}}_{kk_\sigma})} \\
&\times \Bigg[\eta^i_{qk_\tau} \bar{\rho}^i_{kk_\tau} \frac{\partial x^{k_\tau}_\varsigma}{\partial u^q_m} + \theta^i (k_\tau - q - 1) \left((1 - \bar{\rho}^i_{kk_\tau}) \frac{\partial x^{k_\tau - 1}_\varsigma}{\partial u^q_m} + \bar{\rho}^i_{kk_\tau} \frac{\partial x^{k_\tau}_\varsigma}{\partial u^q_m} \right) \Bigg] \\
&+ \bar{h}_i \frac{\partial f_i}{\partial \tilde{u}_m} \Big|_{(s^i_{kk}, \boldsymbol{X}^{k-1}, \delta^i_{kk_\tau} \boldsymbol{X}^{k-1} + (1-\delta^i_{kk_\tau}) \tilde{\boldsymbol{x}}_{kk_\tau}, \boldsymbol{u}^i_{kk}, \tilde{\boldsymbol{u}}_{kk_\sigma})} \big[\eta^i_{qk_\sigma} \tilde{\rho}^i_{kk_\sigma} \\
&+ \eta^i_{(q+1)k_\sigma} (1 - \tilde{\rho}^i_{kk_\sigma})\big] - \sum_{p=1}^{n} \left[\frac{\partial b^k_{ip}}{\partial u^q_m} x^{k-1}_p + b^k_{ip} \frac{\partial x^{k-1}_p}{\partial u^q_m} \right],
\end{aligned}
\tag{6.60}
$$

where, for two given integers j and j', $\eta^i_{jj'}$ is as defined in (6.50); and for a given π, $\theta^i(\pi)$ is as defined in (6.53).

On this basis, we now propose Algorithm 1 for computing the gradients of the cost and constraint functions in Problem (P(h)). From this procedure, we can see that the gradients of the cost and constraint functions with respect to decision variables can be calculated in conjunction with solving fractional time-delay system (6.1) forward in time, simultaneously. In particular, this procedure for calculating the gradients can be integrated with any standard gradient-based nonlinear optimization technique, e.g., SQP, to solve Problem (P(h)) as a nonlinear programming problem.

Algorithm 1 Computing the gradients in Problem (P(h)).

Step 0. Input an integer N and an initial value $\boldsymbol{U} = (\boldsymbol{U}^0, \boldsymbol{U}^1, \ldots, \boldsymbol{U}^N) \in W^{N+1}$. Compute $h = T/N$ and set $k := 1$.

Step 1. Let $t_k = kh$. Then, calculate $\boldsymbol{X}^k$ by (6.20) using $(\boldsymbol{X}^0, \cdots, \boldsymbol{X}^{k-1})$ and $(\boldsymbol{U}^0, \cdots, \boldsymbol{U}^k)$.

Step 2. Calculate $\dfrac{\partial \boldsymbol{B}^k}{\partial u^q_m}$ and $\dfrac{\partial \boldsymbol{C}^k}{\partial u^q_m}$ by (6.48)–(6.49), (6.51)–(6.52), and (6.55)–(6.60) for $m = 1, \ldots, r$, and $q = 0, 1, \ldots, k$.

Step 3. Solve (6.47) for $\dfrac{\partial \boldsymbol{X}^k}{\partial u^q_m}$.

Step 4. If $k < N$, then set $k := k + 1$ and go to Step 1. Otherwise, compute $\dfrac{\partial g^h_j(\boldsymbol{X}^N)}{\partial u^q_m}$, $j = 0, 1, \ldots, N_i$, by (6.45).

6.4 Numerical Examples

We consider four example problems to test the numerical solution algorithm in Sect. 6.3. The toolbox FMINCON in MATLAB is the optimization software to implement SQP method. All computations are performed within MATLAB environment on a PC with 1.99 GHz Intel Core i7-9700T CPU and 16.0 GB RAM.

Example 6.4.1 Consider the following DFOCP:

$$\min\ \tilde{g}_0(u) = \frac{1}{2}\left[10^5 x^2(2) + \int_0^2 u^2(t)dt\right], \tag{6.61}$$

subject to the fractional time-delay system

$$\begin{aligned} {}_0^C\mathrm{D}_t^\alpha\, x(t) &= x(t-1) + u(t), \quad t \in (0, 2], \\ x(t) &= 1, \quad t \in [-1, 0]. \end{aligned}$$

The exact solution for this problem when $\alpha = 1.0$ is

$$u^e(t) = \begin{cases} -2.1 + 1.05t, & t \in [0, 1], \\ -1.05, & t \in [1, 2], \end{cases}$$

$$x^e(t) = \begin{cases} 1 - 1.1t + 0.525t^2, & t \in [0, 1], \\ -0.25 + 1.575t - 1.075t^2 + 0.175t^3, & t \in [1, 2]. \end{cases}$$

Define a new state variable y such that

$$\dot{y}(t) = \frac{1}{2}u^2(t),$$

with

$$y(0) = 0.$$

Then, the cost function (6.61) becomes

$$g_0(u) = \frac{1}{2}10^5 x^2(2) + y(2).$$

We first solve the problem for various values of N when $\alpha = 1.0$. The computed results are listed in Table 6.1. Here, the errors of the control and state are defined as

$$E(x) = \max_{k \in \{0,1,\ldots,N\}} \{|x^e(t_k) - x^a(t_k)|\},$$

and

$$E(u) = \max_{k \in \{0,1,\ldots,N\}} \{|u^e(t_k) - u^a(t_k)|\},$$

where x^a and u^a denote the numerical solutions of x and u, respectively. From Table 6.1, it can be seen that there is an excellent agreement between the approximate and exact solutions of control and state, and the optimal cost decreases as N increases. For comparison, the computed cost, the costs reported in [97, 217, 232] and the cost from the exact solution are listed in Table 6.2. Clearly, our computed cost is comparable or superior to those obtained in [97, 232] and the cost from the exact solution. The cost reported in [217] is lowest. However, the cost reported in [217] is an underestimate of the cost from the exact solution, which should be a lower bound of the cost obtained by any approximation method. We further solve the problem for various values of α when $N = 200$. The computed results are listed in Table 6.3. Furthermore, the corresponding controls and states for various values of α are plotted in Fig. 6.1. From Table 6.3 and Fig. 6.1, we can see that the fractional solutions converge to the integer solution as the fractional order α approaches to the integer order 1.

Example 6.4.2 Consider the following DFOCP:

Table 6.1 Optimal costs, errors and CPU time for Example 6.4.1 when $\alpha = 1.0$

N	20	40	80	160	320
g_0	1.846133	1.841387	1.839330	1.838384	1.837532
$E(x)$	9.0435e–04	2.4351e–04	6.3304e–05	1.6171e–05	1.0504e–05
$E(u)$	3.5411e–02	1.7383e–02	8.6094e–03	4.2828e–03	2.1347e–03
CPU(s)	0.7609	3.5120	2.5810e+01	2.1140e+02	1.7620e+03

Table 6.2 Comparison of optimal costs from our method and those in [97, 217, 232] for Example 6.4.1 when $\alpha = 1.0$

g_0	g_0 in [97]	g_0 in [217]	g_0 in [232]	g_0 from exact solution
1.837532	1.837575	1.837159	1.849370	1.837500

Table 6.3 Optimal costs and CPU time for Example 6.4.1 when $N = 200$

α	1.0	0.95	0.90	0.85	0.8
g_0	1.838201	1.838118	1.834750	1.826941	1.812944
CPU(s)	4.1006e+02	4.1880e+02	4.1861e+02	4.2030e+02	4.1561e+02

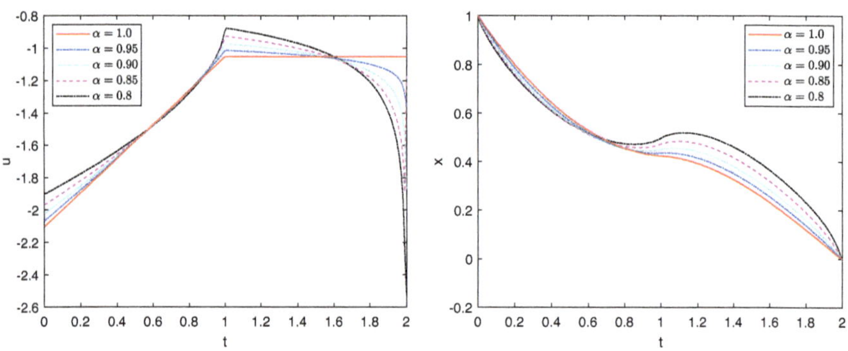

Fig. 6.1 Optimal controls and state trajectories for Example 6.4.1

$$\min\ g_0(u) = \frac{1}{2}\int_0^1 \left[x^2(t) + \frac{1}{2}u^2(t)\right]dt, \tag{6.62}$$

subject to the fractional time-delay system

Table 6.4 Optimal costs and CPU time for Example 6.4.2 when $\alpha = 1.0$

N	20	40	80	160	320
g_0	3.731097	0.373030	0.372910	0.372794	0.372373
CPU(s)	5.1719	11.4817	45.4899	3.4371e+02	2.2625e+03

Table 6.5 Optimal costs and CPU time for Example 6.4.2 when $N = 60$

α	1.0	0.99	0.95	0.9	0.8	0.7
g_0	0.372906	0.371451	0.367783	0.363245	0.354313	0.345538
g_0 in [223]	0.373113	0.372196	0.368551	0.364034	0.355113	0.346307
Error	−0.000207	−0.000745	−0.000768	−0.000789	−0.000800	−0.000769
CPU(s)	25.9688	25.9284	25.8438	25.7500	25.3759	25.0781

$$
\begin{aligned}
{}_0^C\mathrm{D}_t^\alpha x(t) &= -x(t) + x(t - \frac{1}{3}) + u(t) - \frac{1}{2}u(t - \frac{2}{3}), \ t \in (0, 1],\\
x(t) &= 1, \quad t \in [-\frac{1}{3}, 0],\\
u(t) &= 0, \quad t \in [-\frac{2}{3}, 0].
\end{aligned}
$$

Using a similar technique as for Example 6.4.1, the cost function (6.62) can be transformed into the canonical form (6.4) by introducing an additional state variable. We solve the problem for various values of N when $\alpha = 1.0$. The computed optimal costs are listed in Table 6.4, from which we see that the computed optimal cost decreases as N increases. We further solve the problem for various values of α when $N = 60$. The computed costs for various values of α are listed in Table 6.5, where the computed costs are compared with the costs obtained by using Chelyshkov wavelets in [223]. From Table 6.5, we see that the computed cost decreases as α decreases. More importantly, the optimal costs obtained by using our method are all superior to the ones reported in [223]. Figure 6.2 shows the corresponding controls and states for various values of α. From the obtained results, it is clear that the proposed numerical method is effective for solving this DFOCP.

Example 6.4.3 Consider the following nonlinear DFOCP:

$$
\min \ g_0(u) = \int_0^6 \left[x^2(t) + u^2(t)\right]dt, \tag{6.63}
$$

subject to the fractional time-delay system

$$
\begin{aligned}
{}_0^C\mathrm{D}_t^\alpha x(t) &= x(t-1)u(t-2), \ t \in (0, 6],\\
x(t) &= 1, \quad t \in [-1, 0],\\
u(t) &= 0, \quad t \in [-2, 0],
\end{aligned}
$$

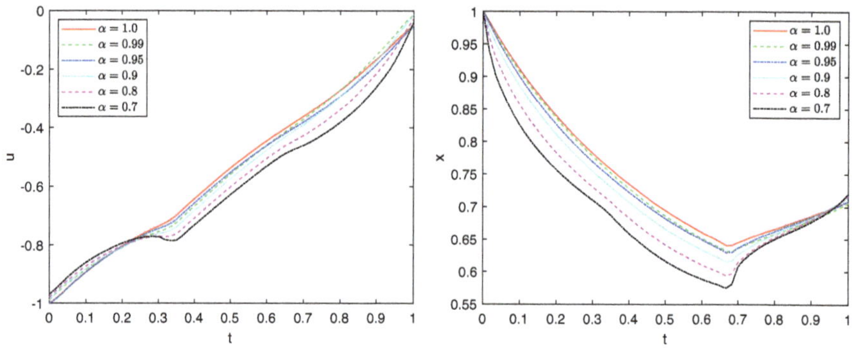

Fig. 6.2 Optimal controls and state trajectories for Example 6.4.2

and the mixed state-control inequality constraint

$$g_1(u) = 0.3 - x(t) - u(t) \leq 0, \quad t \in [0, 6]. \tag{6.64}$$

Note that the constraint (6.64) is a continuous inequality constraint on the state and the control. Using the ϵ–γ approximation method [303], we approximate constraint (6.64) by the following integral constraint:

$$\tilde{g}_1(u) = \int_0^6 \mathcal{L}_\epsilon(0.3 - x(t) - u(t))dt - \gamma \leq 0, \tag{6.65}$$

where $\epsilon > 0$ and $\gamma > 0$ are adjustable parameters; and

$$\mathcal{L}_\epsilon(\vartheta) = \begin{cases} 0, & \text{if } \vartheta < -\epsilon, \\ \dfrac{(\vartheta + \epsilon)^2}{4\epsilon}, & \text{if } -\epsilon \leq \vartheta \leq \epsilon, \\ \vartheta, & \text{if } \vartheta > \epsilon. \end{cases}$$

Furthermore, by applying a technique similar to that used in Example 6.4.1, the integral cost (6.63) and integral constraint (6.65) can be transformed into the canonical forms (6.4) and (6.5) by introducing additional state variables.

We first solve this problem for various values of N when $\alpha = 1.0$. The computed results are listed in Table 6.6. In the computation process, we took $\epsilon = 1.0 \times 10^{-2}$ and $\gamma = 2.5 \times 10^{-3}$ as the initial values for the approximation parameters and adjusting them at each step according to the ϵ–γ process [303]. The process involves reducing γ by a factor of 2 if the optimal solution does not satisfy the constraint (6.64), or reducing both ϵ and γ by a factor of 10 if the optimal solution satisfies (6.64). The ϵ–γ process is terminated when $\epsilon \leq 1.0 \times 10^{-8}$. For comparison, the cost reported in [207], which was obtained by a direct approach based on a hybrid of block-pulse functions and Lagrange interpolation, is also listed in Table 6.6. From Table 6.6, we

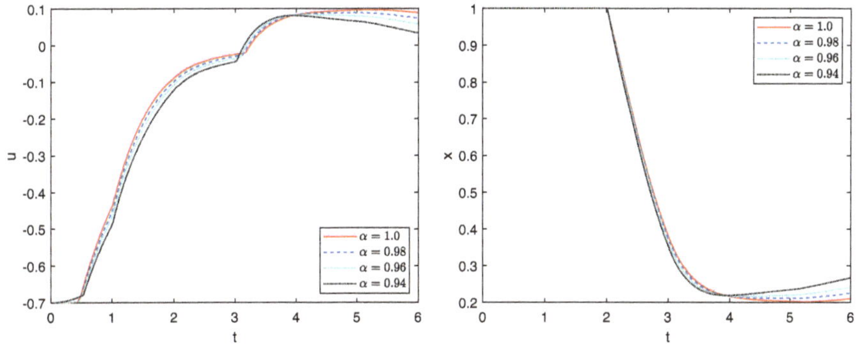

Fig. 6.3 Optimal controls and state trajectories for Example 6.4.3 when $N = 320$

can see that our computed optimal cost decreases as N increases and our computed optimal cost is superior to the one reported in [207] when $N = 320$. We further solve the problem for various values of α when $N = 320$. The obtained optimal costs are listed in Table 6.7, from which we can see that the obtained optimal cost increases as α decreases. The corresponding controls and states for various values of α in Table 6.7 are depicted in Fig. 6.3. In addition, the corresponding mixed state-control constraints are plotted in Fig. 6.4. From Figs. 6.3 and 6.4, it is clear that the fractional solutions converge to the integer solution as the fractional order α approaches to the integer order 1 and the mixed state-control constraint (6.64) is satisfied at every point in the time horizon.

Example 6.4.4 Consider the following nonlinear DFOCP:

$$\min\ g_0(\boldsymbol{u}) = \frac{1}{2}\boldsymbol{x}(2)^\top \boldsymbol{S}\boldsymbol{x}(2) + \frac{1}{2}\int_0^2 \Big[\boldsymbol{x}(t)^\top \boldsymbol{Q}\boldsymbol{x}(t) + \boldsymbol{u}(t)^\top \boldsymbol{R}\boldsymbol{u}(t)\Big]dt, \tag{6.66}$$

where

Table 6.6 Optimal costs and constraint values for Example 6.4.3 when $\alpha = 1.0$

N	20	40	80	160	320	[207]
g_0	3.208586	3.158964	3.133576	3.115869	3.108002	3.108122
g_1	−7.8352e–08	−7.4184e–08	−2.5474e–08	−2.1063e–08	−9.0251e–10	–

Table 6.7 Optimal costs and constraints for Example 6.4.3 when $N = 320$

α	1.0	0.98	0.96	0.94
g_0	3.108002	3.117312	3.124081	3.137930
g_1	−9.0251e–10	7.336e–09	1.073e–08	4.840e–09

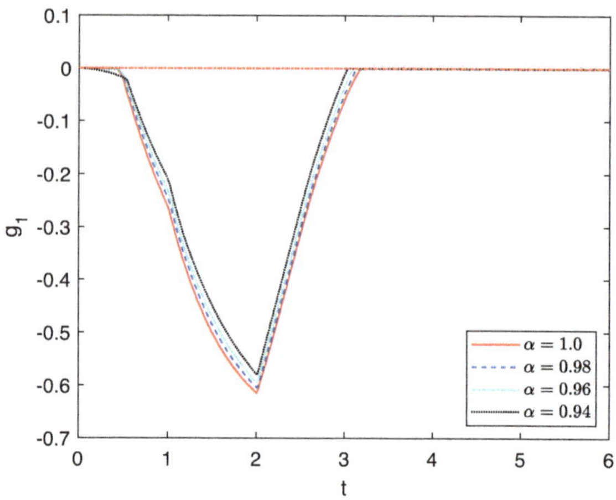

Fig. 6.4 Mixed state-control constraints for Example 6.4.3 when $N = 320$.

$$\boldsymbol{S} = \begin{bmatrix} 1 & 2 & 0 & 0 \\ 2 & 1 & 0 & 0 \\ 0 & 0 & 2 & 1 \\ 0 & 0 & 1 & 1 \end{bmatrix}, \quad \boldsymbol{Q} = \begin{bmatrix} 1 & 0 & 0 & 0 \\ 0 & 2 & 0 & 0 \\ 0 & 0 & 1 & 0 \\ 0 & 0 & 0 & 2 \end{bmatrix}, \quad \boldsymbol{R} = \begin{bmatrix} 1 & 0 & 0 & 0 \\ 0 & 1 & 0 & 0 \\ 0 & 0 & 1 & 0 \\ 0 & 0 & 0 & 1 \end{bmatrix},$$

subject to the fractional time-delay system

$$\begin{aligned}
{}_0^C\mathrm{D}_t^{\alpha_1} x_1(t) &= -2x_1^2(t) + x_1(t)x_2(t-0.1) + 2x_2(t) - u_1(t)u_2(t-0.1),\\
{}_0^C\mathrm{D}_t^{\alpha_2} x_2(t) &= -2x_2^2(t) - x_1(t-0.1) + 2x_3(t) + u_2(t),\\
{}_0^C\mathrm{D}_t^{\alpha_3} x_3(t) &= -x_3^2(t) - x_1(t)x_2(t) - x_2(t-0.1)u_2(t) - u_1(t-0.1) + 2u_3(t),\\
{}_0^C\mathrm{D}_t^{\alpha_4} x_4(t) &= -x_4^2(t) + x_2(t)x_3(t) - 2x_3(t-0.1) + 2u_4(t),
\end{aligned}$$

with the initial conditions

$$\begin{aligned}
x_i(t) &= 1,\ t \in [-0.1, 0],\ i = 1, \ldots, 4,\\
u_i(t) &= 1,\ t \in [-0.1, 0],\ i = 1, \ldots, 4,
\end{aligned}$$

the terminal inequality constraints

$$\begin{aligned}
g_1(\boldsymbol{u}) &= x_1^2(2) + x_2^2(2) + x_3^2(2) + x_4^2(2) - 4 \le 0,\\
g_2(\boldsymbol{u}) &= 0.002 - x_1^2(2) - x_2^2(2) - x_3^2(2) - x_4^2(2) \le 0,
\end{aligned}$$

and the control constraints

$$-3 \le u_i(t) \le 1,\ t \in [0, 2],\ i = 1, \ldots, 4.$$

Table 6.8 Optimal costs and constraint values for Example 6.4.4 when $\alpha = (1.0, 1.0, 1.0, 1.0)^{\top}$

N	100	200	300	400	[334]
g_0	1.113532	1.106782	1.101891	1.100135	1.100820
g_1	−3.954503	−3.958619	−3.959701	−3.958128	−3.959600
g_2	−0.043496	−0.039381	−0.038298	−0.039871	−0.031200

Table 6.9 Optimal costs and constraint values for Example 6.4.4 when $N = 400$

α	$(1.0, 1.0, 1.0, 1.0)^{\top}$	$(0.8, 0.9, 0.6, 0.5)^{\top}$	Error
g_0	1.100135	0.810766	0.289369
g_1	−3.958128	−3.931829	−0.026299
g_2	−0.039871	−0.066171	0.026300

In [334], this problem was solved for $\boldsymbol{\alpha} = (1.0, 1.0, 1.0, 1.0)^{\top}$ by applying the control parameterization technique together with a hybrid time-scaling strategy. The obtained optimal control yields an optimal cost of $g_0(\boldsymbol{u}^*) = 1.10082$, and the corresponding constraint values are

$$g_1(\boldsymbol{u}^*) = -3.9596, \text{ and } g_2(\boldsymbol{u}^*) = -0.0312.$$

For comparison, we first solve the problem using our proposed solution method when $\boldsymbol{\alpha} = (1.0, 1.0, 1.0, 1.0)^{\top}$. Here, the cost function (6.66) is transformed to the canonical form (6.4) by using a similar technique as for Example 6.4.1. The computed results and the ones reported in [334] are listed in Table 6.8. From Table 6.8, we see that the optimal cost improves as N increases and the optimal cost $g_0(\boldsymbol{u}^*) = 1.100135$, which is obtained for the case of $N = 400$, is superior to the one reported in [334].

We now solve this problem when $\boldsymbol{\alpha} = (0.8, 0.9, 0.6, 0.5)^{\top}$ using the uniform mesh with $N = 400$. The computed optimal cost and constraint values are listed in Table 6.9, where, for comparison, the computed optimal cost and constraint values when $\boldsymbol{\alpha} = (1.0, 1.0, 1.0, 1.0)^{\top}$ are also listed in Table 6.9. Furthermore, the corresponding states and controls are shown in Figs. 6.5 and 6.6. As can be seen from the results, the fractional optimal control can achieve a better optimal value than its integer counterpart.

6.5 Conclusions

In this chapter, we have considered numerical solution of delay fractional optimal control problem involving a class of nonlinear time-delay systems with different fractional orders and subject to state inequality constraints. We have designed the

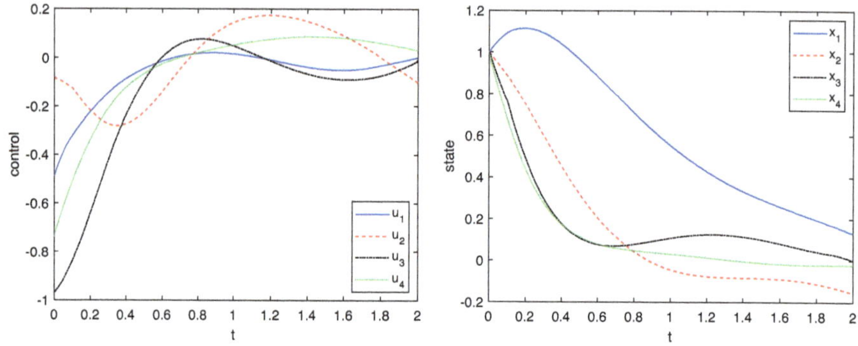

Fig. 6.5 Optimal controls and state trajectories for Example 6.4.4 when $\alpha = (1.0, 1.0, 1.0, 1.0)^\top$ and $N = 400$

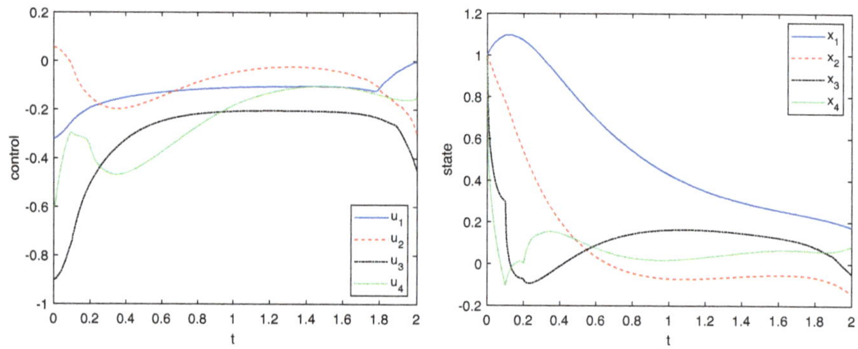

Fig. 6.6 Optimal controls and state trajectories for Example 6.4.4 when $\alpha = (0.8, 0.9, 0.6, 0.5)^\top$ and $N = 400$

explicit numerical scheme for the system based on the Taylor expansion and linear interpolation. This numerical scheme converts the delay fractional optimal control problem into the discrete-time optimal control problem. Then, we have presented the numerical procedure for computing the gradients of the cost and constraint functions with respect to the decision variables. On this basis, the gradient-based optimization approach has been developed to solve the resulting discrete-time optimal control problem. Numerical results obtained from solving several non-trivial example problems show that our method is effective and robust with respect to the orders of the fractional system and the solutions obtained are accurate.

Chapter 7
Free Final-Time Delay Fractional Optimal Control Problems

7.1 Introduction

The terminal time in a fractional optimal control problem (FOCP) can be either fixed or free. This chapter is concerned with FOCPs with free terminal time. FOCPs with free terminal time have been found in various real-world applications such as attitude control [82], HIV treatment [111], resource harvesting [184] and biochemical process [336].

For FOCPs with free terminal time, necessary optimality conditions involving nonlinear systems with both integer-order and fractional-order derivatives are derived in [246]. Numerical algorithms for solving FOCPs with free terminal time are developed in [103, 274], where, by using the well-known time-scaling transformation [303], the FOCPs with free terminal time are first transcribed into the ones with fixed terminal time. In [274], direct transcription methods based on fractional integral approximation formulas are used, whereas a gradient-based optimization method is used to solve the resulting FOCPs in [103]. However, the above theory and numerical solution can not be applied to delay fractional optimal control problems (DFOCPs). By the way, for integer-order delay optimal control problems with free terminal time, a two-stage optimization approach, in which the terminal time is optimized in the outer stage, and the control function is optimized in the inner stage, is developed in [39]. A gradient-based computational approach based on the time-scaling transformation and control parameterization method is discussed in [180]. A fully informed particle swarm optimization together with the control paramterization method is investigated in [51]. However, to our best knowledge, there are no numerical solution methods for DFOCPs with free terminal time. In next section, we state a class of DFOCPs with free terminal time, in which both the terminal time and control vector are to be chosen such that a cost function is minimized subject to fractional time-delay systems and canonical state constraints. The fractional derivatives in this class of problems are described in the Caputo sense and they can be of different orders. By using the time-scaling transformation, we transform the DFOCP with free termi-

C. Liu et al., *Nonlinear Fractional Optimal Control*, Springer Optimization and Its Applications 237, https://doi.org/10.1007/978-981-92-0274-4_7

nal time into an equivalent one with fixed terminal time. This yields a non-standard DFOCP involving fraction-order systems with time-varying delay. In Sect. 7.3, we develop numerical solution approach for the resulting DFOCP. In Sect. 7.3.1, we propose an explicit numerical scheme for solving the transformed fractional time-delay systems, which gives rise to a discrete-time optimal control problem. In Sect. 7.3.2, we derive gradient formulas of the cost and constraint functions with respect to decision variables. Based on this result, we develop a gradient-based optimization approach to solve the resulting discrete-time optimal control problem. In Sect. 7.4, two example problems are solved to demonstrate the effectiveness of our proposed approach. Finally, some conclusions are provided in Sect. 7.5.

The main reference of this chapter is [177].

7.2 Problem Formulation

Consider the following nonlinear fractional time-delay system:

$$ {}_0^C\mathrm{D}_t^{\alpha} \boldsymbol{x}(t) = \boldsymbol{f}(t, \boldsymbol{x}(t), \boldsymbol{x}(t-\sigma), \boldsymbol{u}(t)), \quad t \in (0, T], \tag{7.1} $$

$$ \boldsymbol{x}(t) = \boldsymbol{\phi}(t), \quad t \in [-\sigma, 0], \tag{7.2} $$

where $\boldsymbol{x}(t) = (x_1(t), \ldots, x_n(t))^\top \in \mathbb{R}^n$ is the state vector; $\boldsymbol{u}(t) = (u_1(t), \ldots, u_r(t))^\top \in \mathbb{R}^r$ is the control vector; $T > 0$ is a free terminal time; $\sigma > 0$ is a given time-delay; $\boldsymbol{f} : \mathbb{R} \times \mathbb{R}^n \times \mathbb{R}^n \times \mathbb{R}^r \to \mathbb{R}^n$ and $\boldsymbol{\phi} : \mathbb{R} \to \mathbb{R}^n$ are given functions; and ${}_0^C\mathrm{D}_t^{\alpha} \boldsymbol{x}(t) = ({}_0^C\mathrm{D}_t^{\alpha_1} x_1(t), \ldots, {}_0^C\mathrm{D}_t^{\alpha_n} x_n(t))^\top$ with ${}_0^C\mathrm{D}_t^{\alpha_i} x_i(t)$ denoting the Caputo fractional derivative of $x_i(t)$

The terminal time T in system (7.1) is a decision variable. Define

$$ \mathcal{T} := \{\beta \in \mathbb{R} |\ T_{\min} \leq \beta \leq T_{\max}\}, \tag{7.3} $$

where $T_{\min} > 0$ and $T_{\max} > 0$ are given constants such that $T_{\min} < T_{\max}$. Any $T \in \mathcal{T}$ is called an *admissible terminal time*. Furthermore, define

$$ W := \{w \in \mathbb{R}^r \,|\, a_j \leq w_j \leq b_j,\ j = 1, \ldots, r\}, \tag{7.4} $$

where a_j and b_j are given constants such that $a_j < b_j$. Any piecewise continuous function $\boldsymbol{u} : [0, \infty) \to \mathbb{R}^r$ such that $\boldsymbol{u}(t) \in W$ for almost all $t \geq 0$ is called an *admissible control*. Let $\mathcal{U}$ be the class of all such admissible controls. Accordingly, any pair $(T, \boldsymbol{u}) \in \mathcal{T} \times \mathcal{U}$ is called an *admissible pair* for system (7.1) with (7.2).

We assume throughout this chapter that the following condition is satisfied.

Assumption 7.2.1 The functions $\boldsymbol{f}$ and $\boldsymbol{\phi}$ are twice continuously differentiable with respect to all their arguments.

Let $\boldsymbol{x}(\cdot|\boldsymbol{u})$ denote the continuous solution of (7.1) and (7.2) on $[-\sigma, T]$ corresponding to each $\boldsymbol{u} \in \mathcal{U}$. This solution is called the *state trajectory*. We suppose that the terminal time and state trajectory of system (7.1) with (7.2) are subject to the following canonical constraints:

$$g_k(T, \boldsymbol{u}) = \Phi_k(T, \boldsymbol{x}(T|\boldsymbol{u})) = 0, \quad k = 1, \ldots, N_e, \tag{7.5}$$

$$g_k(T, \boldsymbol{u}) = \Phi_k(T, \boldsymbol{x}(T|\boldsymbol{u})) \geq 0, \quad k = N_e + 1, \ldots, N_e + N_i, \tag{7.6}$$

where N_e is the number of the equality constraints; N_i is the number of the inequality constraints; and $\Phi_k : \mathbb{R} \times \mathbb{R}^n \to \mathbb{R}, k = 1, \ldots, N_e + N_i$, are given continuously differentiable functions. It should be noted that integral constraints can be transformed into the forms of (7.5) and (7.6) by introducing additional differential equations; see, Chap. 6 for details. Thus, there is no loss of generality in ignoring integral terms in the constraint functions (7.5) and (7.6).

We now state our delay fractional optimal control problem as follows.

Problem (P). Find an admissible pair $(T, \boldsymbol{u}) \in \mathcal{T} \times \mathcal{U}$ such that the cost function

$$g_0(T, \boldsymbol{u}) = \Phi_0(T, \boldsymbol{x}(T|\boldsymbol{u})) \tag{7.7}$$

is minimized subject to the canonical constraints (7.5) and (7.6), where $\Phi_0 : \mathbb{R} \times \mathbb{R}^n \to \mathbb{R}$ is a given continuously differentiable function.

Problem (P) is difficult to solve numerically because the time-delay system (7.1) with (7.2) must be integrated over a variable time horizon. To overcome this difficulty, we apply the following time-scaling transformation:

$$t = t(s) = Ts, \tag{7.8}$$

where $s \in [-\sigma T_{\min}^{-1}, 1]$ is a new time variable. Obviously, $s = 0$ corresponds to $t = 0$, and $s = 1$ corresponds to $t = T$. Furthermore, by Theorem 3.13, we note that, for each $(T, \boldsymbol{u}) \in \mathcal{T} \times \mathcal{U}$, Eq. (7.1) is equivalent to the following Volterra integral equation:

$$x_i(t) = \phi_i(0) + \frac{1}{\Gamma(\alpha_i)} \int_0^t (t - \vartheta)^{\alpha_i - 1} f_i(\vartheta, \boldsymbol{x}(\vartheta), \boldsymbol{x}(\vartheta - \sigma), \boldsymbol{u}(\vartheta)) d\vartheta \tag{7.9}$$

for $t \in (0, T]$, $i = 1, 2, \ldots, n$. Let $\tilde{\boldsymbol{x}}(s) = \boldsymbol{x}(t(s))$ and $\tilde{\boldsymbol{u}}(s) = \boldsymbol{u}(t(s))$. Then, by applying change of variables (7.8) to (7.9), we have

$$\tilde{x}_i(s) = \phi_i(0) + \frac{1}{\Gamma(\alpha_i)} \int_0^s (s - \tau)^{\alpha_i - 1} \tilde{f}_i(T\tau, \tilde{\boldsymbol{x}}(\tau), \tilde{\boldsymbol{x}}(\tau - \sigma T^{-1}), \tilde{\boldsymbol{u}}(\tau), T) d\tau \tag{7.10}$$

for $s \in (0, 1]$ and $i = 1, 2, \ldots, n$, where

$$\tilde{f}_i(T\tau, \tilde{\boldsymbol{x}}(\tau), \tilde{\boldsymbol{x}}(\tau - \sigma T^{-1}), \tilde{\boldsymbol{u}}(\tau), T) = T^{\alpha_i} f_i(T\tau, \tilde{\boldsymbol{x}}(\tau), \tilde{\boldsymbol{x}}(\tau - \sigma T^{-1}), \tilde{\boldsymbol{u}}(\tau)). \tag{7.11}$$

Furthermore, the initial condition (7.2) becomes

$$\tilde{\boldsymbol{x}}(s) = \boldsymbol{\phi}(Ts), \quad s \in [-\sigma T_{\min}^{-1}, 0].$$

Thus, the original fractional time-delay system (7.1) with (7.2) is transformed into the following form:

$$ {}_0^C\mathrm{D}_s^{\alpha} \tilde{\boldsymbol{x}}(s) = \tilde{\boldsymbol{f}}(Ts, \tilde{\boldsymbol{x}}(s), \tilde{\boldsymbol{x}}(s - \sigma T^{-1}), \tilde{\boldsymbol{u}}(s), T), \quad s \in (0, 1], \tag{7.12}$$

$$\tilde{\boldsymbol{x}}(s) = \boldsymbol{\phi}(Ts), \quad s \in [-\sigma T_{\min}^{-1}, 0], \tag{7.13}$$

where

$$\tilde{\boldsymbol{f}}(Ts, \tilde{\boldsymbol{x}}(s), \tilde{\boldsymbol{x}}(s - \sigma T^{-1}), \tilde{\boldsymbol{u}}(s), T) = \begin{bmatrix} \tilde{f}_1(Ts, \tilde{\boldsymbol{x}}(s), \tilde{\boldsymbol{x}}(s - \sigma T^{-1}), \tilde{\boldsymbol{u}}(s), T) \\ \vdots \\ \tilde{f}_n(Ts, \tilde{\boldsymbol{x}}(s), \tilde{\boldsymbol{x}}(s - \sigma T^{-1}), \tilde{\boldsymbol{u}}(s), T) \end{bmatrix}. \tag{7.14}$$

Let $\tilde{\mathcal{U}}$ be the class of piecewise continuous functions $\tilde{\boldsymbol{u}} : [0, 1] \to \mathbb{R}^r$ such that $\tilde{\boldsymbol{u}}(s) \in W$ for almost all $s \in [0, 1]$. Any pair $(T, \tilde{\boldsymbol{u}}) \in \mathcal{T} \times \tilde{\mathcal{U}}$ is called an *admissible pair* for system (7.12) with (7.13). Let $\tilde{\boldsymbol{x}}(\cdot|T, \tilde{\boldsymbol{u}})$ denote the solution of system (7.12) with (7.13) on $[-\sigma T_{\min}^{-1}, 1]$ corresponding to a given admissible pair $(T, \tilde{\boldsymbol{u}}) \in \mathcal{T} \times \tilde{\mathcal{U}}$. Then, the canonical constraints (7.5) and (7.6) become

$$\tilde{g}_k(T, \tilde{\boldsymbol{u}}) = \Phi_k(T, \tilde{\boldsymbol{x}}(1, \tilde{\boldsymbol{u}})) = 0, \quad k = 1, \ldots, N_e, \tag{7.15}$$

$$\tilde{g}_k(T, \tilde{\boldsymbol{u}}) = \Phi_k(T, \tilde{\boldsymbol{x}}(1, \tilde{\boldsymbol{u}})) \geq 0, \quad k = N_e + 1, \ldots, N_e + N_i. \tag{7.16}$$

Thus, Problem (P) is equivalent to the following delay fractional optimal control problem with fixed terminal time.

Problem (EP). Find an admissible pair $(T, \tilde{\boldsymbol{u}}) \in \mathcal{T} \times \tilde{\mathcal{U}}$ such that the cost function

$$\tilde{g}_0(T, \tilde{\boldsymbol{u}}) = \Phi_0(T, \tilde{\boldsymbol{x}}(1, \tilde{\boldsymbol{u}})) \tag{7.17}$$

is minimized subject to the canonical constraints (7.15) and (7.16).

7.3 Numerical Solution Approach

7.3.1 Numerical Integration

Let $\{s_l = lh|\ l = 0, 1, \ldots, N\}$ be a partition of $[0, 1]$, where N is an integer and $h = \dfrac{1}{N}$. Then, the integral equation (7.10) can be rewritten as

$$\tilde{x}_i(s_l) = \phi_i(0) \\ + \frac{1}{\Gamma(\alpha_i)} \sum_{q=1}^{l} \int_{(q-1)h}^{qh} (lh - \tau)^{\alpha_i - 1} \tilde{f}_i(T\tau, \tilde{\boldsymbol{x}}(\tau), \tilde{\boldsymbol{x}}(\tau - \sigma T^{-1}), \tilde{\boldsymbol{u}}(\tau), T) d\tau \tag{7.18}$$

for all feasible i and l. We now consider an approximation scheme for the integral in (7.18). By Assumption 7.2.1 and (7.11), $\tilde{f}_i(T\tau, \tilde{\boldsymbol{x}}(\tau), \tilde{\boldsymbol{x}}(\tau - \sigma T^{-1}), \tilde{\boldsymbol{u}}(\tau), T)$ is twice continuously differentiable with respect to all its arguments. Thus, for $q = 1, \ldots, l$, we apply the Taylor expansion to $\tilde{f}_i(T\tau, \tilde{\boldsymbol{x}}(\tau), \tilde{\boldsymbol{x}}(\tau - \sigma T^{-1}), \tilde{\boldsymbol{u}}(\tau), T)$ at any point $\tau_{lq}^i \in ((q-1)h, qh)$ to yield

$$\tilde{f}_i(T\tau, \tilde{\boldsymbol{x}}(\tau), \tilde{\boldsymbol{x}}(\tau - \sigma T^{-1}), \tilde{\boldsymbol{u}}(\tau), T) \\ = \tilde{f}_i(T\tau_{lq}^i, \tilde{\boldsymbol{x}}(\tau_{lq}^i), \tilde{\boldsymbol{x}}(\tau_{lq}^i - \sigma T^{-1}), \tilde{\boldsymbol{u}}(\tau_{lq}^i), T) + K_{lq}^i(\tau - \tau_{lq}^i) + c_{lq}^i(\tau - \tau_{lq}^i)^2, \tag{7.19}$$

where K_{lq}^i and c_{lq}^i are, respectively, the coefficients of the first-order and second-order terms of the expansion at τ_{lq}^i. Then, substituting the right-hand side of (7.19) into the integrand of the last term in (7.18), we obtain, for $q = 1, \ldots, l$,

$$\frac{1}{\Gamma(\alpha_i)} \int_{(q-1)h}^{qh} (lh - \tau)^{\alpha_i - 1} \tilde{f}_i(T\tau, \tilde{\boldsymbol{x}}(\tau), \tilde{\boldsymbol{x}}(\tau - \sigma T^{-1}), \tilde{\boldsymbol{u}}(\tau), T) d\tau \\ = \bar{h}_i \tilde{f}_i(T\tau_{lq}^i, \tilde{\boldsymbol{x}}(\tau_{lq}^i), \tilde{\boldsymbol{x}}(\tau_{lq}^i - \sigma T^{-1}), \tilde{\boldsymbol{u}}(\tau_{lq}^i), T)[(d_{lq} + 1)^{\alpha_i} - d_{lq}^{\alpha_i}] \\ + \frac{K_{lq}^i}{\Gamma(\alpha_i)} \int_{(q-1)h}^{qh} (lh - \tau)^{\alpha_i - 1}(\tau - \tau_{lq}^i) d\tau + R_{lq}^i, \tag{7.20}$$

where $d_{lq} = (l - q)$; $\bar{h}_i = \dfrac{h^{\alpha_i}}{\Gamma(\alpha_i + 1)}$; and $R_{lq}^i = \dfrac{c_{lq}^i}{\Gamma(\alpha_i)} \displaystyle\int_{(q-1)h}^{qh} (lh - \tau)^{\alpha_i - 1}(\tau - \tau_{lq}^i)^2 d\tau$. To achieve the truncation error in (7.20) being R_{lq}^i, we choose the following τ_{lq}^i such that the second term on the right hand-side of (7.20) becomes zero:

$$\tau_{lq}^i = h \frac{[(d_{lq} + 1)^{\alpha_i + 1} - d_{lq}^{\alpha_i + 1}] + (\alpha_i + 1)[(d_{lq} + 1)^{\alpha_i}(q - 1) - d_{lq}^{\alpha_i} q]}{(\alpha_i + 1)[(d_{lq} + 1)^{\alpha_i} - d_{lq}^{\alpha_i}]}. \tag{7.21}$$

Substituting (7.21) into (7.20) and then combining the resulting expression with (7.18), we have, for $l = 1, \ldots, N$,

$$\tilde{x}_i(s_l) = \phi_i(0) + \bar{h}_i \sum_{q=1}^{l} \tilde{f}_i(T\tau_{lq}^i, \tilde{\boldsymbol{x}}(\tau_{lq}^i), \tilde{\boldsymbol{x}}(\tau_{lq}^i - \sigma T^{-1}), \tilde{\boldsymbol{u}}(\tau_{lq}^i), T) \times [(d_{lq}+1)^{\alpha_i} - d_{lq}^{\alpha_i}] + R_l^i, \tag{7.22}$$

where τ_{lq}^i is as given in (7.21); and $R_l^i = \sum_{q=1}^{l} R_{lq}^i$.

From (7.22), it is clear that to compute $\tilde{\boldsymbol{x}}(s_l)$, we need to calculate $\tilde{\boldsymbol{x}}(\tau_{lq}^i)$, $\tilde{\boldsymbol{x}}(\tau_{lq}^i - \sigma T^{-1})$ and $\tilde{\boldsymbol{u}}(\tau_{lq}^i)$, $q = 1, \ldots, l$, which are not available directly from the scheme. Let $\tilde{\boldsymbol{x}}^l$ and $\tilde{\boldsymbol{u}}^l$ be the numerical counterparts of $\tilde{\boldsymbol{x}}(s_l)$ and $\tilde{\boldsymbol{u}}(s_l)$. Then, $\tilde{\boldsymbol{x}}(\tau_{lq}^i)$, $\tilde{\boldsymbol{u}}(\tau_{lq}^i)$ and $\tilde{\boldsymbol{x}}(\tau_{lq}^i - \sigma T^{-1})$ can be approximated by $\tilde{\boldsymbol{x}}_{lq}^i$, $\tilde{\boldsymbol{u}}_{lq}^i$ and $\bar{\boldsymbol{x}}_{lq_\sigma}^i(T)$, respectively, where

$$\tilde{\boldsymbol{x}}_{lq}^i = \tilde{\boldsymbol{x}}^{q-1} + \rho_{lq}^i(\tilde{\boldsymbol{x}}^q - \tilde{\boldsymbol{x}}^{q-1}), \quad \tilde{\boldsymbol{u}}_{lq}^i = \tilde{\boldsymbol{u}}^{q-1} + \rho_{lq}^i(\tilde{\boldsymbol{u}}^q - \tilde{\boldsymbol{u}}^{q-1}), \tag{7.23}$$

and

$$\bar{\boldsymbol{x}}_{lq_\sigma}^i(T) = \begin{cases} \phi(T\tau_{lq}^i - \sigma), & \text{if } \tau_{lq}^i - \sigma T^{-1} \in [-\sigma T_{\min}^{-1}, 0], \\ \tilde{\boldsymbol{x}}^{q_\sigma - 1} + \bar{\rho}_{lq_\sigma}^i(T)(\tilde{\boldsymbol{x}}^{q_\sigma} - \tilde{\boldsymbol{x}}^{q_\sigma - 1}), & \text{if } \tau_{lq}^i - \sigma T^{-1} \in (s_{q_\sigma - 1}, s_{q_\sigma}] \\ & \text{for some } q_\sigma \in \{1, \ldots, q\} \end{cases} \tag{7.24}$$

with $\rho_{lq}^i = \dfrac{\tau_{lq}^i - s_{q-1}}{h} \in (0, 1)$ and $\bar{\rho}_{lq_\sigma}^i(T) = \dfrac{\tau_{lq}^i - \sigma T^{-1} - s_{q_\sigma - 1}}{h} \in (0, 1]$. Furthermore, $\tilde{f}_i(T\tau_{lq}^i, \tilde{\boldsymbol{x}}(\tau_{lq}^i), \tilde{\boldsymbol{x}}(\tau_{lq}^i - \sigma T^{-1}), \tilde{\boldsymbol{u}}(\tau_{lq}^i), T)$ can be approximated by

$$\tilde{f}_i(T\tau_{lq}^i, \tilde{\boldsymbol{x}}(\tau_{lq}^i), \tilde{\boldsymbol{x}}(\tau_{lq}^i - \sigma T^{-1}), \tilde{\boldsymbol{u}}(\tau_{lq}^i), T) = \tilde{f}_i(T\tau_{lq}^i, \tilde{\boldsymbol{x}}_{lq}^i, \bar{\boldsymbol{x}}_{lq_\sigma}^i(T), \tilde{\boldsymbol{u}}_{lq}^i, T). \tag{7.25}$$

Obviously, the truncation error of the above approximations is of order $\mathcal{O}(h^2)$. Thus, we obtain the following time-stepping scheme for (7.18):

$$\tilde{x}_i^l = \tilde{x}_i^0 + \bar{h}_i \sum_{q=1}^{l} \tilde{f}_i(T\tau_{lq}^i, \tilde{\boldsymbol{x}}_{lq}^i, \bar{\boldsymbol{x}}_{lq_\sigma}^i(T), \tilde{\boldsymbol{u}}_{lq}^i, T)[(d_{lq}+1)^{\alpha_i} - d_{lq}^{\alpha_i}] \tag{7.26}$$

for $l = 1, \ldots, N$.

The above scheme is implicit as the last term on the right-hand side of (7.26) contains a nonlinear function of x^l. Although we can use an iteration method such as Newton's method to solve (7.26), it is computationally expensive. Alternatively, we define the following explicit scheme by applying Taylor expansion to approximate the lth term in (7.26):

$$\begin{aligned}
&\tilde{f}_i(T\tau_{ll}^i, \tilde{\boldsymbol{x}}_{ll}^i, \bar{\boldsymbol{x}}_{ll_\sigma}^i(T), \tilde{\boldsymbol{u}}_{ll}^i, T)\\
&= \tilde{f}_i(T\tau_{ll}^i, \tilde{\boldsymbol{x}}^{l-1}, \delta_{ll_\sigma}^i \tilde{\boldsymbol{x}}^{l-1} + (1-\delta_{ll_\sigma}^i)\bar{\boldsymbol{x}}_{ll_\sigma}^i(T), \tilde{\boldsymbol{u}}_{ll}^i, T)\\
&+ \sum_{p=1}^{n}\left[\frac{\partial \tilde{f}_i}{\partial \tilde{x}_p}\Big|_{(T\tau_{ll}^i, \tilde{\boldsymbol{x}}^{l-1}, \delta_{ll_\sigma}^i \tilde{\boldsymbol{x}}^{l-1} + (1-\delta_{ll_\sigma}^i)\bar{\boldsymbol{x}}_{ll_\sigma}^i(T), \tilde{\boldsymbol{u}}_{ll}^i, T)} \rho_{ll}^i(\tilde{x}_p^l - \tilde{x}_p^{l-1})\right]\\
&+ \sum_{p=1}^{n}\left[\frac{\partial \tilde{f}_i}{\partial \bar{x}_p}\Big|_{(T\tau_{ll}^i, \tilde{\boldsymbol{x}}^{l-1}, \delta_{ll_\sigma}^i \tilde{\boldsymbol{x}}^{l-1} + (1-\delta_{ll_\sigma}^i)\bar{\boldsymbol{x}}_{ll_\sigma}^i(T), \tilde{\boldsymbol{u}}_{ll}^i, T)} \delta_{ll_\sigma}^i \tilde{\rho}_{ll_\sigma}^i(T)(\tilde{x}_p^l - \tilde{x}_p^{l-1})\right]\\
&+ \mathcal{O}(h^2),
\end{aligned} \tag{7.27}$$

where $\partial\bar{\boldsymbol{x}}$ denotes the partial differentiation with respect to $\tilde{\boldsymbol{x}}(s - \sigma T^{-1})$; and for two given integers $\imath$ and $\jmath$, $\delta_{\imath\jmath}^i$ is the Kronecker delta function defined by

$$\delta_{\imath\jmath}^i = \begin{cases} 1, & \text{if } \imath = \jmath,\\ 0, & \text{otherwise.}\end{cases} \tag{7.28}$$

Combining (7.27) with (7.26), we obtain the following explicit approximation scheme:

$$\begin{aligned}
\tilde{x}_i^l &= \tilde{x}_i^0 + \bar{h}_i \sum_{q=1}^{l-1} \tilde{f}_i(T\tau_{lq}^i, \tilde{\boldsymbol{x}}_{lq}^i, \bar{\boldsymbol{x}}_{lq_\sigma}^i(T), \tilde{\boldsymbol{u}}_{lq}^i, T)[(d_{lq}+1)^{\alpha_i} - d_{lq}^{\alpha_i}]\\
&+ \bar{h}_i \tilde{f}_i(T\tau_{ll}^i, \tilde{\boldsymbol{x}}^{l-1}, \delta_{ll_\sigma}^i \tilde{\boldsymbol{x}}^{l-1} + (1-\delta_{ll_\sigma}^i)\bar{\boldsymbol{x}}_{ll_\sigma}^i(T), \tilde{\boldsymbol{u}}_{ll}^i, T)\\
&+ \bar{h}_i \sum_{p=1}^{n}\left[\frac{\partial \tilde{f}_i}{\partial \tilde{x}_p}\Big|_{(T\tau_{ll}^i, \tilde{\boldsymbol{x}}^{l-1}, \delta_{ll_\sigma}^i \tilde{\boldsymbol{x}}^{l-1} + (1-\delta_{ll_\sigma}^i)\bar{\boldsymbol{x}}_{ll_\sigma}^i(T), \tilde{\boldsymbol{u}}_{ll}^i, T)} \rho_{ll}^i(\tilde{x}_p^l - \tilde{x}_p^{l-1})\right]\\
&+ \bar{h}_i \sum_{p=1}^{n}\left[\frac{\partial \tilde{f}_i}{\partial \bar{x}_p}\Big|_{(T\tau_{ll}^i, \tilde{\boldsymbol{x}}^{l-1}, \delta_{ll_\sigma}^i \tilde{\boldsymbol{x}}^{l-1} + (1-\delta_{ll_\sigma}^i)\bar{\boldsymbol{x}}_{ll_\sigma}^i(T), \tilde{\boldsymbol{u}}_{ll}^i, T)} \delta_{ll_\sigma}^i \tilde{\rho}_{ll_\sigma}^i(T)(\tilde{x}_p^l - \tilde{x}_p^{l-1})\right]\\
&+ \mathcal{O}(h^{2+\alpha_i})
\end{aligned} \tag{7.29}$$

for $i = 1, \ldots, n$, and $l = 1, \ldots, N$.

Omitting $\mathcal{O}(h^{2+\alpha_i})$ in (7.29) and rearranging the results gives

$$\begin{aligned}
&\boldsymbol{B}^l(\tilde{\boldsymbol{x}}^{l-1}, \{\bar{\boldsymbol{x}}_{ll_\sigma}^i(T)\}_{i=1}^n, \tilde{\boldsymbol{u}}^{l-1}, \tilde{\boldsymbol{u}}^l, T)\tilde{\boldsymbol{x}}^l\\
&= \boldsymbol{C}^l(\{\tilde{\boldsymbol{x}}^\ell\}_{\ell=1}^{l-1}, \{\{\bar{\boldsymbol{x}}_{lq_\sigma}^i(T)\}_{i=1}^n\}_{q=1}^{l-1}, \{\tilde{\boldsymbol{u}}^\jmath\}_{\jmath=0}^l, T)
\end{aligned} \tag{7.30}$$

for $l = 1, \ldots, N$, where $\boldsymbol{B}^l = \boldsymbol{I} - [b^l_{ip}]_{n\times n}$ is an $n \times n$ matrix with

$$\begin{aligned} b^l_{ip} &= \bar{h}_i \rho^i_{ll} \frac{\partial \tilde{f}_i}{\partial \tilde{x}_p}\Big|_{(T\tau^i_{ll}, \tilde{\boldsymbol{x}}^{l-1}, \delta^i_{ll_\sigma}\tilde{\boldsymbol{x}}^{l-1}+(1-\delta^i_{ll_\sigma})\bar{\boldsymbol{x}}^i_{ll_\sigma}(T), \tilde{\boldsymbol{u}}^i_{ll}, T)} \\ &\quad + \bar{h}_i \delta^i_{ll_\sigma} \tilde{\rho}^i_{ll_\sigma}(T) \frac{\partial \tilde{f}_i}{\partial \bar{x}_p}\Big|_{(T\tau^i_{ll}, \tilde{\boldsymbol{x}}^{l-1}, \delta^i_{ll_\sigma}\tilde{\boldsymbol{x}}^{l-1}+(1-\delta^i_{ll_\sigma})\bar{\boldsymbol{x}}^i_{ll_\sigma}(T), \tilde{\boldsymbol{u}}^i_{ll}, T)} \end{aligned} \tag{7.31}$$

for $i = 1, \ldots, n$, $p = 1, \ldots, n$; and $\boldsymbol{C}^l = (c^l_1, \ldots, c^l_n)^\top$ with

$$\begin{aligned} c^l_i &= \tilde{x}^0_i + \bar{h}_i \sum_{q=1}^{l-1} \Big[\tilde{f}_i(T\tau^i_{lq}, \tilde{x}^i_{lq}, \bar{\boldsymbol{x}}^i_{lq_\sigma}(T), \tilde{\boldsymbol{u}}^i_{lq}, T)((d_{lq}+1)^{\alpha_i} - d^{\alpha_i}_{lq}) \Big] \\ &\quad + \bar{h}_i \tilde{f}_i(T\tau^i_{ll}, \tilde{\boldsymbol{x}}^{l-1}, \delta^i_{ll_\sigma}\tilde{\boldsymbol{x}}^{l-1} + (1-\delta^i_{ll_\sigma})\bar{\boldsymbol{x}}^i_{ll_\sigma}(T), \tilde{\boldsymbol{u}}^i_{ll}, T) \\ &\quad - \sum_{p=1}^{n} b^l_{ip} \tilde{x}^{l-1}_p \end{aligned} \tag{7.32}$$

for $i = 1, \ldots, n$. Thus, the approximate solution to (7.18) can be obtained by solving system (7.30) together with (7.13) for each given pair $(T, \tilde{\boldsymbol{u}}) \in \mathcal{T} \times \tilde{\mathcal{U}}$. The following result, which can be proved in a similar manner as given for the proof of Theorem 6.1, gives the convergence rate of the approximation solution.

Theorem 7.1 *Let $\tilde{\boldsymbol{x}}(s)$ be the exact solution of (7.18). Then, there exists an $\hat{h}$ such that $h < \hat{h}$ implies*

$$\|\tilde{\boldsymbol{x}}(s_l) - \tilde{\boldsymbol{x}}^l\|_\infty \le Ch^2, \quad l = 1, \ldots, N, \tag{7.33}$$

where $\|\cdot\|_\infty$ is the infinity norm; $\{\tilde{\boldsymbol{x}}^l\}$ is a sequence generated by (7.30) together with (7.13); and $C > 0$ is a constant independent of h and $\boldsymbol{\alpha}$.

Introduce $\boldsymbol{X}^l = (\tilde{x}^l_1, \ldots, \tilde{x}^l_n)^\top$, and $\boldsymbol{U}^l = (\tilde{u}^l_1, \ldots, \tilde{u}^l_r)^\top$, and let $\boldsymbol{X} = (\boldsymbol{X}^0, \boldsymbol{X}^1, \ldots, \boldsymbol{X}^N) \in R^{n\times(N+1)}$ and $\boldsymbol{U} = (\boldsymbol{U}^0, \boldsymbol{U}^1, \ldots, \boldsymbol{U}^N) \in R^{r\times(N+1)}$. Then, the canonical constrains (7.15) and (7.16) change to

$$\bar{g}_k(T, \boldsymbol{U}) = \Phi_k(T, \boldsymbol{X}^N) = 0, \quad k = 1, \ldots, N_e, \tag{7.34}$$

$$\bar{g}_k(T, \boldsymbol{U}) = \Phi_k(T, \boldsymbol{X}^N) \ge 0, \quad k = N_e + 1, \ldots, N_e + N_i. \tag{7.35}$$

Thus, we propose the following discrete-time optimal control problem:

Problem (EP(h)). Given the discrete-time system (7.30) together with (7.13), find a pair $(T, \boldsymbol{U}) \in \mathcal{T} \times W^{N+1}$ such that the cost function

$$\bar{g}_0(T, \boldsymbol{U}) = \Phi_0(T, \boldsymbol{X}^N) \tag{7.36}$$

is minimized subject to constraints (7.34) and (7.35).

7.3.2 Gradient Computation

The cost function (7.36) has a similar form to constraints (7.34) and (7.35). Hence, the gradients of the cost and constraint functions can be computed in a unified fashion. For $k = 0, 1, \ldots, N_e + N_i$, we have

$$\frac{\partial \bar{g}_k(T, \boldsymbol{U})}{\partial T} = \frac{\partial \Phi_k(T, \boldsymbol{X}^N)}{\partial T} + \sum_{i=1}^{n} \frac{\partial \Phi_k(T, \boldsymbol{X}^N)}{\partial \tilde{x}_i^N} \frac{\partial \tilde{x}_i^N}{\partial T}, \tag{7.37}$$

$$\frac{\partial \bar{g}_k(T, \boldsymbol{U})}{\partial \tilde{u}_m^v} = \sum_{i=1}^{n} \frac{\partial \Phi_k(T, \boldsymbol{X}^N)}{\partial \tilde{x}_i^N} \frac{\partial \tilde{x}_i^N}{\partial \tilde{u}_m^v}, \quad m = 1, \ldots, r, \; v = 0, 1, \ldots, N. \tag{7.38}$$

We now need to determine $\frac{\partial \tilde{x}_i^N}{\partial T}$ and $\frac{\partial \tilde{x}_i^N}{\partial \tilde{u}_m^v}$. Differentiating (7.30) with respect to T and $\tilde{u}_m^v$, and rearranging the results gives

$$\boldsymbol{B}^l \frac{\partial \boldsymbol{X}^l}{\partial T} = \frac{\partial \boldsymbol{C}^l}{\partial T} - \frac{\partial \boldsymbol{B}^l}{\partial T} \boldsymbol{X}^l \tag{7.39}$$

and

$$\boldsymbol{B}^l \frac{\partial \boldsymbol{X}^l}{\partial \tilde{u}_m^v} = \frac{\partial \boldsymbol{C}^l}{\partial \tilde{u}_m^v} - \frac{\partial \boldsymbol{B}^l}{\partial \tilde{u}_m^v} \boldsymbol{X}^l \tag{7.40}$$

for $l = 1, \ldots, N$, where $\frac{\partial \boldsymbol{C}^l}{\partial T} = \left(\frac{\partial c_1^l}{\partial T}, \cdots, \frac{\partial c_n^l}{\partial T} \right)^\top$; $\frac{\partial \boldsymbol{B}^l}{\partial T} = -\left[\frac{\partial b_{ip}^l}{\partial T} \right]_{n \times n}$; $\frac{\partial \boldsymbol{C}^l}{\partial \tilde{u}_m^v} = \left(\frac{\partial c_1^l}{\partial \tilde{u}_m^v}, \cdots, \frac{\partial c_n^l}{\partial \tilde{u}_m^v} \right)^\top$; and $\frac{\partial \boldsymbol{B}^l}{\partial \tilde{u}_m^v} = -\left[\frac{\partial b_{ip}^l}{\partial \tilde{u}_m^v} \right]_{n \times n}$.

The gradient formulas of $\frac{\partial b_{ip}^l}{\partial T}$ and $\frac{\partial c_i^l}{\partial T}$ are given as follows.

1. If $l = 1$, then

$$\begin{aligned}
\frac{\partial b_{ip}^1}{\partial T} &= \bar{h}_i \rho_{11}^i \frac{\partial^2 \tilde{f}_i}{\partial \tilde{x}_p \partial t} \Big|_{(T\tau_{11}^i, \boldsymbol{X}^0, \delta_{11_\sigma}^i \boldsymbol{X}^0 + (1-\delta_{11_\sigma}^i)\bar{\boldsymbol{x}}_{11_\sigma}^i(T), \tilde{\boldsymbol{u}}_{11}^i, T)} \tau_{11}^i \\
&+ \bar{h}_i \delta_{11_\sigma}^i \tilde{\rho}_{11_\sigma}^i(T) \frac{\partial^2 \tilde{f}_i}{\partial \bar{x}_p \partial t} \Big|_{(T\tau_{11}^i, \boldsymbol{X}^0, \delta_{11_\sigma}^i \boldsymbol{X}^0 + (1-\delta_{11_\sigma}^i)\bar{\boldsymbol{x}}_{11_\sigma}^i(T), \tilde{\boldsymbol{u}}_{11}^i, T)} \tau_{11}^i \\
&+ \bar{h}_i \sum_{\varsigma=1}^{n} \Bigg(\rho_{11}^i \frac{\partial^2 \tilde{f}_i}{\partial \tilde{x}_p \partial \bar{x}_\varsigma} \Big|_{(T\tau_{11}^i, \boldsymbol{X}^0, \delta_{11_\sigma}^i \boldsymbol{X}^0 + (1-\delta_{11_\sigma}^i)\bar{\boldsymbol{x}}_{11_\sigma}^i(T), \tilde{\boldsymbol{u}}_{11}^i, T)} \\
&\times (1-\delta_{11_\sigma}^i) \dot{\phi}_\varsigma (T\tau_{11}^i - \sigma) \tau_{11}^i \Bigg) + \bar{h}_i \Bigg(\rho_{11}^i \frac{\partial \tilde{f}_i}{\partial \tilde{x}_p} \\
&+ \delta_{11_\sigma}^i \tilde{\rho}_{11_\sigma}^i(T) \frac{\partial \tilde{f}_i}{\partial \bar{x}_p} \Bigg) \Big|_{(T\tau_{11}^i, \boldsymbol{X}^0, \delta_{11_\sigma}^i \boldsymbol{X}^0 + (1-\delta_{11_\sigma}^i)\bar{\boldsymbol{x}}_{11_\sigma}^i(T), \tilde{\boldsymbol{u}}_{11}^i, T)} \alpha_i T^{-1} \\
&+ \bar{h}_i \delta_{11_\sigma}^i \sigma T^{-2} h^{-1} \frac{\partial \tilde{f}_i}{\partial \bar{x}_p} \Big|_{(T\tau_{11}^i, \boldsymbol{X}^0, \delta_{11_\sigma}^i \boldsymbol{X}^0 + (1-\delta_{11_\sigma}^i)\bar{\boldsymbol{x}}_{11_\sigma}^i(T), \tilde{\boldsymbol{u}}_{11}^i, T)},
\end{aligned} \tag{7.41}$$

$$\begin{aligned}
\frac{\partial c_i^1}{\partial T} &= \bar{h}_i \frac{\partial \tilde{f}_i}{\partial t} \Big|_{(T\tau_{11}^i, \boldsymbol{X}^0, \delta_{11_\sigma}^i \boldsymbol{X}^0 + (1-\delta_{11_\sigma}^i)\bar{\boldsymbol{x}}_{11_\sigma}^i(T), \tilde{\boldsymbol{u}}_{11}^i, T)} \tau_{11}^i \\
&+ \bar{h}_i \sum_{\varsigma=1}^{n} \Bigg(\frac{\partial \tilde{f}_i}{\partial \bar{x}_\varsigma} \Big|_{(T\tau_{11}^i, \boldsymbol{X}^0, \delta_{11_\sigma}^i \boldsymbol{X}^0 + (1-\delta_{11_\sigma}^i)\bar{\boldsymbol{x}}_{11_\sigma}^i(T), \tilde{\boldsymbol{u}}_{11}^i, T)} \\
&\times (1-\delta_{11_\sigma}^i) \dot{\boldsymbol{\phi}}_\varsigma (T\tau_{11}^i - \sigma) \tau_{11}^i \Bigg) \\
&+ \bar{h}_i \alpha_i T^{-1} \tilde{f}_i (T\tau_{11}^i, \boldsymbol{X}^0, \delta_{11_\sigma}^i \boldsymbol{X}^0 + (1-\delta_{11_\sigma}^i)\bar{\boldsymbol{x}}_{11_\sigma}^i(T), \tilde{\boldsymbol{u}}_{11}^i, T) \\
&- \sum_{p=1}^{n} \frac{\partial b_{ip}^1}{\partial T} \tilde{x}_p^0.
\end{aligned} \tag{7.42}$$

If $l > 1$, then

$$\begin{aligned}
&\frac{\partial b_{ip}^l}{\partial T} \\
&= \bar{h}_i \Bigg(\rho_{ll}^i \frac{\partial^2 \tilde{f}_i}{\partial \tilde{x}_p \partial t} + \delta_{ll_\sigma}^i \tilde{\rho}_{ll_\sigma}^i(T) \frac{\partial^2 \tilde{f}_i}{\partial \bar{x}_p \partial t} \Bigg) \Big|_{(T\tau_{ll}^i, \boldsymbol{X}^{l-1}, \delta_{ll_\sigma}^i \boldsymbol{X}^{l-1} + (1-\delta_{ll_\sigma}^i)\bar{\boldsymbol{x}}_{ll_\sigma}^i(T), \tilde{\boldsymbol{u}}_{ll}^i, T)} \tau_{ll}^i \\
&+ \bar{h}_i \sum_{\varsigma=1}^{n} \Bigg[\Bigg(\rho_{ll}^i \frac{\partial^2 \tilde{f}_i}{\partial \tilde{x}_p \partial \tilde{x}_\varsigma} + \rho_{ll}^i \delta_{ll_\sigma}^i \frac{\partial^2 \tilde{f}_i}{\partial \tilde{x}_p \partial \bar{x}_\varsigma} + \delta_{ll_\sigma}^i \tilde{\rho}_{ll_\sigma}^i(T) \frac{\partial^2 \tilde{f}_i}{\partial \bar{x}_p \partial \tilde{x}_\varsigma} \\
&+ (\delta_{ll_\sigma}^i)^2 \tilde{\rho}_{ll_\sigma}^i(T) \frac{\partial^2 \tilde{f}_i}{\partial \bar{x}_p \partial \bar{x}_\varsigma} \Bigg) \Big|_{(T\tau_{ll}^i, \boldsymbol{X}^{l-1}, \delta_{ll_\sigma}^i \boldsymbol{X}^{l-1} + (1-\delta_{ll_\sigma}^i)\bar{\boldsymbol{x}}_{ll_\sigma}^i(T), \tilde{\boldsymbol{u}}_{ll}^i, T)} \frac{\partial \tilde{x}_\varsigma^{l-1}}{\partial T} \Bigg]
\end{aligned}$$

$$+\bar{h}_i \sum_{\varsigma=1}^{n} \Bigg\{ (1-\delta^i_{ll_\sigma})\rho^i_{ll} \frac{\partial^2 \tilde{f}_i}{\partial \tilde{x}_p \partial \tilde{x}_\varsigma} \Big|_{(T\tau^i_{ll}, \boldsymbol{X}^{l-1}, \delta^i_{ll_\sigma}\boldsymbol{X}^{l-1}+(1-\delta^i_{ll_\sigma})\bar{\boldsymbol{x}}^i_{ll_\sigma}(T), \tilde{\boldsymbol{u}}^i_{ll}, T)}$$
$$\times \Big[\theta^i(l_\sigma - 1)\Big((1-\tilde{\rho}^i_{ll_\sigma}(T)) \frac{\partial \tilde{x}^{l_\sigma-1}_\varsigma}{\partial T} + \tilde{\rho}^i_{ll_\sigma}(T)\frac{\partial \tilde{x}^{l_\sigma}_\varsigma}{\partial T}$$
$$+\sigma T^{-2} h^{-1}(\tilde{x}^{l_\sigma}_\varsigma - \tilde{x}^{l_\sigma-1}_\varsigma)\Big) + (1-\theta^i(l_\sigma-1))\dot{\phi}_\varsigma(T\tau^i_{ll}-\sigma)\tau^i_{ll}\Big]\Bigg\}$$
$$+\bar{h}_i\left(\rho^i_{ll}\frac{\partial \tilde{f}_i}{\partial \tilde{x}_p} + \delta^i_{ll_\sigma}\tilde{\rho}^i_{ll_\sigma}(T)\frac{\partial \tilde{f}_i}{\partial \bar{x}_p}\right)\Big|_{(T\tau^i_{ll}, \boldsymbol{X}^{l-1}, \delta^i_{ll_\sigma}\boldsymbol{X}^{l-1}+(1-\delta^i_{ll_\sigma})\bar{\boldsymbol{x}}^i_{ll_\sigma}(T), \tilde{\boldsymbol{u}}^i_{ll}, T)} \alpha_i T^{-1}$$
$$+\bar{h}_i \delta^i_{ll_\sigma}\sigma T^{-2}h^{-1}\frac{\partial \tilde{f}_i}{\partial \bar{x}_p}\Big|_{(T\tau^i_{ll}, \boldsymbol{X}^{l-1}, \delta^i_{ll_\sigma}\boldsymbol{X}^{l-1}+(1-\delta^i_{ll_\sigma})\bar{\boldsymbol{x}}^i_{ll_\sigma}(T), \tilde{\boldsymbol{u}}^i_{ll}, T)}, \tag{7.43}$$

$$\frac{\partial c^l_i}{\partial T} = \bar{h}_i \sum_{q=1}^{l-1}\Bigg\{ \tau^i_{lq}\frac{\partial \tilde{f}_i}{\partial t}\Big|_{(T\tau^i_{lq}, \tilde{\boldsymbol{x}}^i_{lq}, \bar{\boldsymbol{x}}^i_{lq_\sigma}(T), \tilde{\boldsymbol{u}}^i_{lq}, T)} + \sum_{\varsigma=1}^{n}\Bigg[\frac{\partial \tilde{f}_i}{\partial \tilde{x}_\varsigma}\Big|_{(T\tau^i_{lq}, \tilde{\boldsymbol{x}}^i_{lq}, \bar{\boldsymbol{x}}^i_{lq_\sigma}(T), \tilde{\boldsymbol{u}}^i_{lq}, T)}$$
$$\times\left(\rho^i_{lq}\frac{\partial \tilde{x}^q_\varsigma}{\partial T} + (1-\rho^i_{lq})\frac{\partial \tilde{x}^{q-1}_\varsigma}{\partial T}\right) + \frac{\partial \tilde{f}_i}{\partial \bar{x}_\varsigma}\Big|_{(T\tau^i_{lq}, \tilde{\boldsymbol{x}}^i_{lq}, \bar{\boldsymbol{x}}^i_{lq_\sigma}(T), \tilde{\boldsymbol{u}}^i_{lq}, T)}$$
$$\times\Big(\theta^i(q_\sigma-1)((1-\tilde{\rho}^i_{lq_\sigma}(T))\frac{\partial \tilde{x}^{q_\sigma-1}_\varsigma}{\partial T} + \tilde{\rho}^i_{lq_\sigma}(T)\frac{\partial \tilde{x}^{q_\sigma}_\varsigma}{\partial T}$$
$$+\sigma T^{-2}h^{-1}(\tilde{x}^{q_\sigma}_\varsigma - \tilde{x}^{q_\sigma-1}_\varsigma)) + (1-\theta^i(q_\sigma-1))\dot{\phi}_\varsigma(T\tau^i_{lq}-\sigma)\tau^i_{lq}\Big)\Bigg]$$
$$+\alpha_i T^{-1}\tilde{f}_i(T\tau^i_{lq}, \tilde{\boldsymbol{x}}^i_{lq}, \bar{\boldsymbol{x}}^i_{lq_\sigma}(T), \tilde{\boldsymbol{u}}^i_{lq}, T)\Bigg\}\left[(l-q+1)^{\alpha_i} - (l-q)^{\alpha_i}\right]$$
$$+\bar{h}_i\frac{\partial \tilde{f}_i}{\partial t}\Big|_{(T\tau^i_{ll}, \boldsymbol{X}^{l-1}, \delta^i_{ll_\sigma}\boldsymbol{X}^{l-1}+(1-\delta^i_{ll_\sigma})\bar{\boldsymbol{x}}^i_{ll_\sigma}(T), \tilde{\boldsymbol{u}}^i_{ll}, T)}\tau^i_{ll}$$
$$+\bar{h}_i\sum_{\varsigma=1}^{n}\Bigg\{\left(\frac{\partial \tilde{f}_i}{\partial \tilde{x}_\varsigma} + \delta^i_{ll_\sigma}\frac{\partial \tilde{f}_i}{\partial \bar{x}_\varsigma}\right)\Big|_{(T\tau^i_{ll}, \boldsymbol{X}^{l-1}, \delta^i_{ll_\sigma}\boldsymbol{X}^{l-1}+(1-\delta^i_{ll_\sigma})\bar{\boldsymbol{x}}^i_{ll_\sigma}(T), \tilde{\boldsymbol{u}}^i_{ll}, T)}\frac{\partial \tilde{x}^{l-1}_\varsigma}{\partial T}$$
$$+(1-\delta^i_{ll_\sigma})\frac{\partial \tilde{f}_i}{\partial \bar{x}_\varsigma}\Big|_{(T\tau^i_{ll}, \boldsymbol{X}^{l-1}, \delta^i_{ll_\sigma}\boldsymbol{X}^{l-1}+(1-\delta^i_{ll_\sigma})\bar{\boldsymbol{x}}^i_{ll_\sigma}(T), \tilde{\boldsymbol{u}}^i_{ll}, T)}\Big[\theta^i(l_\sigma-1)$$
$$\times\Big((1-\tilde{\rho}^i_{ll_\sigma}(T))\frac{\partial \tilde{x}^{l_\sigma-1}_\varsigma}{\partial T} + \tilde{\rho}^i_{ll_\sigma}(T)\frac{\partial \tilde{x}^{l_\sigma}_\varsigma}{\partial T} + \sigma T^{-2}h^{-1}(\tilde{x}^{l_\sigma}_\varsigma - \tilde{x}^{l_\sigma-1}_\varsigma)\Big)$$
$$+(1-\theta^i(l_\sigma-1))\dot{\phi}_\varsigma(T\tau^i_{ll}-\sigma)\tau^i_{ll}\Big]\Bigg\} - \sum_{p=1}^{n}\left[\frac{\partial b^l_{ip}}{\partial T}\tilde{x}^{l-1}_p + b^l_{ip}\frac{\partial \tilde{x}^{l-1}_p}{\partial T}\right]$$
$$+\bar{h}_i\alpha_i T^{-1}\tilde{f}_i(T\tau^i_{ll}, \boldsymbol{X}^{l-1}, \delta^i_{ll_\sigma}\boldsymbol{X}^{l-1}+(1-\delta^i_{ll_\sigma})\bar{\boldsymbol{x}}^i_{ll_\sigma}(T), \tilde{\boldsymbol{u}}^i_{ll}, T), \tag{7.44}$$

where, for a given π,

$$\theta^i(\pi) = \begin{cases} 1, & \text{if } \pi \geq 0, \\ 0, & \text{otherwise.} \end{cases} \tag{7.45}$$

Furthermore, the gradient formulas of $\frac{\partial b_{ip}^l}{\partial \tilde{u}_m^v}$ and $\frac{\partial c_i^l}{\partial \tilde{u}_m^v}$ are given below:

1. When $v = 0$, if $l = 1$, then

$$\frac{\partial b_{ip}^1}{\partial \tilde{u}_m^0} = \bar{h}_i \delta_{11_\sigma}^i \tilde{\rho}_{11_\sigma}^i(T) \frac{\partial^2 \tilde{f}_i}{\partial \tilde{x}_p \partial \tilde{u}_m} \Big|_{(T\tau_{11}^i, \boldsymbol{X}^0, \delta_{11_\sigma}^i \boldsymbol{X}^0 + (1-\delta_{11_\sigma}^i)\bar{\boldsymbol{x}}_{11_\sigma}^i, \tilde{\boldsymbol{u}}_{11}^i, T)} (1 - \rho_{11}^i)$$
$$+ \bar{h}_i \rho_{11}^i \frac{\partial^2 \tilde{f}_i}{\partial \tilde{x}_p \partial \tilde{u}_m} \Big|_{(T\tau_{11}^i, \boldsymbol{X}^0, \delta_{11_\sigma}^i \boldsymbol{X}^0 + (1-\delta_{11_\sigma}^i)\bar{\boldsymbol{x}}_{11_\sigma}^i, \tilde{\boldsymbol{u}}_{11}^i, T)} (1 - \rho_{11}^i), \tag{7.46}$$

$$\frac{\partial c_i^1}{\partial \tilde{u}_m^0} = \bar{h}_i \frac{\partial \tilde{f}_i}{\partial \tilde{u}_m} \Big|_{(T\tau_{11}^i, \boldsymbol{X}^0, \delta_{11_\sigma}^i \boldsymbol{X}^0 + (1-\delta_{11_\sigma}^i)\bar{\boldsymbol{x}}_{11_\sigma}^i, \tilde{\boldsymbol{u}}_{11}^i, T)} (1 - \rho_{11}^i)$$
$$- \sum_{p=1}^{n} \frac{\partial b_{ip}^1}{\partial \tilde{u}_m^0} \tilde{x}_p^0. \tag{7.47}$$

If $l > 1$, then

$$\frac{\partial b_{ip}^l}{\partial \tilde{u}_m^0} = \bar{h}_i \sum_{\varsigma=1}^{n} \Bigg(\rho_{ll}^i \frac{\partial^2 \tilde{f}_i}{\partial \tilde{x}_p \partial \tilde{x}_\varsigma} + \rho_{ll}^i \delta_{ll_\sigma}^i \frac{\partial^2 \tilde{f}_i}{\partial \tilde{x}_p \partial \bar{x}_\varsigma} + \delta_{ll_\sigma}^i \tilde{\rho}_{ll_\sigma}^i(T) \frac{\partial^2 \tilde{f}_i}{\partial \bar{x}_p \partial \tilde{x}_\varsigma}$$
$$+ (\delta_{ll_\sigma}^i)^2 \tilde{\rho}_{ll_\sigma}^i(T) \frac{\partial^2 \tilde{f}_i}{\partial \bar{x}_p \partial \bar{x}_\varsigma} \Bigg) \Big|_{(T\tau_{ll}^i, \boldsymbol{X}^{l-1}, \delta_{ll_\sigma}^i \boldsymbol{X}^{l-1} + (1-\delta_{ll_\sigma}^i)\bar{\boldsymbol{x}}_{ll_\sigma}^i, \tilde{\boldsymbol{u}}_{ll}^i, T)} \frac{\partial \tilde{x}_\varsigma^{l-1}}{\partial \tilde{u}_m^0}$$
$$+ \bar{h}_i \sum_{\varsigma=1}^{n} \Bigg\{ \rho_{ll}^i \frac{\partial^2 \tilde{f}_i}{\partial \tilde{x}_p \partial \bar{x}_\varsigma} \Big|_{(T\tau_{ll}^i, \boldsymbol{X}^{l-1}, \delta_{ll_\sigma}^i \boldsymbol{X}^{l-1} + (1-\delta_{ll_\sigma}^i)\bar{\boldsymbol{x}}_{ll_\sigma}^i, \tilde{\boldsymbol{u}}_{ll}^i, T)}$$
$$\times (1 - \delta_{ll_\sigma}^i) \theta^i(l_\sigma - 1) \Big[(1 - \tilde{\rho}_{ll_\sigma}^i(T)) \frac{\partial \tilde{x}_\varsigma^{l_\sigma - 1}}{\partial \tilde{u}_m^0} + \tilde{\rho}_{ll_\sigma}^i(T) \frac{\partial \tilde{x}_\varsigma^{l_\sigma}}{\partial \tilde{u}_m^0} \Big] \Bigg\}, \tag{7.48}$$

$$\frac{\partial c_i^l}{\partial \tilde{u}_m^0}$$
$$= \bar{h}_i \sum_{q=1}^{l-1} \Bigg\{ \sum_{\varsigma=1}^{n} \Bigg[\frac{\partial \tilde{f}_i}{\partial \tilde{x}_\varsigma} \Big|_{(T\tau_{lq}^i, \tilde{\boldsymbol{x}}_{lq}^i, \bar{\boldsymbol{x}}_{lq_\sigma}^i, \tilde{\boldsymbol{u}}_{lq}^i, T)} \theta^i(q-1) \Big((1 - \rho_{lq}^i) \frac{\partial \tilde{x}_\varsigma^{q-1}}{\partial \tilde{u}_m^0} + \rho_{lq}^i \frac{\partial \tilde{x}_\varsigma^{q}}{\partial \tilde{u}_m^0} \Big)$$
$$+ \frac{\partial \tilde{f}_i}{\partial \bar{x}_\varsigma} \Big|_{(T\tau_{ll}^i, \tilde{\boldsymbol{x}}_{lq}^i, \bar{\boldsymbol{x}}_{lq_\sigma}^i, \tilde{\boldsymbol{u}}_{lq}^i, T)} \theta^i(q_\sigma - 1) \Big((1 - \tilde{\rho}_{lq_\sigma}^i(T)) \frac{\partial \tilde{x}_\varsigma^{q_\sigma - 1}}{\partial \tilde{u}_m^0} + \tilde{\rho}_{lq_\sigma}^i(T) \frac{\partial \tilde{x}_\varsigma^{q_\sigma}}{\partial \tilde{u}_m^0} \Big) \Bigg]$$
$$+ \frac{\partial \tilde{f}_i}{\partial \tilde{u}_m} \Big|_{(T\tau_{ll}^i, \tilde{\boldsymbol{x}}_{lq}^i, \bar{\boldsymbol{x}}_{lq_\sigma}^i, \tilde{\boldsymbol{u}}_{lq}^i, T)} \delta_{1q}^i (1 - \rho_{lq}^i) \Bigg\} \big[(l - q + 1)^{\alpha_i} - (l - q)^{\alpha_i} \big]$$
$$+ \bar{h}_i \sum_{\varsigma=1}^{n} \Bigg[\Big(\frac{\partial \tilde{f}_i}{\partial \tilde{x}_\varsigma} + \delta_{ll_\sigma}^i \frac{\partial \tilde{f}_i}{\partial \bar{x}_\varsigma} \Big) \Big|_{(T\tau_{ll}^i, \boldsymbol{X}^{l-1}, \delta_{ll_\sigma}^i \boldsymbol{X}^{l-1} + (1-\delta_{ll_\sigma}^i)\bar{\boldsymbol{x}}_{ll_\sigma}^i(T), \tilde{\boldsymbol{u}}_{ll}^i, T)} \frac{\partial \tilde{x}_\varsigma^{l-1}}{\partial \tilde{u}_m^0}$$

$$+\frac{\partial \tilde{f}_i}{\partial \bar{x}_\varsigma}\Big|_{(T\tau^i_{ll},\boldsymbol{X}^{l-1},\delta^i_{ll_\sigma}\boldsymbol{X}^{l-1}+(1-\delta^i_{ll_\sigma})\bar{\boldsymbol{x}}^i_{ll_\sigma}(T),\tilde{\boldsymbol{u}}^i_{ll},T)}(1-\delta^i_{ll_\sigma})\theta^i(l_\sigma-1)$$
$$\times\left((1-\tilde{\rho}^i_{ll_\sigma}(T))\frac{\partial \tilde{x}^{l_\sigma-1}_\varsigma}{\partial \tilde{u}^0_m}+\tilde{\rho}^i_{ll_\sigma}(T)\frac{\partial \tilde{x}^{l_\sigma}_\varsigma}{\partial \tilde{u}^0_m}\right)\Bigg]$$
$$-\sum_{p=1}^{n}\left[\frac{\partial b^l_{ip}}{\partial \tilde{u}^0_m}\tilde{x}^{l-1}_p+b^l_{ip}\frac{\partial \tilde{x}^{l-1}_p}{\partial \tilde{u}^0_m}\right], \tag{7.49}$$

where, for two given integers $\jmath$ and $\jmath'$, $\delta^i_{\jmath\jmath'}$ is as defined in (7.28); and for a given π, $\theta^i(\pi)$ is as defined in (7.45).

2. For $v=1,\ldots,N$, if $v>l$, then

$$\frac{\partial b^l_{ip}}{\partial \tilde{u}^v_m}=0,\ \text{and}\ \frac{\partial c^l_i}{\partial \tilde{u}^v_m}=0. \tag{7.50}$$

If $v=l$, then

$$\frac{\partial b^l_{ip}}{\partial \tilde{u}^v_m}=\bar{h}_i\left(\rho^i_{ll}\frac{\partial^2 \tilde{f}_i}{\partial \tilde{x}_p\partial \tilde{u}_m}+\delta^i_{ll_\sigma}\tilde{\rho}^i_{ll_\sigma}(T)\right.$$
$$\left.\times\frac{\partial^2 \tilde{f}_i}{\partial \bar{x}_p\partial \tilde{u}_m}\right)\Big|_{(T\tau^i_{ll},\boldsymbol{X}^{l-1},\delta^i_{ll_\sigma}\boldsymbol{X}^{l-1}+(1-\delta^i_{ll_\sigma})\bar{\boldsymbol{x}}^i_{ll_\sigma},\tilde{\boldsymbol{u}}^i_{ll},T)}\rho^i_{ll}, \tag{7.51}$$

$$\frac{\partial c^l_i}{\partial \tilde{u}^v_m}=\bar{h}_i\frac{\partial \tilde{f}_i}{\partial \tilde{u}_m}\Big|_{(T\tau^i_{ll},\boldsymbol{X}^{l-1},\delta^i_{ll_\sigma}\boldsymbol{X}^{l-1}+(1-\delta^i_{ll_\sigma})\bar{\boldsymbol{x}}^i_{ll_\sigma},\tilde{\boldsymbol{u}}^i_{ll},T)}\rho^i_{ll}-\sum_{p=1}^{n}\frac{\partial b^l_{ip}}{\partial \tilde{u}^v_m}\tilde{x}^{l-1}_p. \tag{7.52}$$

If $v=l-1$, then

$$\frac{\partial b^l_{ip}}{\partial \tilde{u}^v_m}=\bar{h}_i\sum_{\varsigma=1}^{n}\left(\rho^i_{ll}\frac{\partial^2 \tilde{f}_i}{\partial \tilde{x}_p\partial \tilde{x}_\varsigma}+\rho^i_{ll}\delta^i_{ll_\sigma}\frac{\partial^2 \tilde{f}_i}{\partial \tilde{x}_p\partial \bar{x}_\varsigma}+\delta^i_{ll_\sigma}\tilde{\rho}^i_{ll_\sigma}(T)\frac{\partial^2 \tilde{f}_i}{\partial \bar{x}_p\partial \tilde{x}_\varsigma}\right.$$
$$\left.+(\delta^i_{ll_\sigma})^2\tilde{\rho}^i_{ll_\sigma}(T)\frac{\partial^2 \tilde{f}_i}{\partial \bar{x}_p\partial \bar{x}_\varsigma}\right)\Big|_{(T\tau^i_{ll},\boldsymbol{X}^{l-1},\delta^i_{ll_\sigma}\boldsymbol{X}^{l-1}+(1-\delta^i_{ll_\sigma})\bar{\boldsymbol{x}}^i_{ll_\sigma},\tilde{\boldsymbol{u}}^i_{ll},T)}\frac{\partial \tilde{x}^{l-1}_\varsigma}{\partial \tilde{u}^v_m}$$
$$+\bar{h}_i\sum_{\varsigma=1}^{n}\left[\rho^i_{ll}\frac{\partial^2 \tilde{f}_i}{\partial \tilde{x}_p\partial \bar{x}_\varsigma}\Big|_{(T\tau^i_{ll},\boldsymbol{X}^{l-1},\delta^i_{ll_\sigma}\boldsymbol{X}^{l-1}+(1-\delta^i_{ll_\sigma})\bar{\boldsymbol{x}}^i_{ll_\sigma},\tilde{\boldsymbol{u}}^i_{ll},T)}\right.$$
$$\left.\times(1-\delta^i_{ll_\sigma})\delta^i_{l_\sigma(l-1)}\tilde{\rho}^i_{ll_\sigma}(T)\frac{\partial \tilde{x}^{l_\sigma}_\varsigma}{\partial \tilde{u}^v_m}\right]+\bar{h}_i\left(\rho^i_{ll}\frac{\partial^2 \tilde{f}_i}{\partial \tilde{x}_p\partial \tilde{u}_m}\right.$$
$$\left.+\delta^i_{ll_\sigma}\tilde{\rho}^i_{ll_\sigma}(T)\frac{\partial^2 \tilde{f}_i}{\partial \bar{x}_p\partial \tilde{u}_m}\right)\Big|_{(T\tau^i_{ll},\boldsymbol{X}^{l-1},\delta^i_{ll_\sigma}\boldsymbol{X}^{l-1}+(1-\delta^i_{ll_\sigma})\bar{\boldsymbol{x}}^i_{ll_\sigma},\tilde{\boldsymbol{u}}^i_{ll},T)}(1-\rho^i_{ll}), \tag{7.53}$$

$$\frac{\partial c_i^l}{\partial \tilde{u}_m^v} = \bar{h}_i \Bigg\{ \sum_{\varsigma=1}^{n} \Bigg[\Bigg(\delta^i_{(l-1)_\sigma (l-1)} \tilde{\rho}^i_{l(l-1)_\sigma}(T) \frac{\partial \tilde{f}_i}{\partial \bar{x}_\varsigma} + \rho^i_{l(l-1)} \frac{\partial \tilde{f}_i}{\partial \tilde{x}_\varsigma} \Bigg) \Bigg|_{(T\tau^i_{l(l-1)}, \tilde{\boldsymbol{x}}^i_{l(l-1)}, \bar{\boldsymbol{x}}^i_{l(l-1)_\sigma}, \tilde{\boldsymbol{u}}^i_{l(l-1)}, T)} \frac{\partial \tilde{x}_\varsigma^{l-1}}{\partial \tilde{u}_m^v} \Bigg] + \frac{\partial \tilde{f}_i}{\partial \tilde{u}_m} \Bigg|_{(T\tau^i_{l(l-1)}, \tilde{\boldsymbol{x}}^i_{l(l-1)}, \bar{\boldsymbol{x}}^i_{l(l-1)_\sigma}, \tilde{\boldsymbol{u}}^i_{l(l-1)}, T)} \rho^i_{l(l-1)} \Bigg\} (2^{\alpha_i} - 1)$$
$$+ \bar{h}_i \Bigg\{ \sum_{\varsigma=1}^{n} \Bigg[\Bigg(\delta^i_{ll_\sigma} \frac{\partial \tilde{f}_i}{\partial \bar{x}_\varsigma} + (1-\delta^i_{ll_\sigma}) \delta^i_{(l-1)l_\sigma} \tilde{\rho}^i_{ll_\sigma}(T) \frac{\partial \tilde{f}_i}{\partial \bar{x}_\varsigma} + \frac{\partial \tilde{f}_i}{\partial \tilde{x}_\varsigma} \Bigg) \Bigg|_{(T\tau^i_{ll}, \boldsymbol{X}^{l-1}, \delta^i_{ll_\sigma} \boldsymbol{X}^{l-1} + (1-\delta^i_{ll_\sigma}) \bar{\boldsymbol{x}}^i_{ll_\sigma}, \tilde{\boldsymbol{u}}^i_{ll}, T)} \frac{\partial \tilde{x}_\varsigma^{l-1}}{\partial \tilde{u}_m^v} \Bigg]$$
$$+ \frac{\partial \tilde{f}_i}{\partial \tilde{u}_m} \Bigg|_{(T\tau^i_{ll}, \boldsymbol{X}^{l-1}, \delta^i_{ll_\sigma} \boldsymbol{X}^{l-1} + (1-\delta^i_{ll_\sigma}) \bar{\boldsymbol{x}}^i_{ll_\sigma}, \tilde{\boldsymbol{u}}^i_{ll}, T)} (1 - \rho^i_{ll}) \Bigg\} - \sum_{p=1}^{n} \Bigg[\frac{\partial b^l_{ip}}{\partial \tilde{u}_m^v} \tilde{x}_p^{l-1} + b^l_{ip} \frac{\partial \tilde{x}_p^{l-1}}{\partial \tilde{u}_m^v} \Bigg], \tag{7.54}$$

where, for two given integers $\jmath$ and $\jmath'$, $\delta^i_{\jmath\jmath'}$ is as defined in (7.28).

If $v < l - 1$, then

$$\frac{\partial b^l_{ip}}{\partial \tilde{u}_m^v} = \bar{h}_i \sum_{\varsigma=1}^{n} \Bigg[\Bigg(\rho^i_{ll} \frac{\partial^2 \tilde{f}_i}{\partial \tilde{x}_p \partial \tilde{x}_\varsigma} + \rho^i_{ll} \delta^i_{ll_\sigma} \frac{\partial^2 \tilde{f}_i}{\partial \tilde{x}_p \partial \bar{x}_\varsigma} + \delta^i_{ll_\sigma} \tilde{\rho}^i_{ll_\sigma}(T) \frac{\partial^2 \tilde{f}_i}{\partial \bar{x}_p \partial \tilde{x}_\varsigma} + (\delta^i_{ll_\sigma})^2 \tilde{\rho}^i_{ll_\sigma}(T) \frac{\partial^2 \tilde{f}_i}{\partial \bar{x}_p \partial \bar{x}_\varsigma} \Bigg) \Bigg|_{(T\tau^i_{ll}, \boldsymbol{X}^{l-1}, \delta^i_{ll_\sigma} \boldsymbol{X}^{l-1} + (1-\delta^i_{ll_\sigma}) \bar{\boldsymbol{x}}^i_{ll_\sigma}, \tilde{\boldsymbol{u}}^i_{ll}, T)} \frac{\partial \tilde{x}_\varsigma^{l-1}}{\partial \tilde{u}_m^v} \Bigg]$$
$$+ \bar{h}_i \sum_{\varsigma=1}^{n} \Bigg\{ \rho^i_{ll} \frac{\partial^2 \tilde{f}_i}{\partial \tilde{x}_p \partial \bar{x}_\varsigma} \Bigg|_{(T\tau^i_{ll}, \boldsymbol{X}^{l-1}, \delta^i_{ll_\sigma} \boldsymbol{X}^{l-1} + (1-\delta^i_{ll_\sigma}) \bar{x}^i_{ll_\sigma}, \tilde{\boldsymbol{u}}^i_{ll}, T)} \times (1-\delta^i_{ll_\sigma}) \theta^i(l_\sigma - v) \Bigg[(1 - \tilde{\rho}^i_{ll_\sigma}(T)) \frac{\partial \tilde{x}_\varsigma^{l_\sigma - 1}}{\partial \tilde{u}_m^v} + \tilde{\rho}^i_{ll_\sigma}(T) \frac{\partial \tilde{x}_\varsigma^{l_\sigma}}{\partial \tilde{u}_m^v} \Bigg] \Bigg\}, \tag{7.55}$$

$$\frac{\partial c_i^l}{\partial \tilde{u}_m^v} = \bar{h}_i \sum_{q=v}^{l-1} \Bigg\{ \Bigg[\sum_{\varsigma=1}^{n} \frac{\partial \tilde{f}_i}{\partial \tilde{x}_\varsigma} \Bigg|_{(T\tau^i_{lq}, \tilde{\boldsymbol{x}}^i_{lq}, \bar{\boldsymbol{x}}^i_{lq_\sigma}, \tilde{\boldsymbol{u}}^i_{lq}, T)} \Bigg((1-\rho^i_{lq}) \frac{\partial \tilde{x}_\varsigma^{q-1}}{\partial \tilde{u}_m^v} + \rho^i_{lq} \frac{\partial \tilde{x}_\varsigma^{q}}{\partial \tilde{u}_m^v} \Bigg)$$
$$+ \sum_{\varsigma=1}^{n} \frac{\partial \tilde{f}_i}{\partial \bar{x}_\varsigma} \Bigg|_{(T\tau^i_{lq}, \tilde{\boldsymbol{x}}^i_{lq}, \bar{\boldsymbol{x}}^i_{lq_\sigma}, \tilde{\boldsymbol{u}}^i_{lq}, T)} \theta^i(q_\sigma - v) \Bigg((1 - \tilde{\rho}^i_{lq_\sigma}(T)) \frac{\partial \tilde{x}_\varsigma^{q_\sigma - 1}}{\partial \tilde{u}_m^v} + \tilde{\rho}^i_{lq_\sigma}(T) \frac{\partial \tilde{x}_\varsigma^{q_\sigma}}{\partial \tilde{u}_m^v} \Bigg) + \delta^i_{qv} \frac{\partial \tilde{f}_i}{\partial \tilde{u}_m} \Bigg|_{(T\tau^i_{lq}, \tilde{\boldsymbol{x}}^i_{lq}, \bar{\boldsymbol{x}}^i_{lq_\sigma}, \tilde{\boldsymbol{u}}^i_{lq}, T)} \rho^i_{lq}$$
$$+ \delta^i_{q(v+1)} \frac{\partial \tilde{f}_i}{\partial \tilde{u}_m} \Bigg|_{(T\tau^i_{lq}, \tilde{\boldsymbol{x}}^i_{lq}, \bar{\boldsymbol{x}}^i_{lq_\sigma}, \tilde{\boldsymbol{u}}^i_{lq}, T)} (1 - \rho^i_{lq}) \Bigg] \Big[(l-q+1)^{\alpha_i} - (l-q)^{\alpha_i} \Big] \Bigg\}$$

$$
\begin{aligned}
&+\bar{h}_i \sum_{\varsigma=1}^{n} \Bigg\{ \left(\frac{\partial \tilde{f}_i}{\partial \tilde{x}_\varsigma} + \delta^i_{ll_\sigma} \frac{\partial \tilde{f}_i}{\partial \bar{x}_\varsigma} \right) \Bigg|_{(T\tau^i_{ll}, X^{l-1}, \delta^i_{ll_\sigma} X^{l-1} + (1-\delta^i_{ll_\sigma}) \bar{x}^i_{ll_\sigma}, \tilde{u}^i_{ll}, T)} \frac{\partial \tilde{x}^{l-1}_\varsigma}{\partial \tilde{u}^v_m} \\
&+ \frac{\partial \tilde{f}_i}{\partial \bar{x}_\varsigma} \Bigg|_{(T\tau^i_{ll}, X^{l-1}, \delta^i_{ll_\sigma} X^{l-1} + (1-\delta^i_{ll_\sigma}) \bar{x}^i_{ll_\sigma}, \tilde{u}^i_{ll}, T)} (1 - \delta^i_{ll_\sigma}) \theta^i (l_\sigma - v) \\
&\times \left[(1 - \tilde{\rho}^i_{ll_\sigma}(T)) \frac{\partial \tilde{x}^{l_\sigma - 1}_\varsigma}{\partial \tilde{u}^v_m} + \tilde{\rho}^i_{ll_\sigma}(T) \frac{\partial \tilde{x}^{l_\sigma}_\varsigma}{\partial \tilde{u}^v_m} \right] \Bigg\} \\
&- \sum_{p=1}^{n} \left[\frac{\partial b^l_{ip}}{\partial \tilde{u}^v_m} \tilde{x}^{l-1}_p + b^l_{ip} \frac{\partial \tilde{x}^{l-1}_p}{\partial \tilde{u}^v_m} \right],
\end{aligned}
\tag{7.56}
$$

where, for two given integers $\jmath$ and $\jmath'$, $\delta^i_{\jmath\jmath'}$ is as defined in (7.28); and for a given π, $\theta^i(\pi)$ is as defined in (7.45).

Thus, we propose Algorithm 7.1 for computing the gradients of the cost and constraint functions in Problem (EP(h)). From this procedure, we can see that the gradients of the cost and constraint functions can be calculated together with solving system (7.30) and (7.13) forward in time, simultaneously.

Algorithm 7.1 Computing the gradients in Problem (EP(h)).

Step 0. Input an integer N and an initial value $\boldsymbol{U} = (\boldsymbol{U}^0, \boldsymbol{U}^1, \ldots, \boldsymbol{U}^N) \in W^{N+1}$. Compute $h = 1/N$ and set $l = 1$.

Step 1. Let $s_l = lh$. Then, calculate $\boldsymbol{X}^l$ by (7.30) together with (7.13) using $(\boldsymbol{X}^0, \ldots, \boldsymbol{X}^{l-1})$ and $(\boldsymbol{U}^0, \ldots, \boldsymbol{U}^l)$.

Step 2. Calculate $\frac{\partial \boldsymbol{B}^l}{\partial T}$, $\frac{\partial \boldsymbol{C}^l}{\partial T}$, $\frac{\partial \boldsymbol{B}^l}{\partial \tilde{u}^v_m}$ and $\frac{\partial \boldsymbol{C}^l}{\partial \tilde{u}^v_m}$ by (7.41)–(7.44), (7.46)–(7.56) for $m = 1, \ldots, r$ and $v = 0, 1, \ldots, l$.

Step 3. Using $\boldsymbol{X}^l$, solve (7.39) and (7.40) for $\frac{\partial \boldsymbol{X}^l}{\partial T}$ and $\frac{\partial \boldsymbol{X}^l}{\partial u^v_m}$, $m = 1, \ldots, r$, $v = 0, 1 \ldots, l$.

Step 4. If $l < N$, then set $l := l + 1$ and go to Step 1. Otherwise, compute $\frac{\partial \bar{g}_k(T, \boldsymbol{U})}{\partial T}$ and $\frac{\partial \bar{g}_k(T, \boldsymbol{U})}{\partial u^v_m}$, $k = 0, 1, \ldots, N_e + N_i$, by (7.37) and (7.38).

7.4 Numerical Examples

To illustrate the solution procedure outlined in Sect. 7.3, we solve two example problems. To solve these examples, we write a MATLAB program that combines the gradient computation procedure with the optimization package FMINCON. All computations have been performed on a PC with a 1.99 GHz Intel Core i7-9700T CPU and 16.0 GB RAM.

Example 7.4.1 Consider the following DFOCP with free terminal time:

$$\min\ g_0(T,u) = 0.1T + \int_0^T \left[x^2(t) + u^2(t)\right]dt, \tag{7.57}$$

subject to fractional time-delay system

$$\begin{aligned} {}_0^C\mathrm{D}_t^\alpha x(t) &= x(t) + 0.8x(t-0.15) + 0.9u(t), \quad t \in (0,T], \\ x(t) &= 1, \quad t \in [-0.15, 0], \end{aligned}$$

the terminal equality constraint

$$g_1(T,u) = x(T) = 0,$$

and bound constraint

$$T \in [0.001, 10].$$

In [39], the free time problem was solved for $\alpha = 1.0$ by using two-stage optimization approach based on necessary optimality conditions. The computed optimal cost is $g_0(T^*, u^*) = 5.37$ with $T^* = 1.446$. For comparison, we solve the problem for various values of N when $\alpha = 1.0$. The computed results are listed in Table 7.1, from which we can see that the computed optimal cost approaches $g_0(T^*, u^*)$ as N increases. Furthermore, we solve the problem for various values of α. The obtained results are listed in Table 7.2. Figure 7.1 illustrates the corresponding controls and states for different α. From Fig. 7.1 and Table 7.2, we can see that the optimal cost and terminal time decrease as the fractional order decreases. Also, the fractional solutions converge to the integer solution as fractional order α approach to the integer order 1.

Table 7.1 Optimal costs, terminal time, constraint values and CPU time for Example 7.4.1 with $\alpha = 1.0$

N	20	40	80	160	320
g_0	5.3753	5.3745	5.3740	5.3729	5.3728
T^*	1.4368	1.4592	1.4081	1.4336	1.4435
g_1	1.4691e–06	3.7873e–06	5.8901e–06	1.7860e–06	1.6062e–08
CPU(s)	2.4403e+01	3.7138e+02	4.8100e+02	1.8662e+03	2.9442e+04

Table 7.2 Optimal costs, terminal time and constraint values for Example 7.4.1 with $N = 320$

α	1.0	0.98	0.96	0.94	0.92	0.9
g_0	5.3728	5.3066	5.2385	5.1675	5.0931	5.0151
T^*	1.4435	1.3687	1.2995	1.2384	1.1489	1.0725
g_1	1.6062e–08	1.1213e–06	8.4034e–06	1.0954e–06	9.0533e–08	7.9992e–06

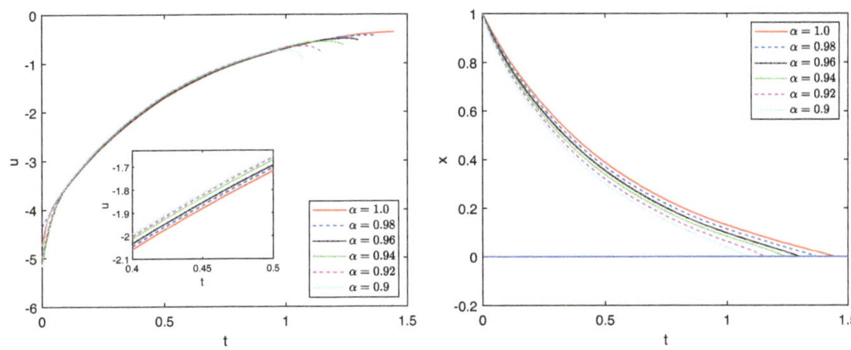

Fig. 7.1 Optimal controls and state trajectories for Example 7.4.1; and the blue line is $x(t) = 0$

Example 7.4.2 Consider the following nonlinear DFOCP with free terminal time:

$$\min \ g_0(T, u) = 0.1T^2 + \int_0^T \exp(-0.05t)(0.2x^{-1}(t)u^3(t) - 2u(t))dt, \quad (7.58)$$

subject to fractional time-delay system

$$\begin{aligned} {}_0^C\mathrm{D}_t^\alpha x(t) &= 3x(t)(1 - 0.2x(t - 0.5)) - u(t), \quad t \in (0, T], \\ x(t) &= 2, \quad t \in [-0.5, 0], \end{aligned}$$

and the continuous inequality constraint

$$x(t) \geq 2, \quad t \in [0, T]. \quad (7.59)$$

Moreover, the control function and terminal time satisfy, respectively, the following bound constraints:

$$u(t) \geq 0, \quad t \in [0, T], \quad (7.60)$$

and

$$T \in [0.2, 20]. \quad (7.61)$$

Note that the constraint (7.59) is a continuous state inequality constraint. Using the ϵ–γ approximation method [303], we approximate constraint (7.59) by the following integral constraint:

$$\gamma + \int_0^T \mathcal{L}_\epsilon(x(t) - 2)dt \geq 0, \quad (7.62)$$

Table 7.3 Optimal costs and terminal time for Example 7.4.2 with constraint (7.59) and $\alpha = 1.0$

N	50	100	150	200	250
T^*	12.0181	12.0272	12.0708	12.1332	12.1644
g_0	–26.0250	–26.1365	–26.1409	–26.14352	–26.1455

where $\epsilon > 0$ and $\gamma > 0$ are adjustable parameters; and

$$\mathcal{L}_\epsilon(\vartheta) = \begin{cases} \vartheta, & \text{if } \vartheta < -\epsilon, \\ -\dfrac{(\vartheta - \epsilon)^2}{4\epsilon}, & \text{if } -\epsilon \leq \vartheta \leq \epsilon, \\ 0, & \text{if } \vartheta > \epsilon. \end{cases} \tag{7.63}$$

Note that the integral term in (7.58) and integral constraint (7.62) can be transformed into the canonical forms (7.5), (7.6) and (7.7) by introducing additional state variables.

In [39], the problem was solved for $\alpha = 1.0$ by using two-stage optimization approach based on necessary optimality conditions. The obtained optimal cost is $\hat{g}_0(T^*, u^*) = -26.146$ with $T^* = 12.24$. For comparison, we first solve the problem for various values of N with $\alpha = 1.0$. The obtained results are listed in Table 7.3. From Table 7.3, we can see that the obtained optimal costs converge to the optimal cost in [39]. Furthermore, we solve the fractional delay optimal control problems with different values of α when $N = 250$. The computed results are listed in Table 7.4 and the corresponding optimal controls and state trajectories when $N = 250$ are plotted in Fig. 7.2. From Fig. 7.2, we can see the continuous state inequality constraint (7.59) is not active, which is consistent with the reported result (for $\alpha = 1.0$) in [39]. Thus, we modify the problem by replacing (7.59) with the following continuous inequality constraint:

$$x(t) \leq 4, \quad t \in [0, T]. \tag{7.64}$$

Using the ϵ-γ approximation method [303], we approximate constraint (7.64) by the following integral constraint:

$$\gamma + \int_0^T \mathcal{L}_\epsilon(4 - x(t))dt \geq 0, \tag{7.65}$$

where $\epsilon > 0$ and $\gamma > 0$ are adjustable parameters; and $\mathcal{L}_\epsilon(\cdot)$ is as defined in (7.63). By running our program with different orders α when $N = 250$, we obtain the optimal costs and terminal time listed in Table 7.4. The corresponding optimal controls and state trajectories are plotted in Fig. 7.3. From Fig. 7.3, we immediately see that the continuous state inequality constraint (7.64) is always satisfied and is active at one point.

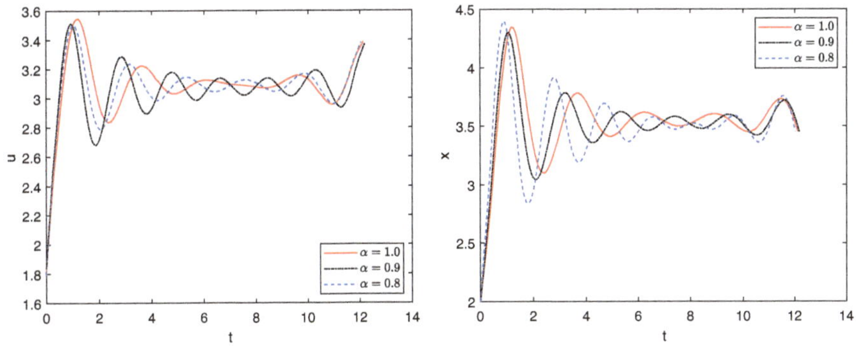

Fig. 7.2 Optimal controls and state trajectories for Example 7.4.2 with constraint (7.59)

Table 7.4 Optimal costs and terminal time for Example 7.4.2 with $N = 250$

α	$1.0^{\dagger}$	$0.9^{\dagger}$	$0.8^{\dagger}$	$1.0^{\ddagger}$	$0.9^{\ddagger}$	$0.8^{\ddagger}$
T^*	12.1664	12.1854	12.0331	12.0717	12.1566	12.1754
g_0	–26.1455	–26.0410	–25.8937	–26.1014	–26.0102	–25.8576

† Fractional orders for example with constraint (7.59)
‡ Fractional orders for example with constraint (7.64)

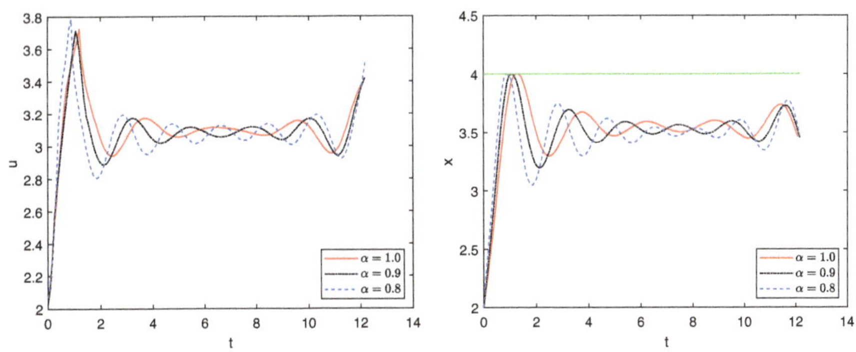

Fig. 7.3 Optimal controls and state trajectories for Example 7.4.2 with $N = 250$ and constraint (7.64); and the green line is $x(t) = 4$

7.5 Conclusions

In this chapter, we have considered numerical solution of free final-time delay fractional optimal control problems. By using the time-scaling transformation, this optimal control problem has been transformed into the equivalent one with fixed terminal time and variable time-delay. Then, we have proposed the explicit numerical scheme for solving the transformed fractional time-delay system, which yields the discrete-time fractional optimal control problem. The gradient-based optimization approach has been developed to solve the resulting discrete-time optimal control problem. Finally, numerical results for two example problems show that this algorithm is highly effective.

Chapter 8
Robust Fractional Optimal Control Problems

8.1 Introduction

In this chapter, we consider robust fractional optimal control with respect to uncertain parameters. As is well-known, parameter uncertainty is inevitably found in almost every system. Moreover, the conventionally obtained optimal open-loop controls in optimal control problems, which make the system highly susceptible to the uncertainties or disturbances. Thus, it is of practical significance to provide a robust optimal control with respect to these uncertainty parameters for fractional optimal control problems.

In robust optimal control, parameter uncertainties are included into an optimal control problem either probabilistically [283, 325] or deterministically [280, 290]. Among these, one of popular robust optimal control methods is minimizing sensitivities of the system output regarding uncertain parameters. The method of sensitivity minimization is presented and applied in [10, 11, 123, 190, 260, 289]. Here, the state system is augmented by the sensitivities which are calculated by either the forward sensitivity or adjoint method. Thus, the discretization method can be used for both the state and sensitivity systems. Although these methods are valid and interesting, they are only applicable to the integer-order optimal control problems. In next section, we describe a robust optimal control problem governed by a nonlinear fractional system with uncertain parameters, where the system sensitivity with respect to the uncertain parameters is explicitly incorporated in the cost function. In Sect. 8.3, we transform the robust fractional optimal control problem to a series of tractable finite-dimension optimization problems. In Sect. 8.3.1, by solving an auxiliary fractional system, we transform the robust fractional optimal control problem to an equivalent standard fractional optimal control. In Sect. 8.3.2, we propose a discretization scheme for the original and auxiliary fractional systems, which results in a finite-dimensional optimization problem. In Sect. 8.4, we develop an optimization algorithm based on

C. Liu et al., *Nonlinear Fractional Optimal Control*, Springer Optimization and Its Applications 237, https://doi.org/10.1007/978-981-92-0274-4_8

the gradients of the cost function to solve the finite-dimensional optimization problem. In Sect. 8.5, two example problems are solved to demonstrate the validity and applicability of the developed algorithm. Finally, the main conclusion is provided in Sect. 8.6.

The main reference of this chapter is [183].

8.2 Problem Formulation

Consider the following nonlinear fractional control system:

$$ {}^{C}_{t_0}\mathrm{D}^{\boldsymbol{\alpha}}_{t}\boldsymbol{x}(t) = \boldsymbol{f}(t, \boldsymbol{x}(t), \boldsymbol{u}(t), \boldsymbol{\sigma}), \quad t \in (t_0, t_f], \tag{8.1} $$

$$ \boldsymbol{x}(t_0) = \boldsymbol{\varphi}(\boldsymbol{\sigma}), \tag{8.2} $$

where t is the time; $\boldsymbol{x}(t) = (x_1(t), x_2(t), \ldots, x_n(t))^\top \in \mathbb{R}^n$ is the state vector; $\boldsymbol{u}(t) = (u_1(t), u_2(t), \ldots, u_r(t))^\top \in \mathbb{R}^r$ is the control vector; $\boldsymbol{\sigma} = (\sigma_1, \sigma_2, \ldots, \sigma_m)^\top \in \mathbb{R}^m$ is the uncertain parameter vector; $t_0 \in \mathbb{R}$ is the initial time; $t_f \in \mathbb{R}$ is the terminal time; $\boldsymbol{f} : \mathbb{R} \times \mathbb{R}^n \times \mathbb{R}^r \times \mathbb{R}^m \to \mathbb{R}^n$ and $\boldsymbol{\varphi} : \mathbb{R}^m \to \mathbb{R}^n$ are given functions; $\boldsymbol{\alpha} = (\alpha_1, \alpha_2, \ldots, \alpha_n)^\top \in (0, 1]^n$ is a given vector of fractional orders; and ${}^{C}_{t_0}\mathrm{D}^{\alpha_i}_{t} x_i(t)$, $i = 1, 2, \ldots, n$, are Caputo fractional derivatives.

Let

$$ W = \{w \in \mathbb{R}^r \mid u_j^{\min} \le w_j \le u_j^{\max},\ j = 1, 2, \ldots, r\}. $$

Here, $u_j^{\min}$ and $u_j^{\max}$, $j = 1, 2, \ldots, r$, are given real numbers such that $u_j^{\min} < u_j^{\max}$. Let $\mathcal{U}$ be the set of Borel measurable functions $\boldsymbol{u} : [t_0, t_f] \to \mathbb{R}^r$ satisfying $\boldsymbol{u}(t) \in W$ for almost all $t \in [t_0, t_f]$.

We assume that fractional system (8.1) with (8.2) satisfies the following conditions.

Assumption 8.2.1 The function $\boldsymbol{\varphi}$ is continuously differentiable.

Assumption 8.2.2 The function $\boldsymbol{f}$ is thrice continuously differentiable with respect to all its arguments.

It follows from Assumptions 8.2.1 and 8.2.2 that, for each given $(\boldsymbol{u}, \boldsymbol{\sigma}) \in \mathcal{U} \times \mathbb{R}^m$, the fractional control system (8.1) with (8.2) has a unique continuous solution $\boldsymbol{x}(\cdot|\boldsymbol{u}, \boldsymbol{\sigma})$. In practice, the values of uncertain parameters are usually not known exactly, and only nominal values can be obtained. Suppose that a nominal parameter vector $\hat{\boldsymbol{\sigma}}$ is given. We consider the cost function as follows:

$$ G(\boldsymbol{u}|\hat{\boldsymbol{\sigma}}) = \Phi(\boldsymbol{x}(t_f|\boldsymbol{u}, \hat{\boldsymbol{\sigma}})), \tag{8.3} $$

where $\Phi : \mathbb{R}^n \to \mathbb{R}$ is a twice continuously differentiable function. Note that any integral term in the cost function can be easily transformed to (8.3) by introducing an additional state variable. In essence, an optimal control can be obtained when minimizes the cost function (8.3). However, this optimal control is determined under the condition that the values of nominal parameters are accurate for the uncertain parameters. It should be noted that the nominal parameter vector is only an approximation of $\boldsymbol{\sigma}$. Thus, for each control $\boldsymbol{u} \in \mathcal{U}$, we further consider the measure of system sensitivity as follows:

$$\left[\frac{\partial G(\boldsymbol{u}|\hat{\boldsymbol{\sigma}})}{\partial \boldsymbol{\sigma}}\right]^{\top} \frac{\partial G(\boldsymbol{u}|\hat{\boldsymbol{\sigma}})}{\partial \boldsymbol{\sigma}}. \tag{8.4}$$

Clearly, a low value of (8.4) implies that the control is robust with respect to $\boldsymbol{\sigma}$.

By considering both (8.3) and (8.4), we state our robust fractional optimal control problem as follows.

Problem (P). Given the nominal parameter vector $\hat{\boldsymbol{\sigma}}$, find a control $\boldsymbol{u} \in \mathcal{U}$ to minimize the following cost function

$$J^{\beta}(\boldsymbol{u}|\hat{\boldsymbol{\sigma}}) = G(\boldsymbol{u}|\hat{\boldsymbol{\sigma}}) + \beta\left[\frac{\partial G(\boldsymbol{u}|\hat{\boldsymbol{\sigma}})}{\partial \boldsymbol{\sigma}}\right]^{\top} \frac{\partial G(\boldsymbol{u}|\hat{\boldsymbol{\sigma}})}{\partial \boldsymbol{\sigma}}, \tag{8.5}$$

where $\beta \geq 0$ is a given weighting factor.

8.3 Problem Transformation

For Problem (P), the cost function (8.5) includes a non-standard system sensitivity term when $\beta > 0$. In this section, we will present an equivalent formulation for Problem (P) and propose a numerical scheme for discretizing the equivalent problem into a finite-dimensional optimization problem.

8.3.1 Computing Sensitivity

Consider the following auxiliary system:

$$\begin{cases} {}_{t_0}^{C}\mathrm{D}_t^{\alpha}\, \boldsymbol{\psi}^k(t) = \left.\dfrac{\partial \boldsymbol{f}}{\partial \boldsymbol{x}}\right|_{(t,\boldsymbol{x}(t),\boldsymbol{u}(t),\hat{\boldsymbol{\sigma}})} \boldsymbol{\psi}^k(t) + \left.\dfrac{\partial \boldsymbol{f}}{\partial \sigma_k}\right|_{(t,\boldsymbol{x}(t),\boldsymbol{u}(t),\hat{\boldsymbol{\sigma}})}, & t \in (t_0, t_f], \\ \boldsymbol{\psi}^k(t_0) = \dfrac{\partial \boldsymbol{\varphi}(\hat{\boldsymbol{\sigma}})}{\partial \sigma_k}, \quad k = 1, 2, \ldots, m. \end{cases} \tag{8.6}$$

Let $\boldsymbol{\psi}^k(\cdot|\boldsymbol{u}, \hat{\boldsymbol{\sigma}})$, $k = 1, 2, \ldots, m$, denote the continuous solutions of the auxiliary system (8.6) corresponding to each $\boldsymbol{u} \in \mathcal{U}$ and vector $\hat{\boldsymbol{\sigma}}$. Then, the following theorem gives an important result.

Theorem 8.1 *Given the nominal parameter vector $\hat{\boldsymbol{\sigma}}$ and for each $\boldsymbol{u} \in \mathcal{U}$,*

$$\frac{\partial \boldsymbol{x}(t|\boldsymbol{u}, \hat{\boldsymbol{\sigma}})}{\partial \sigma_k} = \psi^k(t|\boldsymbol{u}, \hat{\boldsymbol{\sigma}}), \quad t \in [t_0, t_f], \; k = 1, 2, \ldots, m. \tag{8.7}$$

Proof Let ϵ be a sufficiently small real number and let $k \in \{1, 2, \ldots, m\}$ be arbitrary but fixed. For each $\boldsymbol{u} \in \mathcal{U}$, we simply use $\boldsymbol{x}(\cdot)$ instead of $\boldsymbol{x}(\cdot|\boldsymbol{u}, \hat{\boldsymbol{\sigma}})$ and $\boldsymbol{x}^\epsilon(\cdot)$ instead of $\boldsymbol{x}(\cdot|\boldsymbol{u}, \hat{\boldsymbol{\sigma}} + \epsilon \boldsymbol{e}^k)$ to denote the solutions of fractional system (8.1) with (8.2) corresponding to $\hat{\boldsymbol{\sigma}} \in \mathbb{R}^m$ and $\hat{\boldsymbol{\sigma}} + \epsilon \boldsymbol{e}^k \in \mathbb{R}^m$, respectively, where $\boldsymbol{e}^k$ is the kth unit vector in $\mathbb{R}^m$. Then, for $t \in [t_0, t_f]$, and $i = 1, 2, \ldots, n$,

$$x_i(t) = \varphi_i(\hat{\boldsymbol{\sigma}}) + \frac{1}{\Gamma(\alpha_i)} \int_{t_0}^t (t-s)^{\alpha_i - 1} f_i(s, \boldsymbol{x}(s), \boldsymbol{u}(s), \hat{\boldsymbol{\sigma}}) ds,$$

and

$$x_i^\epsilon(t) = \varphi_i(\hat{\boldsymbol{\sigma}} + \epsilon \boldsymbol{e}^k) + \frac{1}{\Gamma(\alpha_i)} \int_{t_0}^t (t-s)^{\alpha_i - 1} f_i(s, \boldsymbol{x}^\epsilon(s), \boldsymbol{u}(s), \hat{\boldsymbol{\sigma}} + \epsilon \boldsymbol{e}^k) ds.$$

Thus,

$$\begin{aligned}
\frac{\partial x_i(t)}{\partial \sigma_k} &= \left. \frac{d x_i^\epsilon(t)}{d\epsilon} \right|_{\epsilon=0} \\
&= \frac{1}{\Gamma(\alpha_i)} \int_{t_0}^t \left\{ (t-s)^{\alpha_i - 1} \left[\sum_{p=1}^n \left. \frac{\partial f_i}{\partial x_p} \right|_{(s, \boldsymbol{x}(s), \boldsymbol{u}(s), \hat{\boldsymbol{\sigma}})} \frac{\partial x_p(s)}{\partial \sigma_k} \right. \right. \\
&\quad \left. \left. + \left. \frac{\partial f_i}{\partial \sigma_k} \right|_{(s, \boldsymbol{x}(s), \boldsymbol{u}(s), \hat{\boldsymbol{\sigma}})} \right] \right\} ds + \frac{\partial \varphi_i(\hat{\boldsymbol{\sigma}})}{\partial \sigma_k}, \quad t \in [t_0, t_f], \; i = 1, 2, \ldots, n.
\end{aligned}$$

Clearly,

$$\begin{cases}
{}_{t_0}^{C}\mathrm{D}_t^\alpha \left\{ \dfrac{\partial \boldsymbol{x}(t)}{\partial \sigma_k} \right\} = \left. \dfrac{\partial \boldsymbol{f}}{\partial \boldsymbol{x}} \right|_{(t, \boldsymbol{x}(t), \boldsymbol{u}(t), \hat{\boldsymbol{\sigma}})} \dfrac{\partial \boldsymbol{x}(t)}{\partial \sigma_k} + \left. \dfrac{\partial \boldsymbol{f}}{\partial \sigma_k} \right|_{(t, \boldsymbol{x}(t), \boldsymbol{u}(t), \hat{\boldsymbol{\sigma}})}, & t \in (t_0, t_f], \\
\dfrac{\partial \boldsymbol{x}(t_0)}{\partial \sigma_k} = \dfrac{\partial \varphi(\hat{\boldsymbol{\sigma}})}{\partial \sigma_k}, &
\end{cases}$$

which implies that for each $\boldsymbol{u} \in \mathcal{U}, \dfrac{\partial \boldsymbol{x}(t)}{\partial \sigma_k}$ is also a solution of auxiliary system (8.6). Therefore, equality (8.7) holds due to the uniqueness of solution for auxiliary system (8.6). The proof is complete. □

Now, we can express the system sensitivity in Problem (P) by the solution of auxiliary system (8.6) as follows.

Theorem 8.2 *Given the nominal parameter vector $\hat{\boldsymbol{\sigma}}$ and for each $\boldsymbol{u} \in \mathcal{U}$,*

$$\left[\frac{\partial G(\boldsymbol{u}|\hat{\boldsymbol{\sigma}})}{\partial \boldsymbol{\sigma}}\right]^\top \frac{\partial G(\boldsymbol{u}|\hat{\boldsymbol{\sigma}})}{\partial \boldsymbol{\sigma}} = \sum_{k=1}^{m}\left[\left(\frac{\partial \Phi(\boldsymbol{x}(t_f|\boldsymbol{u}, \hat{\alpha}))}{\partial \boldsymbol{x}}\right)^\top \boldsymbol{\psi}^k(t_f|\boldsymbol{u}, \hat{\boldsymbol{\sigma}})\right]^2. \tag{8.8}$$

Proof The proof follows from taking the partial differentiation of (8.3) with respect to $\boldsymbol{\sigma}$ and then applying equality (8.7). □

Based on Theorems 8.1 and 8.2, we give an equivalent formulation for Problem (P) as follows.

Problem (EP). Given the nominal parameter vector $\hat{\boldsymbol{\sigma}}$, find a control $\boldsymbol{u} \in \mathcal{U}$ to minimize the cost function

$$J^\beta(\boldsymbol{u}|\hat{\boldsymbol{\sigma}}) = \Phi(\boldsymbol{x}(t_f|\boldsymbol{u}, \hat{\boldsymbol{\sigma}})) + \beta\sum_{k=1}^{m}\left[\left(\frac{\partial \Phi(\boldsymbol{x}(t_f|\boldsymbol{u}, \hat{\boldsymbol{\sigma}}))}{\partial \boldsymbol{x}}\right)^\top \boldsymbol{\psi}^k(t_f|\boldsymbol{u}, \hat{\boldsymbol{\sigma}})\right]^2, \tag{8.9}$$

where $\beta \geq 0$ is a weighting factor.

8.3.2 Discretization Scheme

Given vector $\hat{\boldsymbol{\sigma}}$ and for each $\boldsymbol{u} \in \mathcal{U}$, let

$$\boldsymbol{g}^k(t, \boldsymbol{x}(t), \boldsymbol{\psi}^k(t), \boldsymbol{u}(t), \hat{\boldsymbol{\sigma}}) := \left.\frac{\partial \boldsymbol{f}}{\partial \boldsymbol{x}}\right|_{(t,\boldsymbol{x}(t),\boldsymbol{u}(t),\hat{\boldsymbol{\sigma}})} \boldsymbol{\psi}^k(t) + \left.\frac{\partial \boldsymbol{f}}{\partial \sigma_k}\right|_{(t,\boldsymbol{x}(t),\boldsymbol{u}(t),\hat{\boldsymbol{\sigma}})},$$
$$t \in (t_0, t_f],\ k = 1, 2, \ldots, m. \tag{8.10}$$

Then, it follows from Theorem 3.7 that fractional systems (8.1) with (8.2) and (8.6) can be expressed as

$$x_i(t) = \varphi_i(\hat{\boldsymbol{\sigma}}) + \frac{1}{\Gamma(\alpha_i)}\int_{t_0}^{t}(t-s)^{\alpha_i - 1} f_i(s, \boldsymbol{x}(s), \boldsymbol{u}(s), \hat{\boldsymbol{\sigma}})ds, \tag{8.11}$$

and

$$\psi_i^k(t) = \frac{\partial \varphi_i(\hat{\boldsymbol{\sigma}})}{\partial \sigma_k} + \frac{1}{\Gamma(\alpha_i)}\int_{t_0}^{t}(t-s)^{\alpha_i - 1} g_i^k(s, \boldsymbol{x}(s), \boldsymbol{\psi}^k(s), \boldsymbol{u}(s), \hat{\boldsymbol{\sigma}})ds,$$
$$k = 1, 2, \ldots, m \tag{8.12}$$

for $t \in (t_0, t_f]$, and $i = 1, 2, \ldots, n$, respectively. Let $\{t_l = t_0 + lh : l = 0, 1, \ldots, N\}$ be a uniform grid on $[t_0, t_f]$. Here, N is a given integer; and $h = \frac{t_f - t_0}{N}$. We can rewrite Eqs. (8.11) and (8.12) as

$$x_i^l = \varphi_i(\hat{\boldsymbol{\sigma}}) + \frac{1}{\Gamma(\alpha_i)} \sum_{q=1}^{l} \int_{t_{q-1}}^{t_q} (t^l - s)^{\alpha_i - 1} f_i(s, \boldsymbol{x}(s), \boldsymbol{u}(s), \hat{\boldsymbol{\sigma}}) ds, \tag{8.13}$$

and

$$\psi_i^{k,l} = \frac{\partial \varphi_i(\hat{\boldsymbol{\sigma}})}{\partial \sigma_k} + \frac{1}{\Gamma(\alpha_i)} \sum_{q=1}^{l} \int_{t_{q-1}}^{t_q} (t^l - s)^{\alpha_i - 1} g_i^k(s, \boldsymbol{x}(s), \boldsymbol{\psi}^k(s), \boldsymbol{u}(s), \hat{\boldsymbol{\sigma}}) ds,$$
$$k = 1, 2, \ldots, m \tag{8.14}$$

for $l = 1, 2, \ldots, N$, and $i = 1, 2, \ldots, n$, where $\boldsymbol{x}^l = \boldsymbol{x}(t_l)$; and $\boldsymbol{\psi}^{k,l} = \boldsymbol{\psi}^k(t_l)$. Let

$$s_{ql}^i = t_0 + h \frac{\left[(d_{ql} + 1)^{\alpha_i + 1} - d_{ql}^{\alpha_i + 1}\right] + (\alpha_i + 1)\left[(d_{ql} + 1)^{\alpha_i}(q - 1) - d_{ql}^{\alpha_i} q\right]}{(\alpha_i + 1)\left[(d_{ql} + 1)^{\alpha_i} - d_{ql}^{\alpha_i}\right]}, \tag{8.15}$$

where $d_{ql} = (l - q)$. Obviously, $s_{ql}^i \in (t_{q-1}, t_q]$. By applying Taylor expansion to $f_i(s, \boldsymbol{x}(s), \boldsymbol{u}(s), \hat{\boldsymbol{\sigma}})$ and $g_i^k(s, \boldsymbol{x}(s), \boldsymbol{\psi}^k(s), \boldsymbol{u}(s), \hat{\boldsymbol{\sigma}})$ at the point s_{ql}^i, we have

$$x_i^l = \varphi_i(\hat{\boldsymbol{\sigma}}) + \tilde{h}_i \sum_{q=1}^{l} f_i(s_{ql}^i, \tilde{\boldsymbol{x}}_{ql}^i, \tilde{\boldsymbol{u}}_{ql}^i, \hat{\boldsymbol{\sigma}})\left[(d_{ql} + 1)^{\alpha_i} - d_{ql}^{\alpha_i}\right] + \mathcal{O}(h^2), \tag{8.16}$$

and

$$\psi_i^{k,l} = \frac{\partial \varphi_i(\hat{\boldsymbol{\sigma}})}{\partial \sigma_k} + \tilde{h}_i \sum_{q=1}^{l} g_i^k(s_{ql}^i, \tilde{\boldsymbol{x}}_{ql}^i, \tilde{\psi}_{ql}^{k,i}, \tilde{\boldsymbol{u}}_{ql}^i, \hat{\boldsymbol{\sigma}})\left[(d_{ql} + 1)^{\alpha_i} - d_{ql}^{\alpha_i}\right] + \mathcal{O}(h^2),$$
$$k = 1, 2, \ldots, m \tag{8.17}$$

for $l = 1, 2, \ldots, N$, and $i = 1, 2, \ldots, n$, where $\tilde{h}_i = \frac{h^{\alpha_i}}{\Gamma(\alpha_i + 1)}$; $\tilde{\boldsymbol{x}}_{ql}^i = \boldsymbol{x}(s_{ql}^i)$; $\tilde{\boldsymbol{u}}_{ql}^i = \boldsymbol{u}(s_{ql}^i)$; and $\tilde{\boldsymbol{\psi}}_{ql}^{k,i} = \boldsymbol{\psi}^k(s_{ql}^i)$. However, in Eqs. (8.16) and (8.17), the values of $\tilde{\boldsymbol{x}}_{ql}^i$, $\tilde{\boldsymbol{u}}_{ql}^i$ and $\tilde{\boldsymbol{\psi}}_{ql}^{k,i}$ are not available directly. Thus, we approximate $\tilde{\boldsymbol{x}}_{ql}^i$, $\tilde{\boldsymbol{u}}_{ql}^i$ and $\tilde{\boldsymbol{\psi}}_{ql}^{k,i}$ using the following linear interpolations:

$$\tilde{\boldsymbol{x}}_{ql}^i \approx \boldsymbol{x}^{q-1} + \rho_{ql}^i(\boldsymbol{x}^q - \boldsymbol{x}^{q-1}), \quad \tilde{\boldsymbol{u}}_{ql}^i \approx \boldsymbol{u}^{q-1} + \rho_{ql}^i(\boldsymbol{u}^q - \boldsymbol{u}^{q-1}), \tag{8.18}$$

and

$$\tilde{\psi}_{ql}^{k,i} \approx \psi^{k,q-1} + \rho_{ql}^{i}\big(\psi^{k,q} - \psi^{k,q-1}\big), \tag{8.19}$$

respectively, where $\rho_{ql}^{i} = \dfrac{s_{ql}^{i} - t_{q-1}}{h}$; and $\boldsymbol{u}^{q} = \boldsymbol{u}(t_q)$. Note that approximations (8.18) and (8.19) are of truncation error $\mathcal{O}(h^2)$. Substituting (8.18) and (8.19) into (8.16) and (8.17), we obtain the approximation schemes as follows:

$$x_i^l = \varphi_i(\hat{\boldsymbol{\sigma}}) + \tilde{h}_i \sum_{q=1}^{l} f_i(s_{ql}^i, \tilde{\boldsymbol{x}}_{ql}^i, \tilde{\boldsymbol{u}}_{ql}^i, \hat{\boldsymbol{\sigma}})\big[(d_{ql}+1)^{\alpha_i} - d_{ql}^{\alpha_i}\big], \tag{8.20}$$

and

$$\psi_i^{k,l} = \frac{\partial \varphi_i(\hat{\boldsymbol{\sigma}})}{\partial \sigma_k} + \tilde{h}_i \sum_{q=1}^{l} g_i^k(s_{ql}^i, \tilde{\boldsymbol{x}}_{ql}^i, \tilde{\psi}_{ql}^{k,i}, \tilde{\boldsymbol{u}}_{ql}^i, \hat{\boldsymbol{\sigma}})\big[(d_{ql}+1)^{\alpha_i} - d_{ql}^{\alpha_i}\big],$$
$$k = 1, 2, \ldots, m \tag{8.21}$$

for $l = 1, 2, \ldots, N$, and $i = 1, 2, \ldots, n$.

Note that Eqs. (8.20) and (8.21) are implicit schemes since their right-hand sides are nonlinear functions of $\boldsymbol{x}^l$ and $\psi^{k,l}$. Next, the lth terms at the right-hand sides of (8.20) and (8.21) are approximated by

$$f_i(s_{ll}^i, \tilde{\boldsymbol{x}}_{ll}^i, \tilde{\boldsymbol{u}}_{ll}^i, \hat{\boldsymbol{\sigma}}) = f_i(s_{ll}^i, \boldsymbol{x}^{l-1}, \tilde{\boldsymbol{u}}_{ll}^i, \hat{\boldsymbol{\sigma}}) + \rho_{ll}^i \sum_{p=1}^{n} \left[\frac{\partial f_i}{\partial x_p}\bigg|_{(s_{ll}^i, \boldsymbol{x}^{l-1}, \tilde{\boldsymbol{u}}_{ll}^i, \hat{\boldsymbol{\sigma}})} \big(x_p^l - x_p^{l-1}\big) \right]$$
$$+ \mathcal{O}(h^2), \tag{8.22}$$

and

$$g_i^k(s_{ll}^i, \tilde{\boldsymbol{x}}_{ll}^i, \tilde{\psi}_{ll}^{k,i}, \tilde{\boldsymbol{u}}_{ll}^i, \hat{\boldsymbol{\sigma}}) = g_i^k(s_{ll}^i, \boldsymbol{x}^{l-1}, \psi^{k,l-1}, \tilde{\boldsymbol{u}}_{ll}^i, \hat{\boldsymbol{\sigma}})$$
$$+ \rho_{ll}^i \sum_{p=1}^{n} \Bigg\{ \left[\frac{\partial g_i^k}{\partial x_p}\bigg|_{(s_{ll}^i, \boldsymbol{x}^{l-1}, \psi^{k,l-1}, \tilde{\boldsymbol{u}}_{ll}^i, \hat{\boldsymbol{\sigma}})} \big(x_p^l - x_p^{l-1}\big) \right]$$
$$+ \left[\frac{\partial g_i^k}{\partial \psi_p^k}\bigg|_{(s_{ll}^i, \boldsymbol{x}^{l-1}, \psi^{k,l-1}, \tilde{\boldsymbol{u}}_{ll}^i, \hat{\boldsymbol{\sigma}})} \big(\psi_p^{k,l} - \psi_p^{k,l-1}\big) \right] \Bigg\} + \mathcal{O}(h^2). \tag{8.23}$$

Substituting (8.22) and (8.23) into (8.20) and (8.21) and then omitting the truncation errors give explicit schemes as follows:

$$x_i^l = \varphi_i(\hat{\boldsymbol{\sigma}}) + \tilde{h}_i \sum_{q=1}^{l-1} f_i(s_{ql}^i, \tilde{\boldsymbol{x}}_{ql}^i, \tilde{\boldsymbol{u}}_{ql}^i, \hat{\boldsymbol{\sigma}})\left[(d_{ql}+1)^{\alpha_i} - d_{ql}^{\alpha_i}\right]$$
$$+ \tilde{h}_i f_i(s_{ll}^i, \boldsymbol{x}^{l-1}, \tilde{\boldsymbol{u}}_{ll}^i, \hat{\boldsymbol{\sigma}}) + \tilde{h}_i \rho_{ll}^i \sum_{p=1}^{n} \left[\left.\frac{\partial f_i}{\partial x_p}\right|_{(s_{ll}^i, \boldsymbol{x}^{l-1}, \tilde{\boldsymbol{u}}_{ll}^i, \hat{\boldsymbol{\sigma}})} \left(x_p^l - x_p^{l-1}\right)\right], \tag{8.24}$$

and

$$\psi_i^{k,l} = \frac{\partial \varphi_i(\hat{\boldsymbol{\sigma}})}{\partial \sigma_k} + \tilde{h}_i \sum_{q=1}^{l-1} g_i^k(s_{ql}^i, \tilde{\boldsymbol{x}}_{ql}^i, \tilde{\boldsymbol{\psi}}_{ql}^{k,i}, \tilde{\boldsymbol{u}}_{ql}^i, \hat{\boldsymbol{\sigma}})\left[(d_{ql}+1)^{\alpha_i} - d_{ql}^{\alpha_i}\right]$$
$$+ \tilde{h}_i g_i^k(s_{ll}^i, \boldsymbol{x}^{l-1}, \boldsymbol{\psi}^{k,l-1}, \tilde{\boldsymbol{u}}_{ll}^i, \hat{\boldsymbol{\sigma}}) + \tilde{h}_i \rho_{ll}^i \sum_{p=1}^{n} \left\{ \left[\left.\frac{\partial g_i^k}{\partial x_p}\right|_{(s_{ll}^i, \boldsymbol{x}^{l-1}, \boldsymbol{\psi}^{k,l-1}, \tilde{\boldsymbol{u}}_{ll}^i, \hat{\boldsymbol{\sigma}})} \right.\right.$$
$$\left.\left. \times \left(x_p^l - x_p^{l-1}\right)\right] + \left[\left.\frac{\partial g_i^k}{\partial \psi_p^k}\right|_{(s_{ll}^i, \boldsymbol{x}^{l-1}, \boldsymbol{\psi}^{k,l-1}, \tilde{\boldsymbol{u}}_{ll}^i, \hat{\boldsymbol{\sigma}})} \left(\psi_p^{k,l} - \psi_p^{k,l-1}\right)\right]\right\},$$
$$k = 1, 2, \ldots, m \tag{8.25}$$

for $l = 1, 2, \ldots, N$, and $i = 1, 2, \ldots, n$, where $\tilde{\boldsymbol{x}}_{ql}^i$, $\tilde{\boldsymbol{u}}_{ql}^i$ and $\tilde{\boldsymbol{\psi}}_{ql}^{k,i}$ are defined as in (8.18) and (8.19), respectively. Furthermore, we can rewrite (8.24) and (8.25) as the following matrix forms:

$$(\boldsymbol{I} - \boldsymbol{B}^l)\boldsymbol{x}^l = \boldsymbol{C}^l, \tag{8.26}$$

and

$$-\bar{\boldsymbol{D}}^{k,l}\boldsymbol{x}^l + (\boldsymbol{I} - \tilde{\boldsymbol{D}}^{k,l})\boldsymbol{\psi}^{k,l} = \bar{\boldsymbol{E}}^{k,l}, \quad k = 1, 2, \ldots, m \tag{8.27}$$

for $l = 1, 2, \ldots, N$, where $\boldsymbol{I}$ is the identity matrix; $\boldsymbol{B}^l = \left[b_{ip}^l\right]$, $\bar{\boldsymbol{D}}^{k,l} = \left[\bar{d}_{ip}^{k,l}\right]$ and $\tilde{\boldsymbol{D}}^{k,l} = \left[\tilde{d}_{ip}^{k,l}\right]$ are $n \times n$ matrices whose (i, p)th elements are, respectively, defined as

$$b_{ip}^l = \tilde{h}_i \rho_{ll}^i \left.\frac{\partial f_i}{\partial x_p}\right|_{(s_{ll}^i, \boldsymbol{x}^{l-1}, \tilde{\boldsymbol{u}}_{ll}^i, \hat{\boldsymbol{\sigma}})}, \tag{8.28}$$

$$\bar{d}_{ip}^{k,l} = \tilde{h}_i \rho_{ll}^i \left.\frac{\partial g_i^k}{\partial x_p}\right|_{(s_{ll}^i, \boldsymbol{x}^{l-1}, \boldsymbol{\psi}^{k,l-1}, \tilde{\boldsymbol{u}}_{ll}^i, \hat{\boldsymbol{\sigma}})}, \tag{8.29}$$

and

$$\tilde{d}_{ip}^{k,l} = \tilde{h}_i \rho_{ll}^i \left.\frac{\partial g_i^k}{\partial \psi_p^k}\right|_{(s_{ll}^i, \boldsymbol{x}^{l-1}, \boldsymbol{\psi}^{k,l-1}, \tilde{\boldsymbol{u}}_{ll}^i, \hat{\boldsymbol{\sigma}})}; \tag{8.30}$$

$\boldsymbol{C}^l = (c_1^l, c_2^l, \ldots, c_n^l)^\top$ and $\bar{\boldsymbol{E}}^{k,l} = (\bar{e}_1^{k,l}, \bar{e}_2^{k,l}, \ldots, \bar{e}_n^{k,l})^\top$ are vectors whose ith components are, respectively, defined as

$$c_i^l = \varphi_i(\hat{\boldsymbol{\sigma}}) + \tilde{h}_i \sum_{q=1}^{l-1} f_i(s_{ql}^i, \tilde{\boldsymbol{x}}_{ql}^i, \tilde{\boldsymbol{u}}_{ql}^i, \hat{\boldsymbol{\sigma}})\big[(d_{ql}+1)^{\alpha_i} - d_{ql}^{\alpha_i}\big] + \tilde{h}_i f_i(s_{ll}^i, \boldsymbol{x}^{l-1}, \tilde{\boldsymbol{u}}_{ll}^i, \hat{\boldsymbol{\sigma}})$$
$$- \sum_{p=1}^{n} b_{ip}^l x_p^{l-1}, \tag{8.31}$$

and

$$\bar{e}_i^{k,l} = \frac{\partial \varphi_i(\hat{\boldsymbol{\sigma}})}{\partial \sigma_k} + \tilde{h}_i \sum_{q=1}^{l-1} g_i^k(s_{ql}^i, \tilde{\boldsymbol{x}}_{ql}^i, \tilde{\boldsymbol{\psi}}_{ql}^{k,i}, \tilde{\boldsymbol{u}}_{ql}^i, \hat{\boldsymbol{\sigma}})\big[(d_{ql}+1)^{\alpha_i} - d_{ql}^{\alpha_i}\big]$$
$$+ \tilde{h}_i g_i^k(s_{ll}^i, \boldsymbol{x}^{l-1}, \boldsymbol{\psi}^{k,l-1}, \tilde{\boldsymbol{u}}_{ll}^i, \hat{\boldsymbol{\sigma}}) - \sum_{p=1}^{n} \Big[\bar{d}_{ip}^{k,l} x_p^{l-1} + \tilde{d}_{ip}^{k,l} \psi_p^{k,l-1}\Big]. \tag{8.32}$$

Then, the approximation solutions of fractional systems (8.1) with (8.2) and (8.6) can be obtained by solving systems (8.24) and (8.25) together with the initial conditions $\boldsymbol{x}^0 = \boldsymbol{\varphi}(\hat{\boldsymbol{\sigma}})$ and $\boldsymbol{\psi}^{k,0} = \dfrac{\partial \boldsymbol{\varphi}(\hat{\boldsymbol{\sigma}})}{\partial \sigma_k}$, $k = 1, 2, \ldots, m$.

Thus, we propose a finite-dimensional approximation problem for Problem (EP) as follows.

Problem (EP(h)). Given the nominal parameter vector $\hat{\boldsymbol{\sigma}}$, find a discretized control $\boldsymbol{U} = (\boldsymbol{u}^0, \boldsymbol{u}^1, \ldots, \boldsymbol{u}^N) \in W^{N+1}$ to minimize the cost function

$$\tilde{J}^\beta(\boldsymbol{U}|\hat{\boldsymbol{\sigma}}) = \Phi(\boldsymbol{x}^N) + \beta \sum_{k=1}^{m} \left[\left(\frac{\partial \Phi(\boldsymbol{x}^N)}{\partial \boldsymbol{x}}\right)^\top \boldsymbol{\psi}^{k,N}\right]^2 \tag{8.33}$$

subject to the discretized system

$$\begin{cases} (\boldsymbol{I} - \boldsymbol{B}^l)\boldsymbol{x}^l = \boldsymbol{C}^l, \\ -\bar{\boldsymbol{D}}^{k,l}\boldsymbol{x}^l + (\boldsymbol{I} - \tilde{\boldsymbol{D}}^{k,l})\boldsymbol{\psi}^{k,l} = \bar{\boldsymbol{E}}^{k,l}, \quad l = 1, 2, \ldots, N, \\ \boldsymbol{x}^0 = \boldsymbol{\varphi}(\hat{\boldsymbol{\sigma}}), \\ \boldsymbol{\psi}^{k,0} = \dfrac{\partial \boldsymbol{\varphi}(\hat{\boldsymbol{\sigma}})}{\partial \sigma_k}, \quad k = 1, 2, \ldots, m, \end{cases} \tag{8.34}$$

where $\beta \geq 0$ is a weighting factor.

8.4 Gradient Computation

Given the nominal parameter vector $\hat{\boldsymbol{\sigma}}$, we differentiate the cost function (8.33) with respect to u_j^{π} to give

$$\frac{\partial \tilde{J}^{\beta}(\boldsymbol{U}|\hat{\boldsymbol{\sigma}})}{\partial u_j^{\pi}} = \left(\frac{\partial \Phi(\boldsymbol{x}^N)}{\partial \boldsymbol{x}}\right)^{\top} \frac{\partial \boldsymbol{x}^N}{\partial u_j^{\pi}} + 2\beta \sum_{k=1}^{m} \left\{ \left[\left(\frac{\partial \Phi(\boldsymbol{x}^N)}{\partial \boldsymbol{x}}\right)^{\top} \boldsymbol{\psi}^{k,N} \right] \right.$$
$$\left. \times \left[\left(\frac{\partial \boldsymbol{x}^N}{\partial u_j^{\pi}}\right)^{\top} \frac{\partial^2 \Phi(\boldsymbol{x}^N)}{\partial \boldsymbol{x}^2} \boldsymbol{\psi}^{k,N} + \left(\frac{\partial \Phi(\boldsymbol{x}^N)}{\partial \boldsymbol{x}}\right)^{\top} \frac{\partial \boldsymbol{\psi}^{k,N}}{\partial u_j^{\pi}} \right] \right\},$$
$$j = 1, 2, \ldots, r, \ \pi = 0, 1, 2, \ldots, N. \tag{8.35}$$

However, in (8.35), the values of $\frac{\partial \boldsymbol{x}^N}{\partial u_j^{\pi}}$ and $\frac{\partial \boldsymbol{\psi}^{k,N}}{\partial u_j^{\pi}}$ are required to be determined. For this, differentiating (8.24) and (8.25) with respect to u_j^{π}, $j \in \{1, 2, \ldots, r\}$, $\pi \in \{1, 2, \ldots, N\}$, gives

$$(\boldsymbol{I} - \boldsymbol{B}^l)\frac{\partial \boldsymbol{x}^l}{\partial u_j^{\pi}} = \frac{\partial \boldsymbol{B}^l}{\partial u_j^{\pi}}\boldsymbol{x}^l + \frac{\partial \boldsymbol{C}^l}{\partial u_j^{\pi}}, \tag{8.36}$$

and

$$-\bar{\boldsymbol{D}}^{k,l}\frac{\partial \boldsymbol{x}^l}{\partial u_j^{\pi}} + (\boldsymbol{I} - \tilde{\boldsymbol{D}}^{k,l})\frac{\partial \boldsymbol{\psi}^{k,l}}{\partial u_j^{\pi}} = \frac{\partial \bar{\boldsymbol{D}}^{k,l}}{\partial u_j^{\pi}}\boldsymbol{x}^l + \frac{\partial \tilde{\boldsymbol{D}}^{k,l}}{\partial u_j^{\pi}}\boldsymbol{\psi}^{k,l} + \frac{\partial \bar{\boldsymbol{E}}^{k,l}}{\partial u_j^{\pi}},$$
$$k = 1, 2, \ldots, m \tag{8.37}$$

for $l = 1, 2, \ldots, N$. Thus, to find the values of $\frac{\partial \boldsymbol{x}^N}{\partial u_j^{\pi}}$ and $\frac{\partial \boldsymbol{\psi}^{k,N}}{\partial u_j^{\pi}}$, we need to provide the values of $\frac{\partial \boldsymbol{B}^l}{\partial u_j^{\pi}}$, $\frac{\partial \bar{\boldsymbol{D}}^{k,l}}{\partial u_j^{\pi}}$, $\frac{\partial \tilde{\boldsymbol{D}}^{k,l}}{\partial u_j^{\pi}}$, $\frac{\partial \boldsymbol{C}^l}{\partial u_j^{\pi}}$ and $\frac{\partial \bar{\boldsymbol{E}}^{k,l}}{\partial u_j^{\pi}}$. The following theorems give these required values, whose proofs can be completed by respective differentiating (8.28), (8.31), (8.29), (8.30) and (8.32) with respect to u_j^{π}.

Theorem 8.3 *Let l and $\pi \in \{0, 1, 2, \ldots, N\}$, i and $p \in \{1, 2, \ldots, n\}$, and $j \in \{1, 2, \ldots, r\}$.*
(i) *If $\pi > l$, then*

$$\frac{\partial b_{ip}^l}{\partial u_j^{\pi}} = 0. \tag{8.38}$$

(ii) *If $\pi = l$, then*

$$\frac{\partial b_{ip}^l}{\partial u_j^{\pi}} = \tilde{h}_i(\rho_{ll}^i)^2 \left.\frac{\partial^2 f_i}{\partial x_p \partial u_j}\right|_{(s_{ll}^i, \boldsymbol{x}^{l-1}, \tilde{\boldsymbol{u}}_{ll}^i, \hat{\boldsymbol{\sigma}})}. \tag{8.39}$$

(iii) *If* $\pi = l - 1$, *then*

$$\frac{\partial b_{ip}^{l}}{\partial u_{j}^{\pi}} = \tilde{h}_i \rho_{ll}^{i} \sum_{\varpi=1}^{n} \left[\frac{\partial^2 f_i}{\partial x_p \partial x_{\varpi}} \bigg|_{(s_{ll}^{i}, \boldsymbol{x}^{l-1}, \tilde{\boldsymbol{u}}_{ll}^{i}, \hat{\boldsymbol{\sigma}})} \frac{\partial x_{\varpi}^{l-1}}{\partial u_{j}^{\pi}} \right] + \tilde{h}_i \rho_{ll}^{i} (1 - \rho_{ll}^{i}) \frac{\partial^2 f_i}{\partial x_p \partial u_j} \bigg|_{(s_{ll}^{i}, \boldsymbol{x}^{l-1}, \tilde{\boldsymbol{u}}_{ll}^{i}, \hat{\boldsymbol{\sigma}})}. \tag{8.40}$$

(iv) *If* $\pi < l - 1$, *then*

$$\frac{\partial b_{ip}^{l}}{\partial u_{j}^{\pi}} = \tilde{h}_i \rho_{ll}^{i} \sum_{\varpi=1}^{n} \left[\frac{\partial^2 f_i}{\partial x_p \partial x_{\varpi}} \bigg|_{(s_{ll}^{i}, \boldsymbol{x}^{l-1}, \tilde{\boldsymbol{u}}_{ll}^{i}, \hat{\boldsymbol{\sigma}})} \frac{\partial x_{\varpi}^{l-1}}{\partial u_{j}^{\pi}} \right]. \tag{8.41}$$

Theorem 8.4 *Let* $i \in \{1, 2, \ldots, n\}$, l *and* $\pi \in \{0, 1, 2, \ldots, N\}$, *and* $j \in \{1, 2, \ldots, r\}$.
(i) *If* $\pi > l$, *then*

$$\frac{\partial c_{i}^{l}}{\partial u_{j}^{\pi}} = 0. \tag{8.42}$$

(ii) *If* $\pi = l$, *then*

$$\frac{\partial c_{i}^{l}}{\partial u_{j}^{\pi}} = \tilde{h}_i \rho_{ll}^{i} \frac{\partial f_i}{\partial u_j} \bigg|_{(s_{ll}^{i}, \boldsymbol{x}^{l-1}, \tilde{\boldsymbol{u}}_{ll}^{i}, \hat{\boldsymbol{\sigma}})} - \sum_{p=1}^{n} \frac{\partial b_{ip}^{l}}{\partial u_{j}^{\pi}} x_{p}^{l-1}, \tag{8.43}$$

where $\frac{\partial b_{ip}^{l}}{\partial u_{j}^{\pi}}$ *is as given in Theorem 8.3.*
(iii) *If* $\pi = l - 1$, *then*

$$\begin{aligned} \frac{\partial c_{i}^{l}}{\partial u_{j}^{\pi}} &= \tilde{h}_i \rho_{(l-1)l}^{i} \Bigg[\sum_{\varpi=1}^{n} \frac{\partial f_i}{\partial x_{\varpi}} \bigg|_{(s_{(l-1)l}^{i}, \tilde{\boldsymbol{x}}_{(l-1)l}^{i}, \tilde{\boldsymbol{u}}_{(l-1)l}^{i}, \hat{\boldsymbol{\sigma}})} \frac{\partial x_{\varpi}^{l-1}}{\partial u_{j}^{\pi}} \\ &\quad + \frac{\partial f_i}{\partial u_j} \bigg|_{(s_{(l-1)l}^{i}, \tilde{\boldsymbol{x}}_{(l-1)l}^{i}, \tilde{\boldsymbol{u}}_{(l-1)l}^{i}, \hat{\boldsymbol{\sigma}})} \Bigg] (2^{\alpha_i} - 1) + \tilde{h}_i \Bigg[\sum_{\varpi=1}^{n} \frac{\partial f_i}{\partial x_{\varpi}} \bigg|_{(s_{ll}^{i}, \boldsymbol{x}^{l-1}, \tilde{\boldsymbol{u}}_{ll}^{i}, \hat{\boldsymbol{\sigma}})} \\ &\quad \times \frac{\partial x_{\varpi}^{l-1}}{\partial u_{j}^{\pi}} + (1 - \rho_{ll}^{i}) \frac{\partial f_i}{\partial u_j} \bigg|_{(s_{ll}^{i}, \boldsymbol{x}^{l-1}, \tilde{\boldsymbol{u}}_{ll}^{i}, \hat{\boldsymbol{\sigma}})} \Bigg] - \sum_{p=1}^{n} \left[\frac{\partial b_{ip}^{l}}{\partial u_{j}^{\pi}} x_{p}^{l-1} + b_{ip}^{l} \frac{\partial x_{p}^{l-1}}{\partial u_{j}^{\pi}} \right], \end{aligned} \tag{8.44}$$

where $\frac{\partial b_{ip}^{l}}{\partial u_{j}^{\pi}}$ *is as given in Theorem 8.3.*

(iv) *If* $\pi < l - 1$, *then*

$$\begin{aligned}\frac{\partial c_i^l}{\partial u_j^\pi} &= \tilde{h}_i \sum_{\substack{q=\delta_{\pi 0}+\\(1-\delta_{\pi 0})\pi}}^{l-1} \left\{ \sum_{\varpi=1}^{n} \left[\rho_{ql}^i \frac{\partial f_i}{\partial x_\varpi}\bigg|_{(s_{ql}^i, \tilde{\boldsymbol{x}}_{ql}^i, \tilde{\boldsymbol{u}}_{ql}^i, \hat{\boldsymbol{\sigma}})} \frac{\partial x_\varpi^q}{\partial u_j^\pi} + (1-\rho_{ql}^i) \frac{\partial f_i}{\partial x_\varpi}\bigg|_{(s_{ql}^i, \tilde{\boldsymbol{x}}_{ql}^i, \tilde{\boldsymbol{u}}_{ql}^i, \hat{\boldsymbol{\sigma}})} \right.\right.\\ &\quad \left.\left. \times \frac{\partial x_\varpi^{q-1}}{\partial u_j^\pi} \right] + \left[\rho_{ql}^i \delta_{q\pi} + (1-\rho_{ql}^i)\delta_{q(\pi+1)} \right] \frac{\partial f_i}{\partial u_j}\bigg|_{(s_{ql}^i, \tilde{x}_{ql}^i, \tilde{\boldsymbol{u}}_{ql}^i, \hat{\boldsymbol{\sigma}})} \right\}\\ &\quad \times \left[(d_{ql}+1)^{\alpha_i} - d_{ql}^{\alpha_i} \right] + \tilde{h}_i \sum_{\varpi=1}^{n} \frac{\partial f_i}{\partial x_\varpi}\bigg|_{(s_{ll}^i, \boldsymbol{x}^{l-1}, \tilde{\boldsymbol{u}}_{ll}^i, \hat{\boldsymbol{\sigma}})} \frac{\partial x_\varpi^{l-1}}{\partial u_j^\pi}\\ &\quad - \sum_{p=1}^{n} \left[\frac{\partial b_{ip}^l}{\partial u_j^\pi} x_p^{l-1} + b_{ip}^l \frac{\partial x_p^{l-1}}{\partial u_j^\pi} \right],\end{aligned} \tag{8.45}$$

where $\dfrac{\partial b_{ip}^l}{\partial u_j^\pi}$ *is as given in Theorem 8.3; and* $\delta_{\imath\jmath}$ *is defined as*

$$\delta_{\imath\jmath} = \begin{cases} 1, & \textit{if } \imath = \jmath, \\ 0, & \textit{otherwise.} \end{cases} \tag{8.46}$$

Theorem 8.5 *Let* $k \in \{1, 2, \dots, m\}$, l *and* $\pi \in \{0, 1, 2, \dots, N\}$, i *and* $p \in \{1, 2, \dots, n\}$, *and* $j \in \{1, 2, \dots, r\}$.
(i) *If* $\pi > l$, *then*

$$\frac{\partial \bar{d}_{ip}^{k,l}}{\partial u_j^\pi} = 0. \tag{8.47}$$

(ii) *If* $\pi = l$, *then*

$$\frac{\partial \bar{d}_{ip}^{k,l}}{\partial u_j^\pi} = \tilde{h}_i (\rho_{ll}^i)^2 \frac{\partial^2 g_i^k}{\partial x_p \partial u_j}\bigg|_{(s_{ll}^i, \boldsymbol{x}^{l-1}, \boldsymbol{\psi}^{k,l-1}, \tilde{\boldsymbol{u}}_{ll}^i, \hat{\boldsymbol{\sigma}})}. \tag{8.48}$$

(iii) *If* $\pi = l - 1$, *then*

$$\begin{aligned}\frac{\partial \bar{d}_{ip}^{k,l}}{\partial u_j^\pi} &= \tilde{h}_i \rho_{ll}^i \sum_{\varpi=1}^{n} \left[\frac{\partial^2 g_i^k}{\partial x_p \partial x_\varpi}\bigg|_{(s_{ll}^i, \boldsymbol{x}^{l-1}, \boldsymbol{\psi}^{k,l-1}, \tilde{\boldsymbol{u}}_{ll}^i, \hat{\boldsymbol{\sigma}})} \frac{\partial x_\varpi^{l-1}}{\partial u_j^\pi} \right.\\ &\quad \left. + \frac{\partial^2 g_i^k}{\partial x_p \partial \psi_\varpi^k}\bigg|_{(s_{ll}^i, \boldsymbol{x}^{l-1}, \boldsymbol{\psi}^{k,l-1}, \tilde{\boldsymbol{u}}_{ll}^i, \hat{\boldsymbol{\sigma}})} \frac{\partial \psi_\varpi^{k,l-1}}{\partial u_j^\pi} \right]\\ &\quad + \tilde{h}_i \rho_{ll}^i (1-\rho_{ll}^i) \frac{\partial^2 g_i^k}{\partial x_p \partial u_j}\bigg|_{(s_{ll}^i, \boldsymbol{x}^{l-1}, \boldsymbol{\psi}^{k,l-1}, \tilde{\boldsymbol{u}}_{ll}^i, \hat{\boldsymbol{\sigma}})}.\end{aligned} \tag{8.49}$$

(iv) *If* $\pi < l-1$, *then*

$$\frac{\partial \bar{d}_{ip}^{k,l}}{\partial u_j^{\pi}} = \tilde{h}_i \rho_{ll}^i \sum_{\varpi=1}^{n} \left[\frac{\partial^2 g_i^k}{\partial x_p \partial x_{\varpi}} \bigg|_{(s_{ll}^i, \boldsymbol{x}^{l-1}, \boldsymbol{\psi}^{k,l-1}, \tilde{\boldsymbol{u}}_{ll}^i, \hat{\boldsymbol{\sigma}})} \frac{\partial x_{\varpi}^{l-1}}{\partial u_j^{\pi}} + \frac{\partial^2 g_i^k}{\partial x_p \partial \psi_{\varpi}^k} \bigg|_{(s_{ll}^i, \boldsymbol{x}^{l-1}, \boldsymbol{\psi}^{k,l-1}, \tilde{\boldsymbol{u}}_{ll}^i, \hat{\boldsymbol{\sigma}})} \frac{\partial \psi_{\varpi}^{k,l-1}}{\partial u_j^{\pi}} \right]. \tag{8.50}$$

Theorem 8.6 *Let* $k \in \{1, 2, \ldots, m\}$, l *and* $\pi \in \{0, 1, 2, \ldots, N\}$, i *and* $p \in \{1, 2, \ldots, n\}$, *and* $j \in \{1, 2, \ldots, r\}$.
(i) *If* $\pi > l$, *then*

$$\frac{\partial \tilde{d}_{ip}^{k,l}}{\partial u_j^{\pi}} = 0. \tag{8.51}$$

(ii) *If* $\pi = l$, *then*

$$\frac{\partial \tilde{d}_{ip}^{k,l}}{\partial u_j^{\pi}} = \tilde{h}_i \left(\rho_{ll}^i\right)^2 \frac{\partial^2 g_i^k}{\partial \psi_p^k \partial u_j} \bigg|_{(s_{ll}^i, \boldsymbol{x}^{l-1}, \boldsymbol{\psi}^{k,l-1}, \tilde{\boldsymbol{u}}_{ll}^i, \hat{\boldsymbol{\sigma}})}. \tag{8.52}$$

(iii) *If* $\pi = l-1$, *then*

$$\frac{\partial \tilde{d}_{ip}^{k,l}}{\partial u_j^{\pi}} = \tilde{h}_i \rho_{ll}^i \sum_{\varpi=1}^{n} \left[\frac{\partial^2 g_i^k}{\partial \psi_p^k \partial x_{\varpi}} \bigg|_{(s_{ll}^i, \boldsymbol{x}^{l-1}, \boldsymbol{\psi}^{k,l-1}, \tilde{\boldsymbol{u}}_{ll}^i, \hat{\boldsymbol{\sigma}})} \frac{\partial x_{\varpi}^{l-1}}{\partial u_j^{\pi}} + \frac{\partial^2 g_i^k}{\partial \psi_p^k \partial \psi_{\varpi}^k} \bigg|_{(s_{ll}^i, \boldsymbol{x}^{l-1}, \boldsymbol{\psi}^{k,l-1}, \tilde{\boldsymbol{u}}_{ll}^i, \hat{\boldsymbol{\sigma}})} \frac{\partial \psi_{\varpi}^{k,l-1}}{\partial u_j^{\pi}} \right] + \tilde{h}_i \rho_{ll}^i (1 - \rho_{ll}^i) \frac{\partial^2 g_i^k}{\partial \psi_p^k \partial u_j} \bigg|_{(s_{ll}^i, \boldsymbol{x}^{l-1}, \boldsymbol{\psi}^{k,l-1}, \tilde{\boldsymbol{u}}_{ll}^i, \hat{\boldsymbol{\sigma}})}. \tag{8.53}$$

(iv) *If* $\pi < l-1$, *then*

$$\frac{\partial \tilde{d}_{ip}^{k,l}}{\partial u_j^{\pi}} = \tilde{h}_i \rho_{ll}^i \sum_{\varpi=1}^{n} \left[\frac{\partial^2 g_i^k}{\partial \psi_p^k \partial x_{\varpi}} \bigg|_{(s_{ll}^i, \boldsymbol{x}^{l-1}, \boldsymbol{\psi}^{k,l-1}, \tilde{\boldsymbol{u}}_{ll}^i, \hat{\boldsymbol{\sigma}})} \frac{\partial x_{\varpi}^{l-1}}{\partial u_j^{\pi}} + \frac{\partial^2 g_i^k}{\partial \psi_p^k \partial \psi_{\varpi}^k} \bigg|_{(s_{ll}^i, \boldsymbol{x}^{l-1}, \boldsymbol{\psi}^{k,l-1}, \tilde{\boldsymbol{u}}_{ll}^i, \hat{\boldsymbol{\sigma}})} \frac{\partial \psi_{\varpi}^{k,l-1}}{\partial u_j^{\pi}} \right]. \tag{8.54}$$

Theorem 8.7 *Let* $k \in \{1, 2, \ldots, m\}$, $i \in \{1, 2, \ldots, n\}$, l *and* $\pi \in \{0, 1, 2, \ldots, N\}$, *and* $j \in \{1, 2, \ldots, r\}$.
(i) *If* $\pi > l$, *then*

$$\frac{\partial \bar{e}_i^{k,l}}{\partial u_j^{\pi}} = 0. \tag{8.55}$$

(ii) *If* $\pi = l$, *then*

$$\frac{\partial \bar{e}_i^{k,l}}{\partial u_j^{\pi}} = \tilde{h}_i \rho_{ll}^i \frac{\partial g_i^k}{\partial u_j}\bigg|_{(s_{ll}^i, \boldsymbol{x}^{l-1}, \boldsymbol{\psi}^{k,l-1}, \tilde{\boldsymbol{u}}_{ll}^i, \hat{\boldsymbol{\sigma}})} - \sum_{p=1}^{n}\left[\frac{\partial \bar{d}_{ip}^{k,l}}{\partial u_j^{\pi}} x_p^{l-1} + \frac{\partial \tilde{d}_{ip}^{k,l}}{\partial u_j^{\pi}} \psi_p^{k,l-1}\right], \tag{8.56}$$

where $\frac{\partial \bar{d}_{ip}^{k,l}}{\partial u_j^{\pi}}$ *and* $\frac{\partial \tilde{d}_{ip}^{k,l}}{\partial u_j^{\pi}}$ *are as given in Theorems 8.5 and 8.6, respectively.*

(iii) *If* $\pi = l-1$, *then*

$$\begin{aligned}
\frac{\partial \bar{e}_i^{k,l}}{\partial u_j^{\pi}} &= \tilde{h}_i \rho_{(l-1)l}^i \Bigg\{ \sum_{\varpi=1}^{n} \Bigg[\frac{\partial g_i^k}{\partial x_{\varpi}}\bigg|_{(s_{(l-1)l}^i, \tilde{\boldsymbol{x}}_{(l-1)l}^i, \tilde{\psi}_{(l-1)l}^{k,i}, \tilde{\boldsymbol{u}}_{(l-1)l}^i, \hat{\boldsymbol{\sigma}})} \frac{\partial x_{\varpi}^{l-1}}{\partial u_j^{\pi}} \\
&\quad + \frac{\partial g_i^k}{\partial \psi_{\varpi}^k}\bigg|_{(s_{(l-1)l}^i, \tilde{\boldsymbol{x}}_{(l-1)l}^i, \tilde{\psi}_{(l-1)l}^{k,i}, \tilde{\boldsymbol{u}}_{(l-1)l}^i, \hat{\boldsymbol{\sigma}})} \frac{\partial \psi_{\varpi}^{k,l-1}}{\partial u_j^{\pi}} \Bigg] \\
&\quad + \frac{\partial g_i^k}{\partial u_j}\bigg|_{(s_{(l-1)l}^i, \tilde{\boldsymbol{x}}_{(l-1)l}^i, \tilde{\psi}_{(l-1)l}^{k,i}, \tilde{\boldsymbol{u}}_{(l-1)l}^i, \hat{\boldsymbol{\sigma}})} \Bigg\} (2^{\alpha_i} - 1) \\
&\quad + \tilde{h}_i \Bigg\{ \sum_{\varpi=1}^{n} \Bigg[\frac{\partial g_i^k}{\partial x_{\varpi}}\bigg|_{(s_{ll}^i, \boldsymbol{x}^{l-1}, \boldsymbol{\psi}^{k,l-1}, \tilde{\boldsymbol{u}}_{ll}^i, \hat{\boldsymbol{\sigma}})} \frac{\partial x_{\varpi}^{l-1}}{\partial u_j^{\pi}} + \frac{\partial g_i^k}{\partial \psi_{\varpi}^k}\bigg|_{(s_{ll}^i, \boldsymbol{x}^{l-1}, \boldsymbol{\psi}^{k,l-1}, \tilde{\boldsymbol{u}}_{ll}^i, \hat{\boldsymbol{\sigma}})} \\
&\quad \times \frac{\partial \psi_{\varpi}^{k,l-1}}{\partial u_j^{\pi}} \Bigg] + (1 - \rho_{ll}^i) \frac{\partial g_i^k}{\partial u_j}\bigg|_{(s_{ll}^i, \boldsymbol{x}^{l-1}, \boldsymbol{\psi}^{k,l-1}, \tilde{\boldsymbol{u}}_{ll}^i, \hat{\boldsymbol{\sigma}})} \Bigg\} \\
&\quad - \sum_{p=1}^{n} \Bigg[\frac{\partial \bar{d}_{ip}^{k,l}}{\partial u_j^{\pi}} x_p^{l-1} + \bar{d}_{ip}^{k,l} \frac{\partial x_p^{l-1}}{\partial u_j^{\pi}} + \frac{\partial \tilde{d}_{ip}^{k,l}}{\partial u_j^{\pi}} \psi_p^{k,l-1} + \tilde{d}_{ip}^{k,l} \frac{\partial \psi_p^{k,l-1}}{\partial u_j^{\pi}} \Bigg],
\end{aligned} \tag{8.57}$$

where $\frac{\partial \bar{d}_{ip}^{k,l}}{\partial u_j^{\pi}}$ *and* $\frac{\partial \tilde{d}_{ip}^{k,l}}{\partial u_j^{\pi}}$ *are as given in Theorems 8.5 and 8.6, respectively.*

(iv) *If* $\pi < l-1$, *then*

$$\begin{aligned}
\frac{\partial \bar{e}_i^{k,l}}{\partial u_j^{\pi}} &= \tilde{h}_i \sum_{\substack{q=\delta_{\pi 0}+\\(1-\delta_{\pi 0})\pi}}^{l-1} \Bigg\{ \sum_{\varpi=1}^{n} \Bigg[\rho_{ql}^i \Bigg(\frac{\partial g_i^k}{\partial x_{\varpi}}\bigg|_{(s_{ql}^i, \tilde{\boldsymbol{x}}_{ql}^i, \tilde{\psi}_{ql}^{k,i}, \tilde{\boldsymbol{u}}_{ql}^i, \hat{\boldsymbol{\sigma}})} \frac{\partial x_{\varpi}^q}{\partial u_j^{\pi}} + \frac{\partial g_i^k}{\partial \psi_{\varpi}^k}\bigg|_{(s_{ql}^i, \tilde{\boldsymbol{x}}_{ql}^i, \tilde{\psi}_{ql}^{k,i}, \tilde{\boldsymbol{u}}_{ql}^i, \hat{\boldsymbol{\sigma}})} \\
&\quad \times \frac{\partial \psi_{\varpi}^{k,q}}{\partial u_j^{\pi}} \Bigg) + (1 - \rho_{ql}^i) \Bigg(\frac{\partial g_i^k}{\partial x_{\varpi}}\bigg|_{(s_{ql}^i, \tilde{\boldsymbol{x}}_{ql}^i, \tilde{\psi}_{ql}^{k,i}, \tilde{\boldsymbol{u}}_{ql}^i, \hat{\boldsymbol{\sigma}})} \frac{\partial x_{\varpi}^{q-1}}{\partial u_j^{\pi}} \\
&\quad + \frac{\partial g_i^k}{\partial \psi_{\varpi}^k}\bigg|_{(s_{ql}^i, \tilde{\boldsymbol{x}}_{ql}^i, \tilde{\psi}_{ql}^{k,i}, \tilde{\boldsymbol{u}}_{ql}^i, \hat{\boldsymbol{\sigma}})} \frac{\partial \psi_{\varpi}^{k,q-1}}{\partial u_j^{\pi}} \Bigg) \Bigg] + \left[\rho_{ql}^i \delta_{q\pi} + (1 - \rho_{ql}^i) \delta_{q(\pi+1)} \right] \\
&\quad \times \frac{\partial g_i^k}{\partial u_j}\bigg|_{(s_{ql}^i, \tilde{\boldsymbol{x}}_{ql}^i, \tilde{\psi}_{ql}^{k,i}, \tilde{\boldsymbol{u}}_{ql}^i, \hat{\boldsymbol{\sigma}})} \Bigg\} \left[(d_{ql} + 1)^{\alpha_i} - d_{ql}^{\alpha_i} \right]
\end{aligned}$$

$$+\tilde{h}_i\sum_{\varpi=1}^{n}\left[\frac{\partial g_i^k}{\partial x_\varpi}\bigg|_{(s_{ll}^i,\boldsymbol{x}^{l-1},\boldsymbol{\psi}^{k,l-1},\tilde{\boldsymbol{u}}_{ll}^i,\hat{\boldsymbol{\sigma}})}\frac{\partial x_\varpi^{l-1}}{\partial u_j^\pi}+\frac{\partial g_i^k}{\partial \psi_\varpi^k}\bigg|_{(s_{ll}^i,\boldsymbol{x}^{l-1},\boldsymbol{\psi}^{k,l-1},\tilde{\boldsymbol{u}}_{ll}^i,\hat{\boldsymbol{\sigma}})}\right.$$
$$\left.\times\frac{\partial \psi_\varpi^{k,l-1}}{\partial u_j^\pi}\right]-\sum_{p=1}^{n}\left[\frac{\partial \bar{d}_{ip}^{k,l}}{\partial u_j^\pi}x_p^{l-1}+\bar{d}_{ip}^{k,l}\frac{\partial x_p^{l-1}}{\partial u_j^\pi}+\frac{\partial \tilde{d}_{ip}^{k,l}}{\partial u_j^\pi}\psi_p^{k,l-1}\right.$$
$$\left.+\tilde{d}_{ip}^{k,l}\frac{\partial \psi_p^{k,l-1}}{\partial u_j^\pi}\right], \tag{8.58}$$

where $\frac{\partial \bar{d}_{ip}^{k,l}}{\partial u_j^\pi}$ *and* $\frac{\partial \tilde{d}_{ip}^{k,l}}{\partial u_j^\pi}$ *are as given in Theorems 8.5 and 8.6, respectively; and* $\delta_{\iota j}$ *is as defined as in (8.46).*

On the basis of Theorems 8.3–8.7, we propose Algorithm 8.1 to compute the gradients of the cost function (8.33) in Problem (EP(h)).

Algorithm 8.1 Computing the gradients in Problem (EP(h)).

Step 0. For a given N, choose an initial value $\boldsymbol{U}=(\boldsymbol{u}^0,\boldsymbol{u}^1,\ldots,\boldsymbol{u}^N)$. Set $h=\frac{t_f-t_0}{N}$ and $l:=1$.

Step 1. Let $t_l=t_0+lh$. Use $(\boldsymbol{x}^0,\boldsymbol{x}^1,\cdots,\boldsymbol{x}^{l-1})$, $(\boldsymbol{\psi}^{k,0},\boldsymbol{\psi}^{k,1},\ldots,\boldsymbol{\psi}^{k,l-1})$, $k=1,2,\ldots,m$, and $(\boldsymbol{u}^0,\boldsymbol{u}^1,\cdots,\boldsymbol{u}^l)$ to give $\boldsymbol{x}^l$ and $\boldsymbol{\psi}^{k,l}$ by solving systems (8.26) and (8.27).

Step 2. For $\pi\in\{0,1,\ldots,l\}$ and $j\in\{1,2,\ldots,r\}$, computing $\frac{\partial \boldsymbol{B}^l}{\partial u_j^\pi}$, $\frac{\partial \boldsymbol{C}^l}{\partial u_j^\pi}$, $\frac{\partial \bar{\boldsymbol{D}}^{k,l}}{\partial u_j^\pi}$, $\frac{\partial \tilde{\boldsymbol{D}}^{k,l}}{\partial u_j^\pi}$ and $\frac{\partial \bar{\boldsymbol{E}}^{k,l}}{\partial u_j^\pi}$, $k=1,2,\ldots,m$, by using Theorems 8.3–8.7, respectively.

Step 3. For $\pi\in\{0,1,\ldots,l\}$ and $j\in\{1,2,\ldots,r\}$, solve Eqs. (8.36) and (8.37) to give $\frac{\partial \boldsymbol{x}^l}{\partial u_j^\pi}$ and $\frac{\partial \boldsymbol{\psi}^{k,l}}{\partial u_j^\pi}$, $k=1,2,\ldots,m$.

Step 4. If $l=N$, then compute gradients $\frac{\partial \tilde{J}^\beta(\boldsymbol{U}|\hat{\boldsymbol{\sigma}})}{\partial u_j^\pi}$, $\pi=0,1,2,\ldots,N$, $j=1,2,\ldots,r$, according to (8.35). Otherwise, set $l:=l+1$ and go to Step 1.

8.5 Numerical Examples

To test the performance of the proposed algorithm, we solve two non-trivial example problems in this section. All computations were performed within MATLAB environment on a PC with 3.4 GHz Intel Core i7-6700 and 8.00 GB RAM.

Example 8.5.1 Consider the following robust fractional optimal control problem [190]:

$$\min \quad \tilde{G}(u) = \int_0^{10} \left[x_1^2(t) + u^2(t)\right]dt \tag{8.59}$$

subject to the nonlinear fractional-order system

$$\begin{cases} {}_0^C\mathrm{D}_t^{\alpha_1} x_1(t) = x_2(t), \\ {}_0^C\mathrm{D}_t^{\alpha_2} x_2(t) = \left[1.4 + u(t) - 0.14x_2^2(t)\right]x_2(t) - x_1(t) + 4u(t), \quad t \in (0, 10], \\ \boldsymbol{x}(0) = (-5 + \sigma, -5)^\top, \end{cases}$$

where $u(t)$ is the control function; and σ is an uncertain parameter.

By introducing a new state x_3 satisfying

$$\begin{cases} \dot{x}_3(t) = x_1^2(t) + u^2(t), \quad t \in (0, 10], \\ x_3(0) = 0, \end{cases} \tag{8.60}$$

the cost function (8.59) becomes

$$G(u) = x_3(10). \tag{8.61}$$

Suppose that the nominal value of σ is zero, i.e., $\hat{\sigma} = 0$. Thus, the robust cost function for this example is

$$J^\beta(u|\hat{\sigma}) = G(u|\hat{\sigma}) + \beta\left[\frac{\partial G(u|\hat{\sigma})}{\partial \sigma}\right]^2, \tag{8.62}$$

where $\beta \geq 0$ is a weighting factor.

This example was solved by control parametrization method when $\boldsymbol{\alpha} = (1.0, 1.0)^\top$ in [190]. To give a comparison, this example is solved for different numbers of partition N by using our proposed method when $\boldsymbol{\alpha} = (1.0, 1.0)^\top$. Table 8.1 lists the numerical results for weights $\beta = 0$ and $\beta = 10$. From Table 8.1, it can be seen that as the number of partition increases, the obtained optimal cost decreases. In addition, although the obtained optimal costs for $\beta = 10$ are slightly larger than those for $\beta = 0$, the system sensitivities for weight $\beta = 10$ are much smaller than those for weight $\beta = 0$. More importantly, the obtained optimal costs for both $\beta = 0$ and $\beta = 10$ are all superior to the results reported in [190]. This example is further solved for various fractional orders when $N = 200$. The computed results are displayed in Table 8.2, which shows that the computed optimal costs of fractional orders are smaller than those of integer order. The corresponding controls and states are depicted in Figs. 8.1 and 8.2.

Table 8.1 Numerical results for Example 8.5.1 when $\boldsymbol{\alpha} = (1.0, 1.0)^\top$

N		80	120	160	200	[190]
$\beta = 0$	G	39.8312	39.7702	39.6200	39.4687	44.0099
	$\left[\frac{\partial G}{\partial \sigma}\right]^2$	82.3379	77.1019	91.0635	72.3792	99.6791
$\beta = 10$	G	48.7279	46.9317	44.0511	43.1372	50.0009
	$\left[\frac{\partial G}{\partial \sigma}\right]^2$	1.3083e-06	4.0062e-05	1.6305e-08	2.0389e-06	–

Table 8.2 Numerical results for Example 8.5.1 when $N = 200$

$\boldsymbol{\alpha}$		$(1.0, 1.0)^\top$	$(0.9, 0.9)^\top$	$(0.8, 0.8)^\top$
$\beta = 0$	G	39.4687	33.6205	27.4439
	$\left[\frac{\partial G}{\partial \sigma}\right]^2$	72.3792	69.4536	48.5454
$\beta = 10$	G	43.1372	35.0458	27.5308
	$\left[\frac{\partial G}{\partial \sigma}\right]^2$	2.0389e-06	1.1186e-06	2.5354e-09

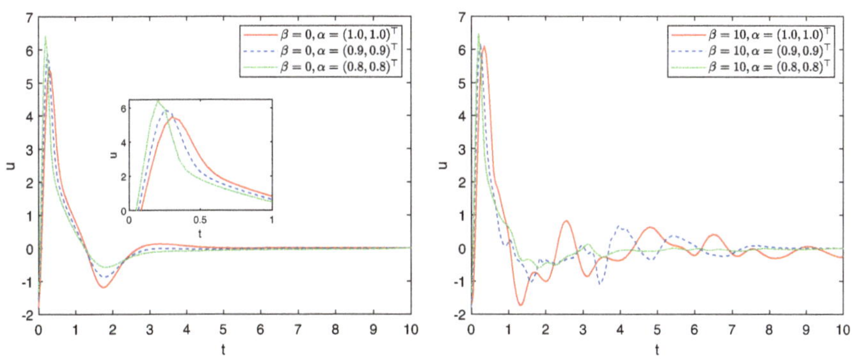

Fig. 8.1 Optimal controls for Example 8.5.1

We test the robustness of the optimal controls in Fig. 8.1 by various disturbances of $\hat{\sigma}$. The obtained system costs are shown in Fig. 8.3, from which we see that the optimal controls obtained through solving this example with $\beta = 10$ are very robust.

Example 8.5.2 Consider the following robust fractional optimal control problem arising in a semi-batch reactor process [288]:

$$\min \quad G(u) = x_1(250)x_3(250) \tag{8.63}$$

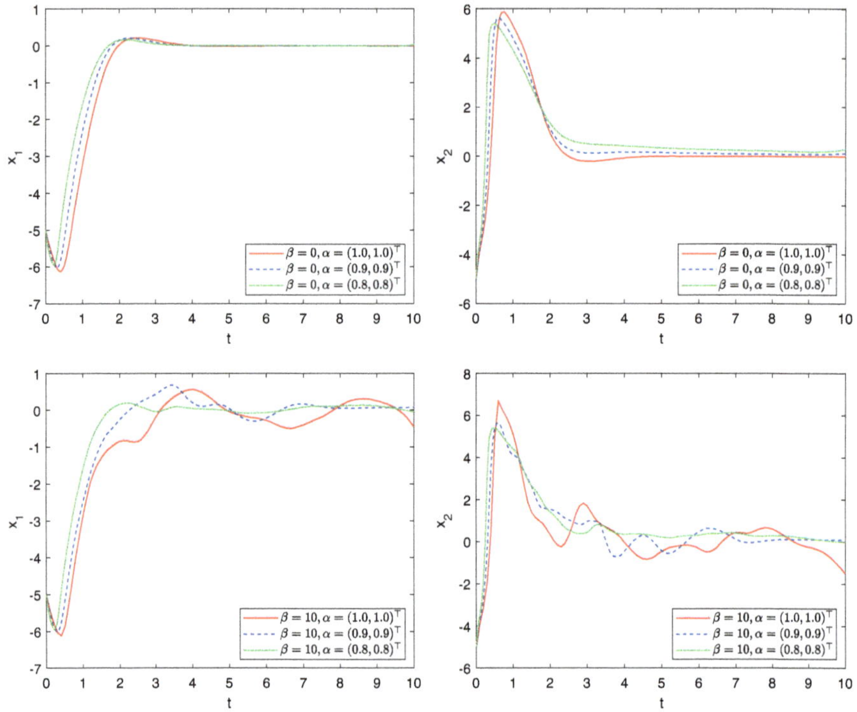

Fig. 8.2 Optimal state trajectories for Example 8.5.1

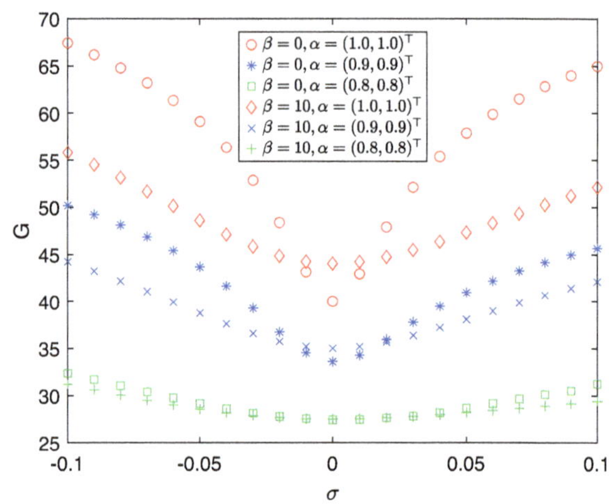

Fig. 8.3 The system costs for different model parameters in Example 8.5.1

subject to the nonlinear fractional system

$$\begin{cases} {}_0^C\mathrm{D}_t^{\alpha_1} x_1(t) = -k_1 x_1(t) x_2(t) - \dfrac{u(t)x_1(t)}{x_3(t)}, \\ {}_0^C\mathrm{D}_t^{\alpha_2} x_2(t) = -k_1 x_1(t) x_2(t) - 2k_2 x_2^2(t) - \dfrac{u(t)(x_2(t)-5)}{x_3(t)}, \quad t \in (0, 250], \\ {}_0^C\mathrm{D}_t^{\alpha_3} x_3(t) = u(t), \\ \boldsymbol{x}(0) = (0.72, 0.05, 1)^\top, \end{cases} \tag{8.64}$$

and

$$0 \le u(t) \le 10^{-3}, \quad t \in [0, 250]. \tag{8.65}$$

Here, $x_1(t)$ and $x_2(t)$ are the concentrations of two species (in molL^{-1}); $x_3(t)$ is the volume of culture fluid (in L); $u(t)$ is the inlet feed rate of reactant (in Lmin^{-1}); and k_1 and k_2 are the kinetic coefficients (in $\mathrm{Lmol}^{-1}\mathrm{min}^{-1}$).

The nominal values $\hat{k}_1 = 0.053$ and $\hat{k}_2 = 0.128$ are given for the parameters k_1 and k_2 in [288]. Therefore, our robust optimal control problem is to choose u such that

$$J^\beta(u|\hat{k}_1, \hat{k}_2) = G(u|\hat{k}_1, \hat{k}_2) + \beta\left[\frac{\partial G(u|\hat{k}_1, \hat{k}_2)}{\partial k_1}\right]^2 + \beta\left[\frac{\partial G(u|\hat{k}_1, \hat{k}_2)}{\partial k_2}\right]^2 \tag{8.66}$$

is minimized and subject to fractional system (8.64) and control constraint (8.65), where $\beta \ge 0$ is a weighting factor.

We solve this example with $N = 250$ by our proposed method. Table 8.3 gives the computed numerical results, from which we can see that the computed optimal costs corresponding to fractional orders are slightly larger than those corresponding to integer order. More importantly, although the computed optimal costs for weight $\beta = 1$ are a little larger than those for weight $\beta = 0$, the system sensitivities for weight $\beta = 1$ are much smaller than those for weight $\beta = 0$. The corresponding controls and states are shown in Figs. 8.4 and 8.5.

Table 8.3 Numerical results for Example 8.5.2 when $N = 250$

α		$(1.0, 1.0, 1.0)^\top$	$(0.95, 0.95, 1.0)^\top$	$(0.9, 0.9, 1.0)^\top$
$\beta = 0$	G	0.4133	0.3122	0.2073
	$\sum_{\ell=1}^{2}\left[\frac{\partial G}{\partial k_\ell}\right]^2$	16.6062	16.0843	13.2241
$\beta = 1$	G	0.6987	0.6965	0.6879
	$\sum_{\ell=1}^{2}\left[\frac{\partial G}{\partial k_\ell}\right]^2$	3.5453e-03	3.3045e-03	2.2362e-03

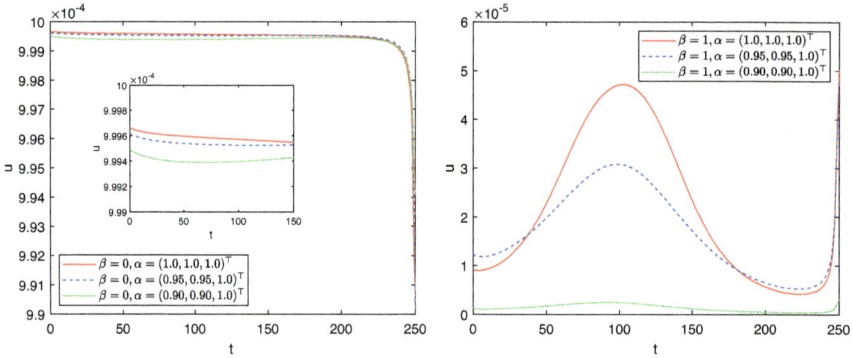

Fig. 8.4 Optimal controls for Example 8.5.2

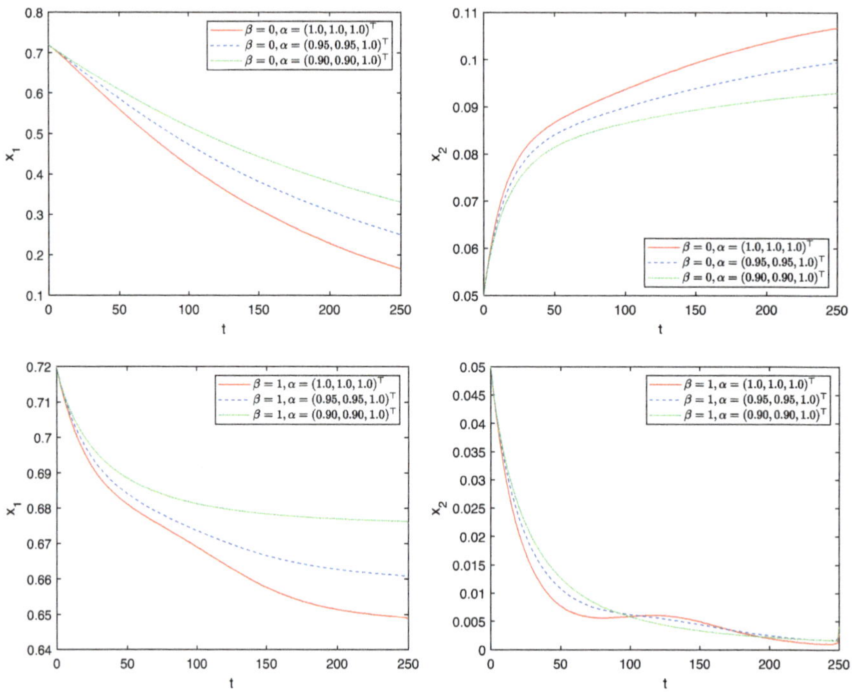

Fig. 8.5 Optimal state trajectories for Example 8.5.2

Table 8.4 System costs for different model parameters in Example 8.5.2

Model parameters	System costs		
	α	$\beta = 0$	$\beta = 1$
$k_1 = 0.0424, k_2 = 0.1024$	$(1.0, 1.0, 1.0)^\top$	0.2342	0.6532
	$(0.95, 0.95, 1.0)^\top$	0.3391	0.6645
	$(0.90, 0.90, 1.0)^\top$	0.4382	0.6772
$k_1 = 0.0424, k_2 = 0.1536$	$(1.0, 1.0, 1.0)^\top$	0.2739	0.6564
	$(0.95, 0.95, 1.0)^\top$	0.3758	0.6672
	$(0.90, 0.90, 1.0)^\top$	0.4698	0.6793
$k_1 = 0.0636, k_2 = 0.1024$	$(1.0, 1.0, 1.0)^\top$	0.1507	0.6497
	$(0.95, 0.95, 1.0)^\top$	0.2555	0.6609
	$(0.90, 0.90, 1.0)^\top$	0.3618	0.6743
$k_1 = 0.0636, k_2 = 0.1536$	$(1.0, 1.0, 1.0)^\top$	0.1861	0.6520
	$(0.95, 0.95, 1.0)^\top$	0.2906	0.6629
	$(0.90, 0.90, 1.0)^\top$	0.3931	0.6759

To test the robustness of controls, we consider various perturbations of parameters $\hat{k}_1$ and $\hat{k}_2$. Table 8.4 lists the numerical results, from which we see that, for both fractional and integer orders, the optimal controls obtained through solving the example with $\beta = 0$ are highly susceptible to the disturbances of parameters $\hat{k}_1$ and $\hat{k}_2$, while the optimal controls through solving the example with $\beta = 1$ are obviously more robust.

8.6 Conclusions

In this chapter, we has considered the robust optimal control problem governed by the nonlinear fractional system with uncertain parameters, in which the system sensitivity with respect to the uncertain parameters is explicitly incorporated in the cost function. By solving the auxiliary fractional system for the system sensitivity, we have proposed the equivalent optimal control problem. We then discretize the equivalent problem into the finite-dimensional optimization problem. Furthermore, we have developed the procedure to compute the gradients of the cost function for solving the finite-dimensional optimization problem. Finally, the effectiveness of the developed algorithm has been illustrated by solving two numerical examples.

Chapter 9
Fractional Optimal Control Problems with Atangana-Baleanu Derivatives

9.1 Introduction

In this chapter, we consider the numerical solution of fractional optimal control problems (FOCPs) with Atangana-Baleanu derivatives. The kernel functions in fractional derivatives can be singular or non-singular. The most well-known fractional derivatives equipped with singular kernels are Riemann-Liouville derivative and Caputo derivative, while those with non-singular kernels are Atangana-Baleanu derivative and Caputo-Fabrizio derivative. FOCPs with Atangana-Baleanu derivatives arise in various real-world applications, including thermistor problem [15], electric circuits [101], and cancer treatment [296].

FOCPs with Atangana-Baleanu derivative have attracted much attention of many researchers since Atangana-Baleanu fractional derivatives are more accurate in characterizing nonlocal dynamics [25]. In [26], the first-order necessary optimality conditions and a finite difference method are discussed. In [119, 130, 297], by expanding states and controls through wavelet or polynomial basis functions, numerical algorithms are developed, where shifted Legendre polynomials [130] and B-spline polynomials [297] are used, while the extended Chebyshev cardinal wavelets are employed in [119]. Nevertheless, it should be noted that the use of high-order basis functions for approximating states as well as controls in FOCPs often gives rise to ill-conditioned and numerically instable algebraic systems. Moreover, it is assumed that the fractional derivatives in FOCPs are of the same order. In next section, we give a class of FOCPs subject to canonical constraints of equality and inequality, where fractional derivatives are described in the Atangana-Baleanu-Caputo sense and they are of different orders. In Sect. 9.3, we present a numerical solution method for the FOCPs. In Sect. 9.3.1, we propose a discretization scheme combining the trapezoidal rule with a novel numerical integration technique for FOCPs with Atangana-Baleanu-Caputo derivatives. In Sect. 9.3.2, we derive the gradients of the cost and constraint functions with respect to the decision variables in the resulting problem. Furthermore, a gradient-based optimization approach is developed to solve the resulting

C. Liu et al., *Nonlinear Fractional Optimal Control*, Springer Optimization and Its Applications 237, https://doi.org/10.1007/978-981-92-0274-4_9

optimization problem. In Sect. 9.4, three non-trivial examples are provided to verify the applicability and effectiveness of the proposed algorithm. Finally, some concluding remarks are provided in Sect. 9.5.

The main reference of this chapter is [182].

9.2 Problem Statement

Consider the following FOCP with Atangana-Baleanu derivatives:

$$\min \quad g_0(\boldsymbol{u}) = \Phi_0(\boldsymbol{x}(t_f), \boldsymbol{u}(t_f)) + \int_{t_0}^{t_f} L_0(t, \boldsymbol{x}(t), \boldsymbol{u}(t))dt, \tag{9.1}$$

over $\mathcal{U}$ and subject to the nonlinear system

$$\begin{cases} {}^{ABC}_{t_0}\mathrm{D}^{\alpha}_{t}\boldsymbol{x}(t) = \boldsymbol{f}(t, \boldsymbol{x}(t), \boldsymbol{u}(t)), \quad t \in (t_0, t_f], \\ \boldsymbol{x}(t_0) = \boldsymbol{x}^0, \end{cases} \tag{9.2}$$

the equality constraints

$$g_i(\boldsymbol{u}) = \Phi_i(\boldsymbol{x}(t_f), \boldsymbol{u}(t_f)) + \int_{t_0}^{t_f} L_i(t, \boldsymbol{x}(t), \boldsymbol{u}(t))dt = 0, \quad i = 1, \ldots, M_e, \tag{9.3}$$

and the inequality constraints

$$g_i(\boldsymbol{u}) = \Phi_i(\boldsymbol{x}(t_f), \boldsymbol{u}(t_f)) + \int_{t_0}^{t_f} L_i(t, \boldsymbol{x}(t), \boldsymbol{u}(t))dt \geq 0, \quad i = M_e + 1, \ldots, M, \tag{9.4}$$

where $\boldsymbol{x}(t) = (x_1(t), \ldots, x_n(t))^\top \in \mathbb{R}^n$ represents the state vector; $\boldsymbol{u}(t) = (u_1(t), \ldots, u_r(t))^\top \in \mathbb{R}^r$ denotes the control vector; $\boldsymbol{x}^0 \in \mathbb{R}^n$ is a given initial state; $t_0 \in \mathbb{R}$ and $t_f \in \mathbb{R}$ are given initial and terminal times, respectively; $\mathcal{U}$ indicates the admissible control set; $\boldsymbol{f} : \mathbb{R} \times \mathbb{R}^n \times \mathbb{R}^r \to \mathbb{R}^n$, $L_i : \mathbb{R} \times \mathbb{R}^n \times \mathbb{R}^r \to \mathbb{R}$ and $\Phi_i : \mathbb{R}^n \times \mathbb{R}^r \to \mathbb{R}$, $i = 0, 1, \ldots, M$, are given twice continuously differentiable functions; and

$${}^{ABC}_{t_0}\mathrm{D}^{\alpha}_{t}\boldsymbol{x}(t) = \left({}^{ABC}_{t_0}\mathrm{D}^{\alpha_1}_{t}x_1(t), \ldots, {}^{ABC}_{t_0}\mathrm{D}^{\alpha_n}_{t}x_n(t)\right)^\top \tag{9.5}$$

with $\alpha_j \in (0, 1]$ and $ABCt_0\mathrm{D}^{\alpha_j}_{t}x_j(t)$ denoting the α_jth Atangana-Baleanu fractional derivative in the Caputo sense of $x_j(t)$.

9.3 Numerical Solution Method

9.3.1 Discretization Scheme

Let N be a positive integer. We divide $(t_0, t_f]$ into N sub-intervals $(t_{k-1}, t_k]$ with partition points $t_k = t_0 + kh, k = 1, 2, \ldots, N$, and $h = \dfrac{t_f - t_0}{N}$. Then, by the trapezoidal rule, the cost function (9.1) and the constraint functions (9.3) and (9.4) can be approximated by

$$
\begin{aligned}
g_0(\boldsymbol{u}) \approx \Phi_0(\boldsymbol{x}^N, \boldsymbol{u}^N) &+ \frac{h}{2}\left[L_0(t_0, \boldsymbol{x}^0, \boldsymbol{u}^0) + L_0(t_N, \boldsymbol{x}^N, \boldsymbol{u}^N)\right] \\
&+ h\sum_{k=1}^{N-1} L_0(t_k, \boldsymbol{x}^k, \boldsymbol{u}^k),
\end{aligned} \tag{9.6}
$$

and

$$
\begin{aligned}
g_i(\boldsymbol{u}) \approx \Phi_i(\boldsymbol{x}^N, \boldsymbol{u}^N) &+ \frac{h}{2}\left[L_i(t_0, \boldsymbol{x}^0, \boldsymbol{u}^0) + L_i(t_N, \boldsymbol{x}^N, \boldsymbol{u}^N)\right] \\
&+ h\sum_{k=1}^{N-1} L_i(t_k, \boldsymbol{x}^k, \boldsymbol{u}^k) \begin{cases} = 0, & i = 1, \ldots, M_e, \\ \geq 0, & i = M_e + 1, \ldots, M, \end{cases}
\end{aligned} \tag{9.7}
$$

respectively, where $\boldsymbol{x}^k = \boldsymbol{x}(t_k)$ and $\boldsymbol{u}^k = \boldsymbol{u}(t_k)$. Furthermore, by Theorem 2.13, we can convert fractional-order system (9.2) to the following equivalent integral form:

$$
\begin{aligned}
x_j^k = x_j^0 &+ \frac{(1-\alpha_j)}{\mathbb{B}(\alpha_j)} f_j(t_k, \boldsymbol{x}^k, \boldsymbol{u}^k) \\
&+ \frac{\alpha_j}{\mathbb{B}(\alpha_j)\Gamma(\alpha_j)} \sum_{l=1}^{k} \int_{t_{l-1}}^{t_l} (t_k - s)^{\alpha_j - 1} f_j(s, \boldsymbol{x}(s), \boldsymbol{u}(s)) ds
\end{aligned} \tag{9.8}
$$

for $k = 1, 2, \ldots, N$ and $j = 1, 2, \ldots, n$. By using Taylor expansion, we expand $f_j(s, \boldsymbol{x}(s), \boldsymbol{u}(s)), s \in (t_{l-1}, t_l]$, at any point $s_{lk}^j \in (t_{l-1}, t_l]$ as

$$
f_j(s, \boldsymbol{x}(s), \boldsymbol{u}(s)) \approx f_j(s_{lk}^j, \boldsymbol{x}_{lk}^j, \boldsymbol{u}_{lk}^j) + \Delta_{lk}^j (s - s_{lk}^j) + \Lambda_{lk}^j (s - s_{lk}^j)^2, \tag{9.9}
$$

where $\boldsymbol{x}_{lk}^j = \boldsymbol{x}(s_{lk}^j)$; $\boldsymbol{u}_{lk}^j = \boldsymbol{u}(s_{lk}^j)$; and Δ_{lk}^j and Λ_{lk}^j are coefficients in the Taylor expansion. Then, substituting (9.9) into (9.8) yields

$$x_j^k = x_j^0 + \frac{(1-\alpha_j)}{\mathbb{B}(\alpha_j)} f_j(t_k, \boldsymbol{x}^k, \boldsymbol{u}^k) + \tilde{h}^j \sum_{l=1}^{k} f_j(s_{lk}^j, \boldsymbol{x}_{lk}^j, \boldsymbol{u}_{lk}^j)\big[(d_{lk}+1)^{\alpha_j} - d_{lk}^{\alpha_j}\big]$$
$$+ \sum_{l=1}^{k} \frac{\alpha_j \Delta_{lk}^j}{\mathbb{B}(\alpha_j)\Gamma(\alpha_j)} \int_{t_{l-1}}^{t_l} (t_k - s)^{\alpha_j - 1}(s - s_{lk}^j)ds + R_k^j \tag{9.10}$$

for $k = 1, 2, \ldots, N$, and $j = 1, 2, \ldots, n$, where $\mathbb{B}(\cdot)$ is the normalized function as in (2.34); $\tilde{h}^j = \dfrac{h^{\alpha_j}}{\mathbb{B}(\alpha_j)\Gamma(\alpha_j)}$; $d_{lk} = (k - l)$; and

$$R_k^j = \sum_{l=1}^{k} \frac{\alpha_j \Lambda_{lk}^j}{\mathbb{B}(\alpha_j)\Gamma(\alpha_j)} \int_{t_{l-1}}^{t_l} (t_k - s)^{\alpha_j - 1}(s - s_{lk}^j)^2 ds.$$

By similar arguments as given in Theorem 3.19, we can show that $R_k^j = \mathcal{O}(h^2)$. To achieve this truncation error, we choose the following s_{lk}^j:

$$s_{lk}^j = t_0 + \frac{\big[(d_{lk}+1)^{\alpha_j+1} - d_{lk}^{\alpha_j+1}\big] + (\alpha_j+1)\big[(d_{lk}+1)^{\alpha_j}(l-1) - d_{lk}^{\alpha_j} l\big]}{(\alpha_j+1)\big[(d_{lk}+1)^{\alpha_j} - d_{lk}^{\alpha_j}\big]} h, \tag{9.11}$$

which makes the fourth term on the right-hand side of (9.10) be equal to zero. Clearly, $s_{lk}^j \in (t_{l-1}, t_l]$. Omitting the truncation error R_{lk}^j in (9.10) gives the following approximation scheme:

$$x_j^k = x_j^0 + \frac{(1-\alpha_j)}{\mathbb{B}(\alpha_j)} f_j(t_k, \boldsymbol{x}^k, \boldsymbol{u}^k) + \tilde{h}^j \sum_{l=1}^{k} f_j(s_{lk}^j, \boldsymbol{x}_{lk}^j, \boldsymbol{u}_{lk}^j)\big[(d_{lk}+1)^{\alpha_j} - d_{lk}^{\alpha_j}\big] \tag{9.12}$$

for $k = 1, 2, \ldots, N$, and $j = 1, 2, \ldots, n$, where s_{lk}^j is given by (9.11). However, in (9.12), the values of $\boldsymbol{x}_{lk}^j$ and $\boldsymbol{u}_{lk}^j$ are not available directly although the s_{lk}^j is given by (9.11). Thus, we approximate $\boldsymbol{x}_{lk}^j$ and $\boldsymbol{u}_{lk}^j$ by the following linear interpolations:

$$\tilde{\boldsymbol{x}}_{lk}^j = \boldsymbol{x}^{l-1} + \rho_{lk}^j(\boldsymbol{x}^l - \boldsymbol{x}^{l-1}) \text{ and } \tilde{\boldsymbol{u}}_{lk}^j = \boldsymbol{u}^{l-1} + \rho_{lk}^j(\boldsymbol{u}^l - \boldsymbol{u}^{l-1}), \tag{9.13}$$

respectively, where $\rho_{lk}^j = \dfrac{s_{lk}^j - t_{l-1}}{h}$. Obviously, the error generated by (9.13) is $\mathcal{O}(h^2)$. Substituting (9.13) into (9.12) gives

$$x_j^k = x_j^0 + \frac{(1-\alpha_j)}{\mathbb{B}(\alpha_j)} f_j(t_k, \boldsymbol{x}^k, \boldsymbol{u}^k) + \tilde{h}^j \sum_{l=1}^{k} f_j(s_{lk}^j, \tilde{\boldsymbol{x}}_{lk}^j, \tilde{\boldsymbol{u}}_{lk}^j)\big[(d_{lk}+1)^{\alpha_j} - d_{lk}^{\alpha_j}\big] \tag{9.14}$$

for $k = 1, 2, \ldots, N$ and $j = 1, 2, \ldots, n$.

Note that equation (9.14) is actually an implicit equation since a nonlinear function of $\boldsymbol{x}^k$ appears in its right-hand side. For deriving an approximation scheme with explicit form, we approximate $f_j(t_k, \boldsymbol{x}^k, \boldsymbol{u}^k)$ and $f_j(s_{kk}^j, \tilde{\boldsymbol{x}}_{kk}^j, \tilde{\boldsymbol{u}}_{kk}^j)$ in (9.14) by

$$f_j(t_k, \boldsymbol{x}^k, \boldsymbol{u}^k) = f_j(t_k, \boldsymbol{x}^{k-1}, \boldsymbol{u}^k) + \left(\frac{\partial f_j}{\partial \boldsymbol{x}}\right)^\top \Bigg|_{(t_k, \boldsymbol{x}^{k-1}, \boldsymbol{u}^k)} (\boldsymbol{x}^k - \boldsymbol{x}^{k-1}) + \mathcal{O}(h^2), \tag{9.15}$$

and

$$f_j(s_{kk}^j, \tilde{\boldsymbol{x}}_{kk}^j, \tilde{\boldsymbol{u}}_{kk}^j) = f_j(s_{kk}^j, \boldsymbol{x}^{k-1}, \tilde{\boldsymbol{u}}_{kk}^j) + \left(\frac{\partial f_j}{\partial \boldsymbol{x}}\right)^\top \Bigg|_{(s_{kk}^j, \boldsymbol{x}^{k-1}, \tilde{\boldsymbol{u}}_{kk}^j)} \rho_{kk}^j (\boldsymbol{x}^k - \boldsymbol{x}^{k-1}) + \mathcal{O}(h^2), \tag{9.16}$$

respectively. Substituting (9.15) and (9.16) into (9.14), we obtain the following explicit approximation scheme for (9.8):

$$\begin{aligned} x_j^k = x_j^0 &+ \frac{(1-\alpha_j)}{\mathbb{B}(\alpha_j)} f_j(t_k, \boldsymbol{x}^{k-1}, \boldsymbol{u}^k) + \tilde{h}^j f_j(s_{kk}^j, \boldsymbol{x}^{k-1}, \tilde{\boldsymbol{u}}_{kk}^j) \\ &+ \frac{(1-\alpha_j)}{\mathbb{B}(\alpha_j)} \left(\frac{\partial f_j}{\partial \boldsymbol{x}}\right)^\top \Bigg|_{(t_k, \boldsymbol{x}^{k-1}, \boldsymbol{u}^k)} (\boldsymbol{x}^k - \boldsymbol{x}^{k-1}) + \tilde{h}^j \left(\frac{\partial f_j}{\partial \boldsymbol{x}}\right)^\top \Bigg|_{(s_{kk}^j, \boldsymbol{x}^{k-1}, \tilde{\boldsymbol{u}}_{kk}^j)} \\ &\times \rho_{kk}^j (\boldsymbol{x}^k - \boldsymbol{x}^{k-1}) + \tilde{h}^j \sum_{l=1}^{k-1} f_j(s_{lk}^j, \tilde{\boldsymbol{x}}_{lk}^j, \tilde{\boldsymbol{u}}_{lk}^j)\big[(d_{lk}+1)^{\alpha_j} - d_{lk}^{\alpha_j}\big] \end{aligned} \tag{9.17}$$

for $k = 1, 2, \ldots, N$ and $j = 1, 2 \ldots, n$. We can further rewrite (9.17) as

$$(\boldsymbol{I} - \boldsymbol{Q}^k(\boldsymbol{x}^{k-1}, \boldsymbol{u}^{k-1}, \boldsymbol{u}^k))\boldsymbol{x}^k = \boldsymbol{C}^k(\boldsymbol{x}^0, \ldots, \boldsymbol{x}^{k-1}, \boldsymbol{u}^0, \boldsymbol{u}^1, \ldots, \boldsymbol{u}^k), \quad k = 1, 2, \ldots, N. \tag{9.18}$$

Here, $\boldsymbol{I}$ is the identity matrix; $\boldsymbol{Q}^k = [q_{jp}^k]_{n\times n}$ denotes a matrix with the elements as

$$q_{jp}^k = \frac{(1-\alpha_j)}{\mathbb{B}(\alpha_j)} \frac{\partial f_j}{\partial x_p}\Bigg|_{(t_k, \boldsymbol{x}^{k-1}, \boldsymbol{u}^k)} + \tilde{h}^j \rho_{kk}^j \frac{\partial f_j}{\partial x_p}\Bigg|_{(s_{kk}^j, \boldsymbol{x}^{k-1}, \tilde{\boldsymbol{u}}_{kk}^j)}; \tag{9.19}$$

and $\boldsymbol{C}^k = (c_1^k, \ldots, c_n^k)^\top$ is a vector with components as

$$\begin{aligned} c_j^k = x_j^0 &+ \frac{(1-\alpha_j)}{\mathbb{B}(\alpha_j)} f_j(t_k, \boldsymbol{x}^{k-1}, \boldsymbol{u}^k) + \tilde{h}^j f_j(s_{kk}^j, \boldsymbol{x}^{k-1}, \tilde{\boldsymbol{u}}_{kk}^j) \\ &+ \tilde{h}^j \sum_{l=1}^{k-1} f_j(s_{lk}^j, \tilde{\boldsymbol{x}}_{lk}^j, \tilde{\boldsymbol{u}}_{lk}^j)\big[(d_{lk}+1)^{\alpha_j} - d_{lk}^{\alpha_j}\big] - \sum_{p=1}^{n} q_{jp}^k x_p^{k-1}. \end{aligned} \tag{9.20}$$

The following result, which can be proved in a similar manner to Theorem 3.20, gives the convergence order of the above discretization method.

Theorem 9.1 *Let $\boldsymbol{x}(t)$ be the exact solution of (9.2). Then, there exists an $\hat{h}$ such that $h < \hat{h}$ implies*

$$\|\boldsymbol{x}(t_k) - \boldsymbol{x}^k\|_\infty \le Ch^2, \quad k = 1, 2, \dots, N, \tag{9.21}$$

where $\|\cdot\|_\infty$ is the infinity norm; $\{\boldsymbol{x}^k\}$ is the sequence generated by (9.18); and $C > 0$ is a constant independent of h and α.

Let $\boldsymbol{U} = (\boldsymbol{u}^0, \boldsymbol{u}^1, \dots, \boldsymbol{u}^N) \in \mathbb{R}^{r\times(N+1)}$. Then, combining (9.18) with (9.6) and (9.7) yields the discretized problem as follows:

$$\begin{aligned} \min \quad \tilde{g}_0(\boldsymbol{U}) = {} & \Phi_0(\boldsymbol{x}^N, \boldsymbol{u}^N) + \frac{h}{2}\big[L_0(t_0, \boldsymbol{x}^0, \boldsymbol{u}^0) + L_0(t_N, \boldsymbol{x}^N, \boldsymbol{u}^N)\big] \\ & + h\sum_{k=1}^{N-1} L_0(t_k, \boldsymbol{x}^k, \boldsymbol{u}^k), \end{aligned} \tag{9.22}$$

over $\mathbb{R}^{r\times(N+1)}$ and subject to the discretized system

$$\begin{cases} (\boldsymbol{I} - \boldsymbol{Q}^k(\boldsymbol{x}^{k-1}, \boldsymbol{u}^{k-1}, \boldsymbol{u}^k))\boldsymbol{x}^k = \boldsymbol{C}^k(\boldsymbol{x}^0, \dots, \boldsymbol{x}^{k-1}, \boldsymbol{u}^0, \boldsymbol{u}^1, \dots, \boldsymbol{u}^k), \\ \qquad\qquad\qquad\qquad\qquad\qquad k = 1, 2, \dots, N, \\ \boldsymbol{x}(t_0) = \boldsymbol{x}^0, \end{cases} \tag{9.23}$$

and constraints

$$\begin{aligned} \tilde{g}_i(\boldsymbol{U}) = {} & \Phi_i(\boldsymbol{x}^N, \boldsymbol{u}^N) + \frac{h}{2}\big[L_i(t_0, \boldsymbol{x}^0, \boldsymbol{u}^0) + L_i(t_N, \boldsymbol{x}^N, \boldsymbol{u}^N)\big] \\ & + h\sum_{k=1}^{N-1} L_i(t_k, \boldsymbol{x}^k, \boldsymbol{u}^k) \begin{cases} = 0, & i = 1, 2, \dots, M_e, \\ \ge 0, & i = M_e + 1, \dots, M. \end{cases} \end{aligned} \tag{9.24}$$

9.3.2 Optimization Technique

In view of the similarity of cost function (9.22) and constraint function (9.24), we can compute the required gradients in a unified way. For $i = 0, 1, \dots, M$, differentiating $\tilde{g}_i(\boldsymbol{U})$ with respect to u_ς^π gives

$$\begin{aligned}\frac{\partial \tilde{g}_i(\boldsymbol{U})}{\partial u_\varsigma^\pi} &= \delta_{\pi 0}\frac{h}{2}\frac{\partial L_i}{\partial u_\varsigma^\pi}\bigg|_{(t_0,\boldsymbol{x}^0,\boldsymbol{u}^0)} + \left[\frac{\partial \Phi_i}{\partial \boldsymbol{x}^N}\bigg|_{(\boldsymbol{x}^N,\boldsymbol{u}^N)} + \frac{h}{2}\frac{\partial L_i}{\partial \boldsymbol{x}^N}\bigg|_{(t_N,\boldsymbol{x}^N,\boldsymbol{u}^N)}\right]^\top \frac{\partial \boldsymbol{x}^N}{\partial u_\varsigma^\pi}\\ &+ \delta_{\pi N}\left[\frac{\partial \Phi_i}{\partial u_\varsigma^\pi}\bigg|_{(\boldsymbol{x}^N,\boldsymbol{u}^N)} + \frac{h}{2}\frac{\partial L_i}{\partial u_\varsigma^\pi}\bigg|_{(t_N,\boldsymbol{x}^N,\boldsymbol{u}^N)}\right]\\ &+ h\sum_{k=1}^{N-1}\left[\left(\frac{\partial L_i}{\partial \boldsymbol{x}^k}\right)^\top\bigg|_{(t_k,\boldsymbol{x}^k,\boldsymbol{u}^k)}\frac{\partial \boldsymbol{x}^k}{\partial u_\varsigma^\pi} + \frac{\partial L_i}{\partial u_\varsigma^\pi}\bigg|_{(t_k,\boldsymbol{x}^k,\boldsymbol{u}^k)}\right],\\ &\qquad \varsigma \in \{1,2,\ldots,r\},\ \pi \in \{0,1,\ldots,N\},\end{aligned} \tag{9.25}$$

where $\delta_{\imath\jmath}$ is the indicator function defined by

$$\delta_{\imath\jmath} = \begin{cases} 1, & \text{if } \imath = \jmath,\\ 0, & \text{otherwise.}\end{cases} \tag{9.26}$$

Using $\dfrac{\partial \boldsymbol{x}^k}{\partial u_\varsigma^\pi} = 0$, $\varsigma \in \{1,2,\ldots,r\}$, for $k < \pi$, we can rewrite (9.25) as

$$\begin{aligned}\frac{\partial \tilde{g}_i(\boldsymbol{U})}{\partial u_\varsigma^\pi} &= \delta_{\pi 0}\frac{h}{2}\frac{\partial L_i}{\partial u_\varsigma^\pi}\bigg|_{(t_0,\boldsymbol{x}^0,\boldsymbol{u}^0)} + \delta_{\pi N}\left[\frac{\partial \Phi_i}{\partial u_\varsigma^\pi}\bigg|_{(\boldsymbol{x}^N,\boldsymbol{u}^N)} + \frac{h}{2}\frac{\partial L_i}{\partial u_\varsigma^\pi}\bigg|_{(t_N,\boldsymbol{x}^N,\boldsymbol{u}^N)}\right]\\ &+ (1-\delta_{\pi 0})(1-\delta_{\pi N})h\frac{\partial L_i}{\partial u_\varsigma^\pi}\bigg|_{(t_\pi,\boldsymbol{x}^\pi,\boldsymbol{u}^\pi)} + \left[\frac{\partial \Phi_i}{\partial \boldsymbol{x}^N}\bigg|_{(\boldsymbol{x}^N,\boldsymbol{u}^N)}\right.\\ &\left.+ \frac{h}{2}\frac{\partial L_i}{\partial \boldsymbol{x}^N}\bigg|_{(t_N,\boldsymbol{x}^N,\boldsymbol{u}^N)}\right]^\top \frac{\partial \boldsymbol{x}^N}{\partial u_\varsigma^\pi} + h\sum_{k=\pi}^{N-1}\left(\frac{\partial L_i}{\partial \boldsymbol{x}^k}\right)^\top\bigg|_{(t_k,\boldsymbol{x}^k,\boldsymbol{u}^k)}\frac{\partial \boldsymbol{x}^k}{\partial u_\varsigma^\pi}\\ &\qquad \varsigma \in \{1,2,\ldots,r\},\ \pi \in \{0,1,\ldots,N\},\end{aligned} \tag{9.27}$$

where $\delta_{\imath\jmath}$ is as defined in (9.26). However, in (9.27), the partial derivatives $\dfrac{\partial \boldsymbol{x}^k}{\partial u_\varsigma^\pi}$ are required to be determined. From (9.23), we have

$$(\boldsymbol{I} - \boldsymbol{Q}^k)\frac{\partial \boldsymbol{x}^k}{\partial u_\varsigma^\pi} = \frac{\partial \boldsymbol{Q}^k}{\partial u_\varsigma^\pi}\boldsymbol{x}^k + \frac{\partial \boldsymbol{C}^k}{\partial u_\varsigma^\pi} \tag{9.28}$$

for $k = 1,2,\ldots,N$. Thus, to compute $\dfrac{\partial \boldsymbol{x}^k}{\partial u_\varsigma^\pi}$, we need to find the values of $\dfrac{\partial \boldsymbol{Q}^k}{\partial u_\varsigma^\pi}$ and $\dfrac{\partial \boldsymbol{C}^k}{\partial u_\varsigma^\pi}$. The following theorems give the calculations of $\dfrac{\partial \boldsymbol{Q}^k}{\partial u_\varsigma^\pi}$ and $\dfrac{\partial \boldsymbol{C}^k}{\partial u_\varsigma^\pi}$.

Theorem 9.2 *Suppose that* $\varsigma \in \{1,2,\ldots,r\}$, p *and* $j \in \{1,2,\ldots,n\}$, *and* π *and* $k \in \{0,1,\ldots,N\}$.
(i) *If* $\pi > k$, *then*

$$\frac{\partial q_{jp}^k}{\partial u_\varsigma^\pi} = 0. \tag{9.29}$$

(ii) *If* $\pi = k$, *then*

$$\frac{\partial q_{jp}^k}{\partial u_\varsigma^\pi} = \frac{(1-\alpha_j)}{\mathbb{B}(\alpha_j)} \frac{\partial^2 f_j}{\partial x_p \partial u_\varsigma}\bigg|_{(t_k, \boldsymbol{x}^{k-1}, \boldsymbol{u}^k)} + \tilde{h}^j (\rho_{kk}^j)^2 \frac{\partial^2 f_j}{\partial x_p \partial u_\varsigma}\bigg|_{(s_{kk}^j, \boldsymbol{x}^{k-1}, \tilde{\boldsymbol{u}}_{kk}^j)}. \tag{9.30}$$

(iii) *If* $\pi = k-1$, *then*

$$\frac{\partial q_{jp}^k}{\partial u_\varsigma^\pi} = \sum_{\varpi=1}^{n} \left[\frac{(1-\alpha_j)}{\mathbb{B}(\alpha_j)} \frac{\partial^2 f_j}{\partial x_p \partial x_\varpi}\bigg|_{(t_k, \boldsymbol{x}^{k-1}, \boldsymbol{u}^k)} + \tilde{h}^j \rho_{kk}^j \frac{\partial^2 f_j}{\partial x_p \partial x_\varpi}\bigg|_{(s_{kk}^j, \boldsymbol{x}^{k-1}, \tilde{\boldsymbol{u}}_{kk}^j)} \right]$$
$$\times \frac{\partial \boldsymbol{x}_\varpi^{k-1}}{\partial u_\varsigma^\pi} + \tilde{h}^j \rho_{kk}^j (1-\rho_{kk}^j) \frac{\partial^2 f_j}{\partial x_p \partial u_\varsigma}\bigg|_{(s_{kk}^j, \boldsymbol{x}^{k-1}, \tilde{\boldsymbol{u}}_{kk}^j)}. \tag{9.31}$$

(iv) *If* $\pi < k-1$, *then*

$$\frac{\partial q_{jp}^k}{\partial u_\varsigma^\pi} = \sum_{\varpi=1}^{n} \left[\frac{(1-\alpha_j)}{\mathbb{B}(\alpha_j)} \frac{\partial^2 f_j}{\partial x_p \partial x_\varpi}\bigg|_{(t_k, \boldsymbol{x}^{k-1}, \boldsymbol{u}^k)} + \tilde{h}^j \rho_{kk}^j \frac{\partial^2 f_j}{\partial x_p \partial x_\varpi}\bigg|_{(s_{kk}^j, \boldsymbol{x}^{k-1}, \tilde{\boldsymbol{u}}_{kk}^j)} \right]$$
$$\times \frac{\partial \boldsymbol{x}_\varpi^{k-1}}{\partial u_\varsigma^\pi}. \tag{9.32}$$

Proof The proof follows from differentiating (9.19) and making use of the chain rule. □

Theorem 9.3 *Suppose that* $\varsigma \in \{1, 2, \ldots, r\}$, $j \in \{1, 2, \ldots, n\}$ *and* π *and* $k \in \{0, 1, \ldots, N\}$.
(i) *If* $\pi > k$, *then*

$$\frac{\partial c_j^k}{\partial u_\varsigma^\pi} = 0. \tag{9.33}$$

(ii) *If* $\pi = k$, *then*

$$\frac{\partial c_j^k}{\partial u_\varsigma^\pi} = \frac{(1-\alpha_j)}{\mathbb{B}(\alpha_j)} \frac{\partial f_j}{\partial u_\varsigma}\bigg|_{(t_k, \boldsymbol{x}^{k-1}, \boldsymbol{u}^k)} + \tilde{h}^j \rho_{kk}^j \frac{\partial f_j}{\partial u_\varsigma}\bigg|_{(s_{kk}^j, \boldsymbol{x}^{k-1}, \tilde{\boldsymbol{u}}_{kk}^j)} - \sum_{p=1}^{n} \frac{\partial q_{jp}^k}{\partial u_\varsigma^\pi} x_p^{k-1}, \tag{9.34}$$

where $\frac{\partial q_{jp}^k}{\partial u_\varsigma^\pi}$ *is as given in Theorem 9.2.*
(iii) *If* $\pi = k-1$, *then*

$$\frac{\partial c_j^k}{\partial u_\varsigma^\pi} = \sum_{\varpi=1}^{n} \left[\frac{(1-\alpha_j)}{\mathbb{B}(\alpha_j)} \frac{\partial f_j}{\partial x_\varpi}\bigg|_{(t_k, \boldsymbol{x}^{k-1}, \boldsymbol{u}^k)} + \tilde{h}^j \frac{\partial f_j}{\partial x_\varpi}\bigg|_{(s_{kk}^j, \boldsymbol{x}^{k-1}, \tilde{\boldsymbol{u}}_{kk}^j)} \right] \frac{\partial x_\varpi^{k-1}}{\partial u_\varsigma^\pi}$$
$$+ \tilde{h}^j (1-\rho_{kk}^j) \frac{\partial f_j}{\partial u_\varsigma}\bigg|_{(s_{kk}^j, \boldsymbol{x}^{k-1}, \tilde{\boldsymbol{u}}_{kk}^j)} + \left[\tilde{h}^j \rho_{(k-1)k}^j \frac{\partial f_j}{\partial u_\varsigma}\bigg|_{(s_{(k-1)k}^j, \tilde{\boldsymbol{x}}_{(k-1)k}^j, \tilde{\boldsymbol{u}}_{(k-1)k}^j)} \right.$$
$$\left. + \sum_{\varpi=1}^{n} \tilde{h}^j \rho_{(k-1)k}^j \frac{\partial f_j}{\partial x_\varpi}\bigg|_{(s_{(k-1)k}^j, \tilde{\boldsymbol{x}}_{(k-1)k}^j, \tilde{\boldsymbol{u}}_{(k-1)k}^j)} \frac{\partial x_\varpi^{k-1}}{\partial u_\varsigma^\pi} \right] [2^{\alpha_j} - 1]$$
$$- \sum_{p=1}^{n} \left[\frac{\partial q_{jp}^k}{\partial u_\varsigma^\pi} x_p^{k-1} + q_{jp}^k \frac{\partial x_p^{k-1}}{\partial u_\varsigma^\pi} \right], \tag{9.35}$$

where q_{jp}^k *is as defined in (9.19) and* $\dfrac{\partial q_{jp}^k}{\partial u_\varsigma^\pi}$ *is as given in Theorem 9.2.*

(iv) *If* $\pi < k-1$, *then*

$$\frac{\partial c_j^k}{\partial u_\varsigma^\pi} = \sum_{\varpi=1}^{n} \left[\frac{(1-\alpha_j)}{\mathbb{B}(\alpha_j)} \frac{\partial f_j}{\partial x_\varpi}\bigg|_{(t_k, \boldsymbol{x}^{k-1}, \boldsymbol{u}^k)} + \tilde{h}^j \frac{\partial f_j}{\partial x_\varpi}\bigg|_{(s_{kk}^j, \boldsymbol{x}^{k-1}, \tilde{\boldsymbol{u}}_{kk}^j)} \right] \frac{\partial x_\varpi^{k-1}}{\partial u_\varsigma^\pi}$$
$$+ \tilde{h}^j \sum_{\substack{l=\delta_{\pi 0}+ \\ (1-\delta_{\pi 0})\pi}}^{k-1} \left\{ \sum_{\varpi=1}^{n} \left[\delta_{l\pi} \rho_{lk}^j \frac{\partial f_j}{\partial x_\varpi}\bigg|_{(s_{lk}^j, \tilde{\boldsymbol{x}}_{lk}^j, \tilde{\boldsymbol{u}}_{lk}^j)} \frac{\partial x_\varpi^l}{\partial u_\varsigma^\pi} + (1-\delta_{l\pi}) \right. \right.$$
$$\left. \times \left(\rho_{lk}^j \frac{\partial f_j}{\partial x_\varpi}\bigg|_{(s_{lk}^j, \tilde{\boldsymbol{x}}_{lk}^j, \tilde{\boldsymbol{u}}_{lk}^j)} \frac{\partial x_\varpi^l}{\partial u_\varsigma^\pi} + (1-\rho_{lk}^j) \frac{\partial f_j}{\partial x_\varpi}\bigg|_{(s_{lk}^j, \tilde{\boldsymbol{x}}_{lk}^j, \tilde{\boldsymbol{u}}_{lk}^j)} \frac{\partial x_\varpi^{l-1}}{\partial u_\varsigma^\pi} \right) \right]$$
$$\left. + \delta_{l\pi} \rho_{lk}^j \frac{\partial f_j}{\partial u_\varsigma}\bigg|_{(s_{lk}^j, \tilde{\boldsymbol{x}}_{lk}^j, \tilde{\boldsymbol{u}}_{lk}^j)} + \delta_{l(\pi+1)} (1-\rho_{lk}^j) \frac{\partial f_j}{\partial u_\varsigma}\bigg|_{(s_{lk}^j, \tilde{\boldsymbol{x}}_{lk}^j, \tilde{\boldsymbol{u}}_{lk}^j)} \right\}$$
$$\times \left[(d_{lk}+1)^{\alpha_j} - d_{lk}^{\alpha_j} \right] - \sum_{p=1}^{n} \left[\frac{\partial q_{jp}^k}{\partial u_\varsigma^\pi} x_p^{k-1} + q_{jp}^k \frac{\partial x_p^{k-1}}{\partial u_\varsigma^\pi} \right], \tag{9.36}$$

where q_{jp}^k *and* $\delta_{\iota\jmath}$ *are as defined in (9.19) and (9.26), respectively, and* $\dfrac{\partial q_{jp}^k}{\partial u_\varsigma^\pi}$ *is as given in Theorem 9.2.*

Proof The proof follows from differentiating (9.20) and making use of the chain rule. □

Based on (9.27), (9.28) and Theorems 9.2 and 9.3, we propose Algorithm 9.1 to solve problems (9.22)–(9.24). Note that any standard gradient optimization technique, e.g., sequential quadratic programming, will automatically implement Steps 5–7 in Algorithm 9.1.

Algorithm 9.1 Gradient optimization algorithm for problem (9.22)-(9.24).

Step 0. Select the positive integer N and the initial value $\boldsymbol{U} = (\boldsymbol{u}^0, \boldsymbol{u}^1, \ldots, \boldsymbol{u}^N)$, and set $h = \frac{t_f - t_0}{N}$ and $k := 1$.
Step 1. Set $t_k = t_0 + kh$ and use $(\boldsymbol{x}^0, \boldsymbol{x}^1, \cdots, \boldsymbol{x}^{k-1})$ and $(\boldsymbol{u}^0, \boldsymbol{u}^1, \cdots, \boldsymbol{u}^k)$ to give $\boldsymbol{x}^k$ by solving system (9.23).
Step 2. For $\varsigma \in \{1, 2, \ldots, r\}$, and $\pi \in \{0, 1, \ldots, k\}$, computing $\frac{\partial \boldsymbol{Q}^k}{\partial u_\varsigma^\pi}$ and $\frac{\partial \boldsymbol{C}^k}{\partial u_\varsigma^\pi}$ by Theorems 9.2 and 9.3.
Step 3. For $\varsigma \in \{1, 2, \ldots, r\}$, and $\pi \in \{0, 1, \ldots, k\}$, solve (9.28) for $\frac{\partial \boldsymbol{x}^k}{\partial u_\varsigma^\pi}$ by using $\boldsymbol{x}^k$.
Step 4. If $k = N$, then for $i \in \{0, 1, \ldots, M\}$, $\varsigma \in \{1, 2, \ldots, r\}$, and $\pi \in \{0, 1, \ldots, N\}$, compute $\frac{\partial \tilde{g}_i(\boldsymbol{U})}{\partial u_\varsigma^\pi}$ by (9.27). Otherwise, set $k := k + 1$ and go to Step 1.
Step 5. Checking the optimality of $\boldsymbol{U}$ by the computed gradients from Step 4. If $\boldsymbol{U}$ is an optimum, then output $\boldsymbol{U}$ and stop.
Step 6. Calculate a search direction and perform a line search to give $\tilde{\boldsymbol{U}}$.
Step 7. Set $k := 1$ and $\boldsymbol{U} := \tilde{\boldsymbol{U}}$, and go to Step 1.

9.4 Numerical Examples

Three example problems are solved to test Algorithm 9.1 in this section. All computations are carried out within MATLAB environment and on a personal computer with 16.0 GB RAM and 2.90 GHz Intel Core i7-10700F CPU.

Example 9.4.1 Consider the following FOCP with Atangana-Baleanu derivatives:

$$\min \quad g_0(u) = \frac{1}{2}\int_0^1 \left[x_1^2(t) + u^2(t)\right]dt,$$

subject to the linear system

$$\begin{cases} {}^{ABC}_{\ \ 0}\mathrm{D}_t^{\alpha_1} x_1(t) = x_2(t), \\ {}^{ABC}_{\ \ 0}\mathrm{D}_t^{\alpha_2} x_2(t) = -x_2(t) + u(t), \quad t \in (0, 1], \\ \boldsymbol{x}(0) = (0, 10)^\top, \end{cases}$$

and the constraint

$$-1 \le u(t) \le 1, \quad t \in [0, 1].$$

For $\boldsymbol{\alpha} = (1.0, 1.0)^\top$, this example was solved by the Chebyshev cardinal wavelets [119] and the Boubaker hybrid function scheme [249]. Thus, we first solve it when $\boldsymbol{\alpha} = (1.0, 1.0)^\top$ for comparison. Table 9.1 lists the computed optimal costs for a variety of positive integers N. It can be seen from Table 9.1 that the computed optimal cost decreases as integer N increases. More importantly, we obtain optimal costs $g_0^* = 8.070335$ when $N = 160$ and $g_0^* = 8.069133$ when $N = 320$, which are superior to those reported in [119] and [249]. Furthermore, this example with

Table 9.1 Computational results when $\alpha = (1.0, 1.0)^\top$ for Example 9.4.1

N	40	80	160	320	[119]	[249]
g_0	8.070505	8.070448	8.070335	8.069133	8.070430	8.07059
CPU (s)	4.2607e+01	7.0074e+01	5.3892e+02	5.4675e+03	–	–

Table 9.2 Computational results when $N = 320$ for Example 9.4.1

α	$(1.0, 1.0)^\top$	$(0.9, 0.9)^\top$	$(0.8, 0.8)^\top$	$(0.7, 0.7)^\top$	$(0.6, 0.6)^\top$	$(0.5, 0.5)^\top$
g_0	8.069133	8.815717	9.657076	10.49462	11.19725	11.70935
g_0 in [119]	8.070430	8.816074	9.668431	10.49575	11.19966	11.71328
CPU (s)	5.4675e+03	1.0893e+03	1.2277e+03	6.4909e+02	1.2959e+03	4.7292e+02

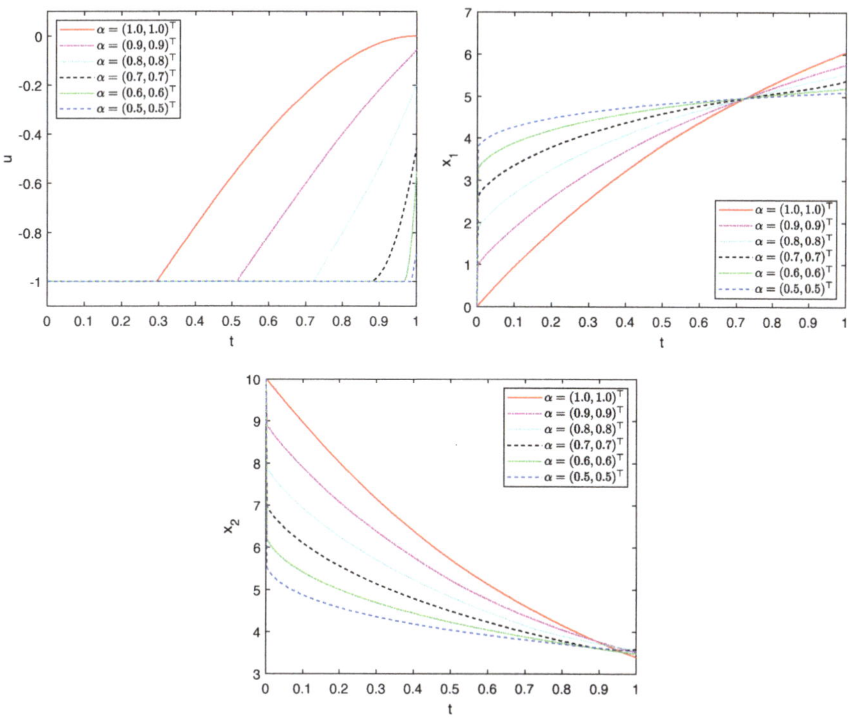

Fig. 9.1 Optimal controls and states when $N = 320$ for Example 9.4.1

various fractional orders is solved when $N = 320$. Table 9.2 shows the computed optimal costs by our proposed method. From Table 9.2, we can see that the optimal cost increases as the fractional order α decreases. In particular, all computed optimal costs are superior to those reported in [119]. Figure 9.1 depicts the behaviors of controls and states corresponding to the listed optimal costs in Table 9.2.

Example 9.4.2 Consider the following FOCP with Atangana-Baleanu derivatives:

$$\min \quad g_0(u) = \int_0^{4.5} \left[x_1^2(t) + u^2(t)\right]dt,$$

subject to the nonlinear system

$$\begin{cases} {}^{ABC}_{\ \ 0}\mathrm{D}_t^{\alpha_1} x_1(t) = x_2(t), \\ {}^{ABC}_{\ \ 0}\mathrm{D}_t^{\alpha_2} x_2(t) = -x_1(t) + x_2(t)(1.4 - 0.14x_2^2(t)) + 4u(t), \quad t \in (0, 4.5], \\ \boldsymbol{x}(0) = (-5, -5)^\top, \end{cases}$$

and the mixed control and state constraint

$$g_1(u) = -u(t) - \frac{1}{6}x_1(t) \geq 0, \quad t \in [0, 4.5]. \tag{9.37}$$

It should be noted that (9.37) is a path constraint. By using the constraint transcription technique in [303], constraint (9.37) can be approximated by

$$\tilde{g}_1(u) = \gamma + \int_0^{4.5} \psi_\epsilon(-u(t) - \frac{1}{6}x_1(t))dt \geq 0.$$

Here, ϵ and γ denote two positive parameters, and

$$\psi_\epsilon(\eta) = \begin{cases} \eta, & \text{if } \eta < -\epsilon, \\ -\dfrac{(\eta - \epsilon)^2}{4\epsilon}, & \text{if } -\epsilon \leq \eta \leq \epsilon, \\ 0, & \text{if } \eta > \epsilon. \end{cases}$$

In [167], this example when $\boldsymbol{\alpha} = (1.0, 1.0)^\top$ was solved by the exact penalty method and an optimal cost $g_0^* = 45.804838$ was obtained. For comparison, we first solve this example when $\boldsymbol{\alpha} = (1.0, 1.0)^\top$ for a variety of integers N. Table 9.3 lists the computational results by our proposed method. It can be seen from Table 9.3 that the computed optimal cost decreases as integer N increases, and all of our computed optimal costs are superior to those reported in [167]. We further solve this example when $N = 200$ for a variety of fractional orders. Table 9.4 shows the computational results by our proposed method. It can be seen from Table 9.4 that smaller optimal costs for fractional orders have been obtained than those for integer order $\boldsymbol{\alpha} = (1.0, 1.0)^\top$. Figure 9.2 plots the behaviors of controls, states, and constraint functions corresponding to the computed optimal costs listed in Table 9.4. From Fig. 9.2, we can see that the constraint (9.37) is always satisfied on the time horizon.

Example 9.4.3 Consider the following FOCP with Atangana-Baleanu derivatives in continuous stirred tank reactor [192]:

Table 9.3 Computational results when $\alpha = (1.0, 1.0)^{\top}$ for Example 9.4.2

N	50	100	150	200	[167]
g_0	44.808857	44.804411	44.803991	44.777272	45.804838

Table 9.4 Computational results when $N = 200$ for Example 9.4.2

α	$(1.0, 1.0)^{\top}$	$(0.95, 0.95)^{\top}$	$(0.92, 0.93)^{\top}$	$(0.9, 0.9)^{\top}$
g_0	44.777272	37.036915	36.975642	34.255879

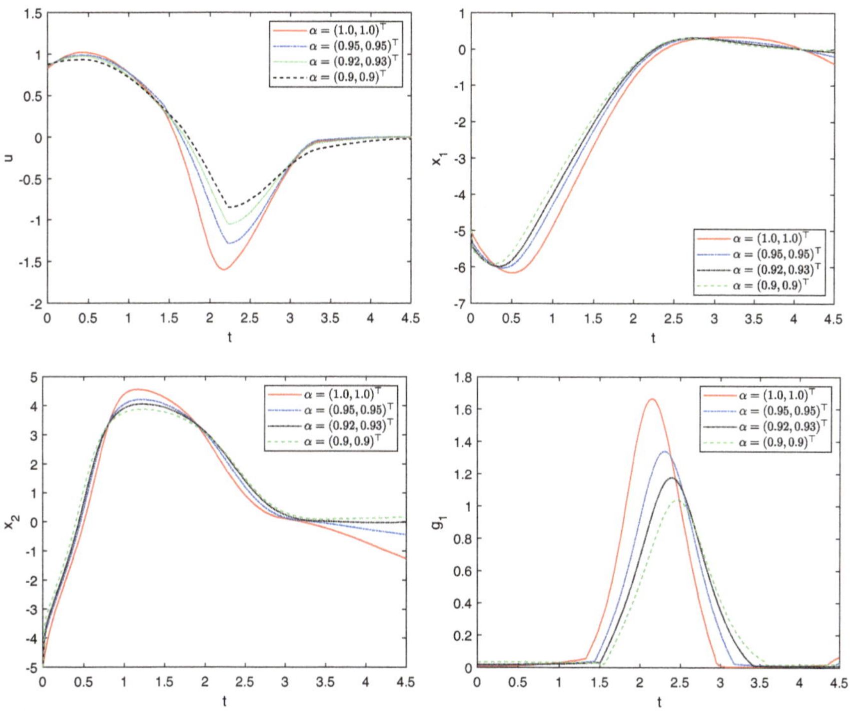

Fig. 9.2 Optimal controls, states, and constraints when $N = 200$ for Example 9.4.2

$$\min \quad g_0(u) = \int_0^{0.78} \left[x_1^2(t) + x_2^2(t) + 0.1u^2(t)\right]dt,$$

subject to the nonlinear system

Table 9.5 Computational results when $\boldsymbol{\alpha} = (1.0, 1.0)^\top$ for Example 9.4.3

N	100	200	300	400	[192]
g_0	0.133206	0.133125	0.133109	0.133095	0.133094
CPU (s)	2.5629e+01	1.6975e+02	5.6533e+02	1.2382e+03	–

Table 9.6 Computational results when $N = 320$ for Example 9.4.3

$\boldsymbol{\alpha}$	$(1.0, 1.0)^\top$	$(0.99, 0.99)^\top$	$(0.98, 0.98)^\top$	$(0.97, 0.97)^\top$
g_0	0.133102	0.122543	0.112973	0.104366
CPU (s)	6.9545e+02	8.5951e+02	9.5569e+02	1.3234e+03

$$\begin{cases} {}^{ABC}_{\ 0}\mathrm{D}_t^{\alpha_1} x_1(t) = -(1+u(t))(x_1(t)+0.25) + (x_2(t)+0.5)\exp\left(\dfrac{25x_1(t)}{x_1(t)+2}\right) \\ \qquad -(x_1(t)+0.25), \\ {}^{ABC}_{\ 0}\mathrm{D}_t^{\alpha_2} x_2(t) = 0.5 - (x_2(t)+0.5)\exp\left(\dfrac{25x_1(t)}{x_1(t)+2}\right) - x_2(t), \quad t \in (0, 0.78], \\ \boldsymbol{x}(0) = (0.09, 0.09)^\top, \end{cases}$$

where $\boldsymbol{x}(t) = (x_1(t), x_2(t))^\top$; $x_1(t)$ is deviation from the dimensionless steady-state temperature; $x_2(t)$ represents deviation from dimensionless concentration; and $u(t)$ denotes the effect of coolant flow on the chemical reaction.

As shown in [192], the global minimum of this example when $\boldsymbol{\alpha} = (1.0, 1.0)^\top$ is $g_0^* = 0.133094$, which was obtained by using Pontryagin's maximum principle. We first solve this example when $\boldsymbol{\alpha} = (1.0, 1.0)^\top$ for a variety of integers N. Table 9.5 lists the computational results by our proposed method. From Table 9.5, we can see that the obtained optimal cost decreases as integer N increases, and when $N = 400$, the optimal cost computed by our proposed method is comparable to the global minimum in [192]. Furthermore, we solve this example when $N = 320$ for a variety of fractional orders. Table 9.6 presents the computational results by our proposed method. It can be seen from Table 9.6 that smaller optimal costs for fractional orders are computed than the one for integer order. Figure 9.3 illustrates the behaviors of controls and states for the optimal costs listed in Table 9.6.

9.5 Conclusions

In this chapter, we have studied numerical solution of fractional optimal control problems with Atangana-Baleanu-Caputo derivatives. We have proposed the discretization scheme to convert the fractional optimal control problem into a sequence of discretized problems. The gradient optimization technique has been presented to

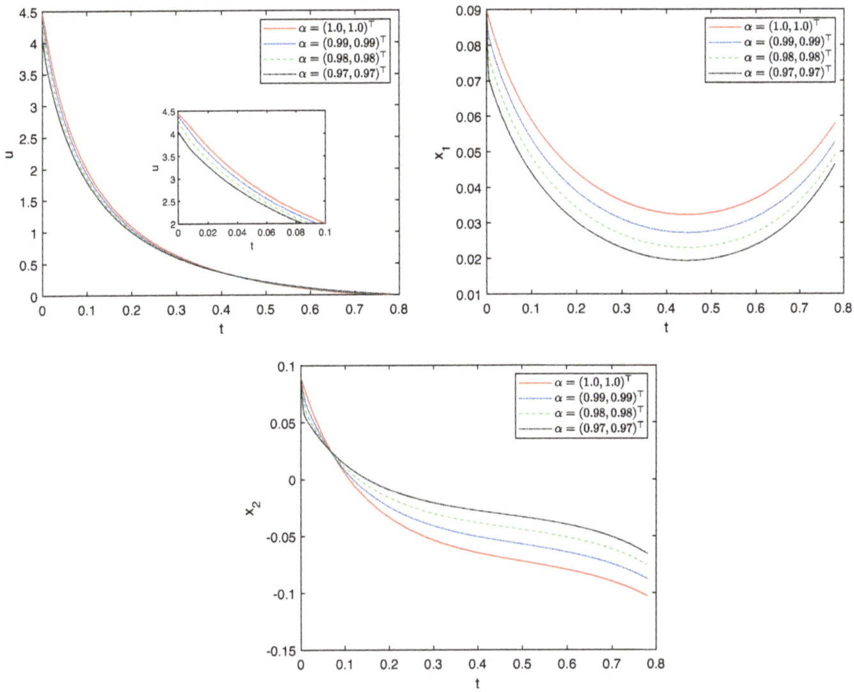

Fig. 9.3 Optimal controls and states when $N = 320$ for Example 9.4.3

solve the resulting problem. Numerical results obtained by solving three non-trivial examples have been provided to verify the applicability and effectiveness of the proposed algorithm.

Chapter 10
Fractional Optimal Control Problems with Conformable Derivatives

10.1 Introduction

In this chapter, we consider numerical solution of fractional optimal control problems (FOCPs) with conformable derivatives. There exist several classical definitions of fractional-order derivatives including Caputo, Riemann-Liouville, and Atangana-Baleanu derivatives. However, some challenges arise with these fractional derivatives in practical applications because they generally do not obey product rule, chain rule, quotient rule, etc. Thus, the conformable fractional derivative is proposed in [147], which follows almost all important properties like the integer-order derivative. It has been found a variety of applications in different fields such as Newton mechanics [62], epidemic model [115], and prey-predator system [143].

FOCPs can be defined in accordance with diverse fractional-order derivatives. In the past decades, FOCPs have attracted extensive research attention from both the theoretical and the computational perspectives [23, 67, 229, 311]. However, theory and solution techniques for FOCPs with conformable derivatives are scarce in the literature. In [163], necessary conditions for optimality are discussed. In [60], an analytical solution approach based on differential transform is developed. In [61], a close-form solution technique is presented. However, these analytical techniques have two limitations: (i) they are only applicable to FOCPs governed by fractional-order linear systems in the conformable sense; and (ii) they can not be used to solve FOCPs with general constraints. In next section, we present a class of nonlinear FOCPs with the conformable fractional derivatives and subject to the equality and inequality constraints. In Sect. 10.3, we develop a numerical solution method for the FOCPs. In Sect. 10.3.1, we propose a discretization technique for the governed nonlinear fractional systems as well as the functions of cost and constraint. This gives rise to a discretized approximate problem. In Sect. 10.3.2, we derive the gradient formulas of the cost and constraint functions in regard to the discretized controls. On this basis, we develop a numerical algorithm to solve the resulting optimization problem. In Sect. 10.4, numerical results of three non-trivial examples are provided

C. Liu et al., *Nonlinear Fractional Optimal Control*, Springer Optimization and Its Applications 237, https://doi.org/10.1007/978-981-92-0274-4_10

to demonstrate the applicability and effectiveness of the proposed algorithm. Finally, we provide conclusions of this chapter in Sect. 10.5.

The main reference of this chapter is [104].

10.2 Problem Formulation

Consider the following control system with the conformable fractional derivatives:

$$\frac{d_{t_0}^{\boldsymbol{\alpha}}}{dt_{t_0}^{\boldsymbol{\alpha}}}\boldsymbol{x}(t) = \boldsymbol{f}(t, \boldsymbol{x}(t), \boldsymbol{u}(t)), \quad t \in (t_0, t_f], \tag{10.1}$$

$$\boldsymbol{x}(t_0) = \boldsymbol{x}^0, \tag{10.2}$$

where $\boldsymbol{\alpha} \in (0, 1]^n$, $\boldsymbol{x}(t) \in \mathbb{R}^n$ and $\boldsymbol{u}(t) \in \mathbb{R}^m$ are the vectors of fractional orders, states and controls, respectively; $t_f \in \mathbb{R}$ denotes the final time; $\boldsymbol{x}^0 \in \mathbb{R}^n$ represents the initial state; $\boldsymbol{f} : \mathbb{R} \times \mathbb{R}^n \times \mathbb{R}^m \to \mathbb{R}^n$ indicates a twice continuously differentiable function; and

$$\frac{d_{t_0}^{\boldsymbol{\alpha}}}{dt_{t_0}^{\boldsymbol{\alpha}}}\boldsymbol{x}(t) = \left(\frac{d_{t_0}^{\alpha_1}}{dt_{t_0}^{\alpha_1}}x_1(t), \ldots, \frac{d_{t_0}^{\alpha_1}}{dt_{t_0}^{\alpha_n}}x_n(t)\right)^{\top}$$

with $\frac{d_{t_0}^{\alpha_i}}{dt_{t_0}^{\alpha_i}}x_i(t)$ denoting the conformable fractional derivative of $x_i(t)$.

Define

$$W = \{\boldsymbol{w} \in \mathbb{R}^m \mid a_j \le w_j \le b_j, \ j = 1, 2, \ldots, m\},$$

where a_j and b_j are real numbers satisfying $a_j \le b_j$. Any Borel measurable function $\boldsymbol{u} : [t_0, t_f] \to \mathbb{R}^m$ such that $\boldsymbol{u}(t) \in W$ for almost all $t \in [t_0, t_f]$ is called an admissible control. Let $\mathcal{U}$ be the class of all such admissible controls. Then, system (10.1) with (10.2) has a continuous solution, denoted by $\boldsymbol{x}(\cdot|\boldsymbol{u})$, for each $\boldsymbol{u} \in \mathcal{U}$. Furthermore, the following constraints of equality and inequality are imposed on the solution $\boldsymbol{x}(\cdot|\boldsymbol{u})$:

$$g_k(\boldsymbol{u}) = \Phi_k(\boldsymbol{x}(t_f|\boldsymbol{u}), \boldsymbol{u}(t_f)) + \mathrm{I}_{t_0}^{\gamma_k} L_k(t_f, \boldsymbol{x}(t_f|\boldsymbol{u}), \boldsymbol{u}(t_f)) = 0, \ k \in \mathcal{E}, \tag{10.3}$$

$$g_k(\boldsymbol{u}) = \Phi_k(\boldsymbol{x}(t_f|\boldsymbol{u}), \boldsymbol{u}(t_f)) + \mathrm{I}_{t_0}^{\gamma_k} L_k(t_f, \boldsymbol{x}(t_f|\boldsymbol{u}), \boldsymbol{u}(t_f)) \le 0, \ k \in \mathcal{I}. \tag{10.4}$$

Here, $\gamma_k \in (0, 1]$; $\mathrm{I}_{t_0}^{\gamma_k}$ is the conformable fractional integral of order γ_k; $\mathcal{E}$ and $\mathcal{I}$ are, respectively, the index sets of the equality and inequality constraints; $\Phi_k : \mathbb{R}^n \times \mathbb{R}^m \to \mathbb{R}$, $k \in \mathcal{E} \cup \mathcal{I}$, is a continuously differentiable function; and $L_k : \mathbb{R} \times \mathbb{R}^n \times \mathbb{R}^m \to \mathbb{R}$, $k \in \mathcal{E} \cup \mathcal{I}$, indicates a twice continuously differentiable function. It should be noted that both the terminal term and the conformable fractional integral term appear in the constraint functions (10.3) and (10.4).

Thus, we state the following FOCP in the conformable sense:

Problem (P). *Choose an admissible control* $\boldsymbol{u} \in \mathcal{U}$ *such that the cost function*

$$g_0(\boldsymbol{u}) = \Phi_0(\boldsymbol{x}(t_f|\boldsymbol{u}), \boldsymbol{u}(t_f)) + \mathrm{I}_{t_0}^{\gamma_0} L_0(t_f, \boldsymbol{x}(t_f|\boldsymbol{u}), \boldsymbol{u}(t_f))$$

is minimized subject to the equality and inequality constraints (10.3) *and* (10.4)*, where* $\gamma_0 \in (0, 1]$ *denotes the fractional order;* $\Phi_0 : \mathbb{R}^n \times \mathbb{R}^m \to \mathbb{R}$ *is a continuously differentiable function; and* $L_0 : \mathbb{R} \times \mathbb{R}^n \times \mathbb{R}^m \to \mathbb{R}$ *indicates a twice continuously differentiable function.*

10.3 Numerical Solution Method

10.3.1 Discretization Scheme

Let time interval $(t_0, t_f]$ be partitioned into N equal parts with the mesh points $t_l = t_0 + lh, l = 1, \ldots, N$. Here, N denotes a positive integer; and $h = \dfrac{t_f - t_0}{N}$. Then, it follows from Theorem 2.17 that system (10.1) with (10.2) can be transformed to the following equivalent integral equations:

$$x_i^l = x_i^0 + \sum_{q=1}^{l} \int_{t_{q-1}}^{t_q} (s - t_0)^{\alpha_i - 1} f_i(s, \boldsymbol{x}(s), \boldsymbol{u}(s))ds \tag{10.5}$$

for $i = 1, \ldots, n$, $l = 1, \ldots, N$, where $\boldsymbol{x}^l = \boldsymbol{x}(t_l)$. Furthermore, $f_i(s, \boldsymbol{x}(s), \boldsymbol{u}(s))$, $s \in (t_{q-1}, t_q)$, in (10.5) can be expanded as

$$f_i(s, \boldsymbol{x}(s), \boldsymbol{u}(s)) \approx f_i(s_q^i, \boldsymbol{x}_q^i, \boldsymbol{u}_q^i) + \Delta_q^i (s - s_q^i) + \Lambda_q^i (s - s_q^i)^2, \tag{10.6}$$

where $s_q^i \in (t_{q-1}, t_q)$; $\boldsymbol{x}_q^i = \boldsymbol{x}(s_q^i)$; $\boldsymbol{u}_q^i = \boldsymbol{u}(s_q^i)$; and Δ_q^i and Λ_q^i are coefficients in the Taylor expansion. Substituting (10.6) into (10.5) gives

$$\begin{aligned} x_i^l = x_i^0 &+ \tilde{h}^i \sum_{q=1}^{l} f_i(s_q^i, \boldsymbol{x}_q^i, \boldsymbol{u}_q^i)\left[q^{\alpha_i} - (q-1)^{\alpha_i}\right] \\ &+ \sum_{q=1}^{l} \Delta_q^i \int_{t_{q-1}}^{t_q} (s - t_0)^{\alpha_i - 1}(s - s_q^i)ds + R_i^l \end{aligned} \tag{10.7}$$

for $i = 1, \ldots, n$, and $l = 1, \ldots, N$, where $\tilde{h}^i = \dfrac{h^{\alpha_i}}{\alpha_i}$; and $R_i^l = \sum_{q=1}^{l} \Lambda_q^i \int_{t_{q-1}}^{t_q} (s - t_0)^{\alpha_i - 1}(s - s_q^i)^2 ds$. In order to ensure the truncation error of (10.7) being R_i^l, we choose the point s_q^i in such a way that the third term at the right-hand side of (10.7) is equal to zero. This choice yields

$$s_q^i = t_{q-1} + \frac{(\alpha_i + 1)q^{\alpha_i} - q^{\alpha_i+1} + (q-1)^{\alpha_i+1}}{(\alpha_i+1)\big[q^{\alpha_i} - (q-1)^{\alpha_i}\big]}h. \tag{10.8}$$

It should be noted that $s_q^i \in (t_{q-1}, t_q)$ and it is independent of the value of f_i. Moreover, $R_i^l = \mathcal{O}(h^2)$, whose proof is similar to that given for Theorem 3.19. Substituting (10.8) into (10.7) and then omitting the truncation error R_i^l, we have

$$x_i^l = x_i^0 + \tilde{h}^i \sum_{q=1}^{l} f_i(s_q^i, \boldsymbol{x}_q^i, \boldsymbol{u}_q^i)\big[q^{\alpha_i} - (q-1)^{\alpha_i}\big] \tag{10.9}$$

for $i = 1, \ldots, n$, and $l = 1, \ldots, N$, where s_q^i is as given in (10.8). Nevertheless, $\boldsymbol{x}_q^i$ and $\boldsymbol{u}_q^i$ in (10.9) are not available directly. Thus, we respectively approximate $\boldsymbol{x}_q^i$ and $\boldsymbol{u}_q^i$ by $\tilde{\boldsymbol{x}}_q^i$ and $\tilde{\boldsymbol{u}}_q^i$, which are defined as

$$\tilde{\boldsymbol{x}}_q^i = \boldsymbol{x}^{q-1} + \rho_q^i(\boldsymbol{x}^q - \boldsymbol{x}^{q-1}), \text{ and } \tilde{\boldsymbol{u}}_q^i = \boldsymbol{u}^{q-1} + \rho_q^i(\boldsymbol{u}^q - \boldsymbol{u}^{q-1}), \tag{10.10}$$

where $\rho_q^i = \dfrac{s_q^i - t_{q-1}}{h}$. Obviously, the truncation error of (10.10) is also $\mathcal{O}(h^2)$. Combining (10.9) with (10.10), we propose the following implicit scheme for solving system (10.1) with (10.2):

$$x_i^l = x_i^0 + \tilde{h}^i \sum_{q=1}^{l} f_i(s_q^i, \tilde{\boldsymbol{x}}_q^i, \tilde{\boldsymbol{u}}_q^i)\big[q^{\alpha_i} - (q-1)^{\alpha_i}\big] \tag{10.11}$$

for $i = 1, \ldots, n$, and $l = 1, \ldots, N$. However, it is known that the implicit schemes are not stable numerically in solving differential equations. Thus, to derive an explicit scheme for solving system (10.1) with (10.2), we further approximate $f_i(s_l^i, \tilde{\boldsymbol{x}}_l^i, \tilde{\boldsymbol{u}}_l^i)$ by

$$f_i(s_l^i, \tilde{\boldsymbol{x}}_l^i, \tilde{\boldsymbol{u}}_l^i) = f_i(s_l^i, \boldsymbol{x}^{l-1}, \tilde{\boldsymbol{u}}_l^i) + \left(\frac{\partial f_i}{\partial \boldsymbol{x}}\right)^{\top}\Bigg|_{(s_l^i, \boldsymbol{x}^{l-1}, \tilde{\boldsymbol{u}}_l^i)} \rho_l^i(\boldsymbol{x}^l - \boldsymbol{x}^{l-1}) + \mathcal{O}(h^2). \tag{10.12}$$

Substituting (10.12) into (10.11) gives

$$x_i^l = x_i^0 + \tilde{h}^i \sum_{q=1}^{l-1} f_i(s_q^i, \tilde{\boldsymbol{x}}_q^i, \tilde{\boldsymbol{u}}_q^i)\big[q^{\alpha_i} - (q-1)^{\alpha_i}\big] + \tilde{h}^i \Bigg[f_i(s_l^i, \boldsymbol{x}^{l-1}, \tilde{\boldsymbol{u}}_l^i)$$
$$+ \left(\frac{\partial f_i}{\partial \boldsymbol{x}}\right)^{\top}\Bigg|_{(s_l^i, \boldsymbol{x}^{l-1}, \tilde{\boldsymbol{u}}_l^i)} \rho_l^i (\boldsymbol{x}^l - \boldsymbol{x}^{l-1})\Bigg]\big[l^{\alpha_i} - (l-1)^{\alpha_i}\big] \tag{10.13}$$

for $i = 1, \ldots, n$, and $l = 1, \ldots, N$. We further reorganize (10.13) as the following explicit scheme:

$$(\boldsymbol{I} - \boldsymbol{D}^l(\boldsymbol{x}^{l-1}, \boldsymbol{u}^{l-1}, \boldsymbol{u}^l))\boldsymbol{x}^l = \boldsymbol{C}^l(\boldsymbol{x}^0, \ldots, \boldsymbol{x}^{l-1}, \boldsymbol{u}^0, \ldots, \boldsymbol{u}^l),\ l = 1, \ldots, N. \tag{10.14}$$

Here, $\boldsymbol{I}$ denotes the identity matrix; $\boldsymbol{D}^l = [d_{ip}^l]$ represents a matrix whose elements are defined by

$$d_{ip}^l = \tilde{h}^i \rho_l^i \frac{\partial f_i}{\partial x_p}\Bigg|_{(s_l^i, \boldsymbol{x}^{l-1}, \tilde{\boldsymbol{u}}_l^i)} \big[l^{\alpha_i} - (l-1)^{\alpha_i}\big]; \tag{10.15}$$

and $\boldsymbol{C}^l = (c_1^l, \ldots, c_n^l)^{\top}$ indicates a vector with components defined by

$$c_i^l = x_i^0 + \tilde{h}^i \sum_{q=1}^{l-1} f_i(s_q^i, \tilde{\boldsymbol{x}}_q^i, \tilde{\boldsymbol{u}}_q^i)\big[q^{\alpha_i} - (q-1)^{\alpha_i}\big]$$
$$+ \tilde{h}^i f_i(s_l^i, \boldsymbol{x}^{l-1}, \tilde{\boldsymbol{u}}_l^i)\big[l^{\alpha_i} - (l-1)^{\alpha_i}\big] - \sum_{p=1}^{n} d_{ip}^l x_p^{l-1}. \tag{10.16}$$

Hence, the solution of system (10.1) with (10.2) can be computed by solving the discretized system (10.14) in conjunction with $\boldsymbol{x}(t_0) = \boldsymbol{x}^0$. The following theorem, which can be proved by using a similar proof as given for Theorem 3.20, states the error of the explicit scheme (10.14).

Theorem 10.1 *Let $\boldsymbol{x}(t)$ be the exact solution of (10.1) with (10.2). Then, there exists an $\hat{h}$ such that $h < \hat{h}$ implies*

$$\|\boldsymbol{x}(t_l) - \boldsymbol{x}^l\|_\infty \le Ch^2, \quad l = 1, \ldots, N,$$

where $\|\cdot\|_\infty$ is the infinity norm; $\{\boldsymbol{x}^l\}$ is a sequence generated by (10.14); and $C > 0$ is a constant independent of h and $\boldsymbol{\alpha}$.

Recall that the conformal fractional integrals appear in the functions of cost and constraint. Then, for each $k \in \{0\} \cup \mathcal{E} \cup \mathcal{I}$,

$$\mathrm{I}_{t_0}^{\gamma_k} L_k(t_f, \boldsymbol{x}(t_f|\boldsymbol{u}), \boldsymbol{u}(t_f)) = \sum_{l=1}^{N} \int_{t_{l-1}}^{t_l} (\tau - t_0)^{\gamma_k - 1} L_k(\tau, \boldsymbol{x}(\tau|\boldsymbol{u}), \boldsymbol{u}(\tau)) d\tau. \tag{10.17}$$

By using similar derivations as for (10.6)–(10.11), we can approximate the conformable fractional integral (10.17) by

$$\mathrm{I}_{t_0}^{\gamma_k} L_k(t_f, \boldsymbol{x}(t_f|\boldsymbol{u}), \boldsymbol{u}(t_f)) = \hat{h}^k \sum_{l=1}^{N} L_k(\tau_l^k, \hat{\boldsymbol{x}}_l^k, \hat{\boldsymbol{u}}_l^k)\big[l^{\gamma_k} - (l-1)^{\gamma_k}\big] + \mathcal{O}(h^2),$$

where $\hat{h}^k = \dfrac{h^{\gamma_k}}{\gamma_k}$; $\tau_l^k = t_{l-1} + \dfrac{(\gamma_k+1)l^{\gamma_k} - l^{\gamma_k+1} + (l-1)^{\gamma_k+1}}{(\gamma_k+1)\big[l^{\gamma_k} - (l-1)^{\gamma_k}\big]}h$; $\hat{\boldsymbol{x}}_l^k = \boldsymbol{x}^{l-1} + \hat{\rho}_l^k(\boldsymbol{x}^l - \boldsymbol{x}^{l-1})$; $\hat{\boldsymbol{u}}_l^k = \boldsymbol{u}^{l-1} + \hat{\rho}_l^k(\boldsymbol{u}^l - \boldsymbol{u}^{l-1})$; and $\hat{\rho}_l^k = \dfrac{\tau_l^k - t_{l-1}}{h}$.

Then, we propose the following discretized approximate problem for Problem (P):

Problem (P(h)). *Find a discretized control* $\boldsymbol{U} = (\boldsymbol{u}^0, \boldsymbol{u}^1, \ldots, \boldsymbol{u}^N) \in W^{N+1}$ *to minimize the cost function*

$$\hat{g}_0(\boldsymbol{U}) = \Phi_0(\boldsymbol{x}^N, \boldsymbol{u}^N) + \hat{h}^0 \sum_{l=1}^{N} L_0(\tau_l^0, \hat{\boldsymbol{x}}_l^0, \hat{\boldsymbol{u}}_l^0)\big[l^{\gamma_0} - (l-1)^{\gamma_0}\big], \tag{10.18}$$

subject to the discretized system

$$\begin{cases} (\boldsymbol{I} - \boldsymbol{D}^l(\boldsymbol{x}^{l-1}, \boldsymbol{u}^{l-1}, \boldsymbol{u}^l))\boldsymbol{x}^l = \boldsymbol{C}^l(\boldsymbol{x}^0, \ldots, \boldsymbol{x}^{l-1}, \boldsymbol{u}^0, \ldots, \boldsymbol{u}^l), \quad l = 1, \ldots, N, \\ \boldsymbol{x}(t_0) = \boldsymbol{x}^0, \end{cases}$$

and the equality and inequality constraints

$$\hat{g}_k(\boldsymbol{U}) = \Phi_k(\boldsymbol{x}^N, \boldsymbol{u}^N) + \hat{h}^k \sum_{l=1}^{N} L_k(\tau_l^k, \hat{\boldsymbol{x}}_l^k, \hat{\boldsymbol{u}}_l^k)\big[l^{\gamma_k} - (l-1)^{\gamma_k}\big] = 0, \ k \in \mathcal{E}, \tag{10.19}$$

$$\hat{g}_k(\boldsymbol{U}) = \Phi_k(\boldsymbol{x}^N, \boldsymbol{u}^N) + \hat{h}^k \sum_{l=1}^{N} L_k(\tau_l^k, \hat{\boldsymbol{x}}_l^k, \hat{\boldsymbol{u}}_l^k)\big[l^{\gamma_k} - (l-1)^{\gamma_k}\big] \le 0, \ k \in \mathcal{I}. \tag{10.20}$$

10.3.2 Optimization Algorithm

The required gradients of cost and constraint functions for solving Problem (P(h)) will be derived in this section.

For each $k \in \{0\} \cup \mathcal{E} \cup \mathcal{I}$, we have

$$\begin{aligned}
\frac{\partial \hat{g}_k(\boldsymbol{U})}{\partial u_j^{\pi}} = & \left(\frac{\partial \Phi_k}{\partial \boldsymbol{x}}\right)^{\top}\Bigg|_{(\boldsymbol{x}^N,\boldsymbol{u}^N)} \frac{\partial \boldsymbol{x}^N}{\partial u_j^{\pi}} + \delta_{\pi N} \frac{\partial \Phi_k}{\partial u_j^{\pi}}\Bigg|_{(\boldsymbol{x}^N,\boldsymbol{u}^N)} \\
& + \hat{h}^k \sum_{\substack{l=\delta_{\pi 0}+ \\ (1-\delta_{\pi 0})\pi}}^{N-1} \left\{ \hat{\rho}_l^k \left(\frac{\partial L_k}{\partial \boldsymbol{x}}\right)^{\top}\Bigg|_{(\tau_l^k,\hat{\boldsymbol{x}}_l^k,\hat{\boldsymbol{u}}_l^k)} \left[l^{\gamma_k} - (l-1)^{\gamma_k}\right] \right. \\
& \left. + (1-\hat{\rho}_{l+1}^k)\left(\frac{\partial L_k}{\partial \boldsymbol{x}}\right)^{\top}\Bigg|_{(\tau_{l+1}^k,\hat{\boldsymbol{x}}_{l+1}^k,\hat{\boldsymbol{u}}_{l+1}^k)} \left[(l+1)^{\gamma_k} - l^{\gamma_k}\right] \right\} \frac{\partial \boldsymbol{x}^l}{\partial u_j^{\pi}} \\
& + \hat{h}^k \hat{\rho}_N^k \left(\frac{\partial L_k}{\partial \boldsymbol{x}}\right)^{\top}\Bigg|_{(\tau_N^k,\hat{\boldsymbol{x}}_N^k,\hat{\boldsymbol{u}}_N^k)} \frac{\partial \boldsymbol{x}^N}{\partial u_j^{\pi}} \left[N^{\gamma_k} - (N-1)^{\gamma_k}\right] \\
& + (1-\delta_{\pi 0}) \hat{h}^k \hat{\rho}_{\pi}^k \frac{\partial L_k}{\partial u_j}\Bigg|_{(\tau_{\pi}^k,\hat{\boldsymbol{x}}_{\pi}^k,\hat{\boldsymbol{u}}_{\pi}^k)} \left[\pi^{\gamma_k} - (\pi-1)^{\gamma_k}\right] \\
& + (1-\delta_{\pi N}) \hat{h}^k (1-\hat{\rho}_{\pi+1}^k) \frac{\partial L_k}{\partial u_j}\Bigg|_{(\tau_{\pi+1}^k,\hat{\boldsymbol{x}}_{\pi+1}^k,\hat{\boldsymbol{u}}_{\pi+1}^k)} \left[(\pi+1)^{\gamma_k} - \pi^{\gamma_k}\right], \\
& \qquad\qquad j = 1, \ldots, m, \ \pi = 1, \ldots, N. \qquad (10.21)
\end{aligned}$$

Here, for given $\imath$ and $\jmath$, $\delta_{\imath\jmath}$ is defined as

$$\delta_{\imath\jmath} = \begin{cases} 1, & \text{if } \imath = \jmath, \\ 0, & \text{otherwise.} \end{cases} \qquad (10.22)$$

In (10.21), we need to evaluate $\dfrac{\partial \boldsymbol{x}^l}{\partial \boldsymbol{u}^{\pi}}$, $\pi \in \{0, \ldots, N\}$, $l \in \{1, \ldots, N\}$. To this end, differentiating (10.14) with respect to u_j^{π} gives

$$(\boldsymbol{I} - \boldsymbol{D}^l) \frac{\partial \boldsymbol{x}^l}{\partial u_j^{\pi}} = \frac{\partial \boldsymbol{D}^l}{\partial u_j^{\pi}} \boldsymbol{x}^l + \frac{\partial \boldsymbol{C}^l}{\partial u_j^{\pi}}, \quad l = 1, \ldots, N. \qquad (10.23)$$

Hence, to solve (10.23) for $\dfrac{\partial \boldsymbol{x}^l}{\partial \boldsymbol{u}^{\pi}}$, we need to find the values of $\dfrac{\partial \boldsymbol{D}^l}{\partial u_j^{\pi}}$ and $\dfrac{\partial \boldsymbol{C}^l}{\partial u_j^{\pi}}$. The following theorems present the expressions of $\dfrac{\partial \boldsymbol{D}^l}{\partial u_j^{\pi}}$ and $\dfrac{\partial \boldsymbol{C}^l}{\partial u_j^{\pi}}$.

Theorem 10.2 *Let $l \in \{1, \ldots, N\}$, i and $p \in \{1, \ldots, n\}$, $j \in \{1, \ldots, m\}$ and $\pi \in \{0, \ldots, N\}$.*
(i) *If $\pi < l - 1$, then*

$$\frac{\partial d_{ip}^l}{\partial u_j^\pi} = \tilde{h}^i \rho_l^i \Bigg(\sum_{\varpi=1}^{n} \frac{\partial^2 f_i}{\partial x_p \partial x_\varpi}\bigg|_{(s_l^i, \boldsymbol{x}^{l-1}, \tilde{\boldsymbol{u}}_l^i)} \frac{\partial \boldsymbol{x}_\varpi^{l-1}}{\partial u_j^\pi} \Bigg) \big[l^{\alpha_i} - (l-1)^{\alpha_i} \big]. \tag{10.24}$$

(ii) *If* $\pi = l - 1$, *then*

$$\begin{aligned} \frac{\partial d_{ip}^l}{\partial u_j^\pi} = \tilde{h}^i \rho_l^i \Bigg(& \sum_{\varpi=1}^{n} \frac{\partial^2 f_i}{\partial x_p \partial x_\varpi}\bigg|_{(s_l^i, \boldsymbol{x}^{l-1}, \tilde{\boldsymbol{u}}_l^i)} \frac{\partial x_\varpi^{l-1}}{\partial u_j^\pi} \\ & + (1-\rho_l^i) \frac{\partial^2 f_i}{\partial x_p \partial u_j}\bigg|_{(s_l^i, \boldsymbol{x}^{l-1}, \tilde{\boldsymbol{u}}_l^i)} \Bigg) \big[l^{\alpha_i} - (l-1)^{\alpha_i} \big]. \end{aligned} \tag{10.25}$$

(iii) *If* $\pi = l$, *then*

$$\frac{\partial d_{ip}^l}{\partial u_j^\pi} = \tilde{h}^i (\rho_l^i)^2 \frac{\partial^2 f_i}{\partial x_p \partial u_j}\bigg|_{(s_l^i, \boldsymbol{x}^{l-1}, \tilde{\boldsymbol{u}}_l^i)} \big[l^{\alpha_i} - (l-1)^{\alpha_i} \big]. \tag{10.26}$$

(iv) *If* $\pi > l$, *then*

$$\frac{\partial d_{ip}^l}{\partial u_j^\pi} = 0. \tag{10.27}$$

Proof Differentiating (10.15) with respect to u_j^π and using the chain rule complete the proof. □

Theorem 10.3 *Let* $l \in \{1, \ldots, N\}$, $i, p \in \{1, \ldots, n\}$, $j \in \{1, \ldots, m\}$ *and* $\pi \in \{0, \ldots, N\}$.
(i) *If* $\pi < l - 1$, *then*

$$\begin{aligned} \frac{\partial c_i^l}{\partial u_j^\pi} = \tilde{h}^i \sum_{\substack{q=\delta_{\pi 0}+ \\ (1-\delta_{\pi 0})\pi}}^{l-1} & \Bigg\{ \sum_{\varpi=1}^{n} \frac{\partial f_i}{\partial x_\varpi}\bigg|_{(s_q^i, \tilde{\boldsymbol{x}}_q^i, \tilde{\boldsymbol{u}}_q^i)} \Bigg[\rho_q^i \frac{\partial x_\varpi^q}{\partial u_j^\pi} + (1-\rho_q^i) \frac{\partial x_\varpi^{q-1}}{\partial u_j^\pi} \Bigg] \\ & + \delta_{q\pi} \rho_q^i \frac{\partial f_i}{\partial u_j}\bigg|_{(s_q^i, \tilde{\boldsymbol{x}}_q^i, \tilde{\boldsymbol{u}}_q^i)} + \delta_{q(\pi+1)} (1-\rho_q^i) \frac{\partial f_i}{\partial u_j}\bigg|_{(s_q^i, \tilde{\boldsymbol{x}}_q^i, \tilde{\boldsymbol{u}}_q^i)} \Bigg\} \\ & \times \big[q^{\alpha_i} - (q-1)^{\alpha_i} \big] + \tilde{h}^i \sum_{\varpi=1}^{n} \frac{\partial f_i}{\partial x_\varpi}\bigg|_{(s_l^i, \boldsymbol{x}^{l-1}, \tilde{\boldsymbol{u}}_l^i)} \frac{\partial x_\varpi^{l-1}}{\partial u_j^\pi} \big[l^{\alpha_i} - (l-1)^{\alpha_i} \big] \\ & - \sum_{p=1}^{n} \Bigg[\frac{\partial d_{ip}^l}{\partial u_j^\pi} x_p^{l-1} + d_{ip}^l \frac{\partial x_p^{l-1}}{\partial u_j^\pi} \Bigg], \end{aligned} \tag{10.28}$$

where δ_{ij} *is as defined in (10.22); and* $\dfrac{\partial d_{ip}^l}{\partial u_j^\pi}$ *can be computed by Theorem 10.2.*

(ii) *If* $\pi = l - 1$, *then*

$$\begin{aligned}\frac{\partial c_i^l}{\partial u_j^\pi} &= \tilde{h}^i \rho_{(l-1)l}^i \Bigg(\sum_{\varpi=1}^n \frac{\partial f_i}{\partial x_\varpi}\Bigg|_{(s_{(l-1)l}^i, \tilde{\boldsymbol{x}}_{(l-1)l}^i, \tilde{\boldsymbol{u}}_{(l-1)l}^i)} \frac{\partial x_\varpi^{l-1}}{\partial u_j^\pi} \\ &\quad + \frac{\partial f_i}{\partial u_j}\Bigg|_{(s_{(l-1)l}^i, \tilde{\boldsymbol{x}}_{(l-1)l}^i, \tilde{\boldsymbol{u}}_{(l-1)l}^i)} \Bigg) [(l-1)^{\alpha_i} - (l-2)^{\alpha_i}] \\ &\quad + \tilde{h}^i \Bigg(\sum_{\varpi=1}^n \frac{\partial f_i}{\partial x_\varpi}\Bigg|_{(s_l^i, \boldsymbol{x}^{l-1}, \tilde{\boldsymbol{u}}_l^i)} \frac{\partial x_\varpi^{l-1}}{\partial u_j^\pi} + (1 - \rho_l^i) \frac{\partial f_i}{\partial u_j}\Bigg|_{(s_l^i, \boldsymbol{x}^{l-1}, \tilde{\boldsymbol{u}}_l^i)} \Bigg) \\ &\quad \times \left[l^{\alpha_i} - (l-1)^{\alpha_i} \right] - \sum_{p=1}^n \left[\frac{\partial d_{ip}^l}{\partial u_j^\pi} x_p^{l-1} + d_{ip}^l \frac{\partial x_p^{l-1}}{\partial u_j^\pi} \right], \end{aligned} \tag{10.29}$$

where $\frac{\partial d_{ip}^l}{\partial u_j^\pi}$ *can be computed by Theorem 10.2.*

(iii) *If* $\pi = l$, *then*

$$\frac{\partial c_i^l}{\partial u_j^\pi} = \tilde{h}^i \rho_l^i \frac{\partial f_i}{\partial u_j}\Bigg|_{(s_l^i, \boldsymbol{x}^{l-1}, \tilde{\boldsymbol{u}}_l^i)} \left[l^{\alpha_i} - (l-1)^{\alpha_i} \right] - \sum_{p=1}^n \frac{\partial d_{ip}^l}{\partial u_j^\pi} x_p^{l-1}, \tag{10.30}$$

where $\frac{\partial d_{ip}^l}{\partial u_j^\pi}$ *can be computed by Theorem 10.2.*

(iv) *If* $\pi > l$, *then*

$$\frac{\partial c_i^l}{\partial u_j^\pi} = 0. \tag{10.31}$$

Proof Differentiating (10.16) with respect to u_j^π and using the chain rule complete the proof. □

Based on Theorems 10.2 and 10.3, the cost and constraint functions as well as their gradients at the discretized control $\boldsymbol{U} = (\boldsymbol{u}^0, \boldsymbol{u}^1, \ldots, \boldsymbol{u}^N) \in W^{N+1}$ can be computed by Algorithm 10.1.

10.4 Numerical Examples

We solve three examples to illustrate the effectiveness of numerical solution approach in the previous section. All computations are implemented using MATLAB on a personal computer equipped with 16.0 GB RAM and 2.9 GHz CPU.

Algorithm 10.1 Computing the gradients in Problem (P(h)).

Step 0. Select a $\boldsymbol{U} = (\boldsymbol{u}^0, \boldsymbol{u}^1, \ldots, \boldsymbol{u}^N) \in W^{N+1}$ as the initial control, and set $l := 1$ and $h = \frac{t_f - t_0}{N}$.

Step 1. Let $t_l = t_0 + lh$. Use $(\boldsymbol{x}^0, \cdots, \boldsymbol{x}^{l-1})$ and $(\boldsymbol{u}^0, \cdots, \boldsymbol{u}^l)$ to calculate $\boldsymbol{x}^l$ by solving (10.14) with $\boldsymbol{x}(t_0) = \boldsymbol{x}^0$.

Step 2. Compute $\frac{\partial \boldsymbol{D}^l}{\partial u_j^\pi}$ and $\frac{\partial \boldsymbol{C}^l}{\partial u_j^\pi}$ for $j = 1, \ldots, m$, and $\pi = 0, \ldots, N$, via (10.24)–(10.27) and (10.28)–(10.31) in Theorems 10.2 and 10.3.

Step 3. Solve (10.23) to give $\frac{\partial \boldsymbol{x}^l}{\partial u_j^\pi}$ for $j = 1, \ldots, m$, and $\pi = 0, \ldots, N$.

Step 4. If $l < N$, then set $l := l + 1$ and go to Step 1. Otherwise, for $k \in \{0\} \cup \mathcal{E} \cup \mathcal{I}$, calculate the cost and constraint functions $\hat{g}_k(\boldsymbol{U})$ via (10.18)–(10.20), and their corresponding gradients $\frac{\partial \hat{g}_k(\boldsymbol{U})}{\partial u_j^\pi}$ for $j = 1, \ldots, m$, and $\pi = 0, \ldots, N$, via (10.21).

Example 10.4.1 Consider the following FOCP with conformable derivative [61]:

$$\min \quad g_0(u) = \frac{1}{2}\, \mathrm{I}_0^{\gamma_0}(x^2(1) + u^2(1)),$$

subject to fractional-order linear system

$$\begin{cases} \dfrac{d_0^{\gamma_0}}{dt_0^{\gamma_0}} x(t) = -x(t) + u(t), \quad t \in (0, 1], \\ x(0) = 1, \end{cases}$$

and terminal constraint

$$g_1(u) = x(1) = 0.$$

This example for $\gamma_0 = 1.0$ was solved by the analytical technique and an optimal value of $g_0^* = 0.2596$ was reported in [61]. As a comparison, this problem with $\gamma_0 = 1.0$ is first solved. The optimal costs, constraints and CPU time for various of N are shown in Table 10.1. It is observed from Table 10.1 that the computed optimal costs decrease by increasing the partition number N. In particular, when $N = 400$, the obtained optimal cost approaches to the analytical optimal cost reported in [61]. It also can be seen from Table 10.1 that the CPU time increases sharply by increasing the partition number N. This is because the computational time mainly lies in the time-steps of numerically solving fractional systems and the number of decision variables in the discretized approximate problem, which are all proportional to the partition number N. Furthermore, when $N = 400$, we solve this problem with several fractional orders γ_0. The numerical results for various fractional orders are presented in Table 10.2. It is observed from Table 10.2 that as the fractional orders decrease, the obtained optimal costs decrease. More importantly, the computed optimal costs are

Table 10.1 Optimal costs, constraints and CPU time of Example 10.4.1 for $\gamma_0 = 1.0$

N	100	200	300	400	[61]
g_0	0.295941	0.295844	0.295750	0.295694	0.2956
g_1	7.28254e–19	4.21651e–16	2.92343e–17	2.15436e–16	–
CPU(s)	2.5423e+01	4.0472e+02	4.6842e+02	2.7268e+03	–

Table 10.2 Optimal costs, constraints and CPU time of Example 10.4.1 when $N = 400$

γ_0	0.5	0.7	0.85	0.95	1.0
g_0	0.211740	0.232459	0.259739	0.282998	0.295694
g_0 in [61]	0.2117	0.2321	0.2594	–	0.2956
g_1	2.92132e–17	4.23345e–18	7.31676e–17	3.37142e–17	2.15436e–16
CPU(s)	3.4506e+03	2.3885e+03	2.8316e+03	3.8891e+03	2.7268e+03

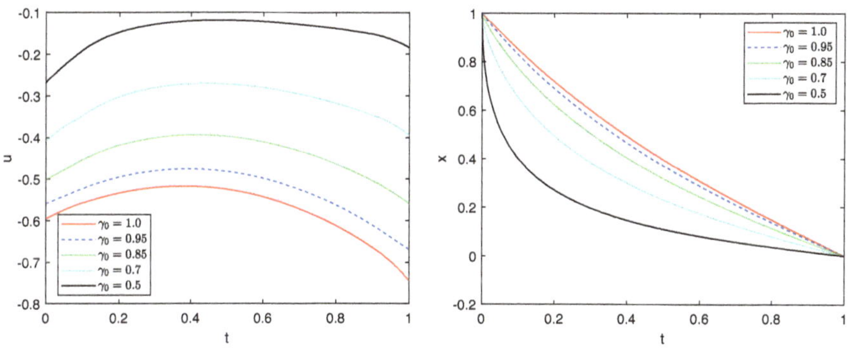

Fig. 10.1 Optimal controls and optimal states of Example 10.4.1

comparable to the optimal costs in [61]. The corresponding optimal control policies and optimal states are depicted in Fig. 10.1.

Example 10.4.2 Consider the following FOCP with conformable derivatives [194]:

$$\min \quad g_0(u) = \int_0^{0.78} \left[x_1^2(t) + x_2^2(t) + 0.1u^2(t)\right]dt,$$

subject to the fractional-order nonlinear system

Table 10.3 Optimal costs, constraints and CPU time of Example 10.4.2 with $\boldsymbol{\alpha} = (1.0, 1.0)^\top$

N	100	200	300	400	[194]
g_0	0.143038	0.143037	0.143036	0.143025	0.14490
g_1	7.53969e–10	-5.02461e–09	1.48213e–09	-2.03313e–10	-0.00371
g_2	-6.75931e–10	5.91540e–09	-1.80222e–09	2.25984e–10	-0.00368
CPU(s)	1.5561e+02	1.0157e+03	4.0742e+03	6.6193e+03	–

Table 10.4 Optimal costs, constraints and CPU time of Example 10.4.2 when $N = 320$

α	$(1.0, 1.0)^\top$	$(0.98, 0.98)^\top$	$(0.96, 0.96)^\top$	$(0.94, 0.94)^\top$
g_0	0.143033	0.133265	0.123604	0.114125
g_1	-1.00422e–09	1.67374e–09	6.95813e–09	1.54491e–09
g_2	1.28568e–09	-2.65701e–09	-8.99980e–09	-1.77454e–09
CPU(s)	2.5158e+03	3.3087e+03	3.5985e+03	5.7789e+03

$$
\begin{cases}
\dfrac{d_0^{\alpha_1}}{dt_0^{\alpha_1}} x_1(t) = -(x_1(t) + 0.25) - (1 + u(t))(x_1(t) + 0.25) \\
\qquad\qquad + (x_2(t) + 0.5) \exp\left[\dfrac{25x_1(t)}{x_1(t) + 2}\right], \quad t \in (0, 0.78], \\
\dfrac{d_0^{\alpha_2}}{dt_0^{\alpha_2}} x_2(t) = 0.5 - x_2(t) - (x_2(t) + 0.5) \exp\left[\dfrac{25x_1(t)}{x_1(t) + 2}\right], \\
\boldsymbol{x}(0) = (0.09, 0.09)^\top,
\end{cases}
$$

and the terminal constraints:

$$
g_1(u) = x_1(0.78) = 0, \quad g_2(u) = x_2(0.78) = 0.
$$

This problem with $\boldsymbol{\alpha} = (1.0, 1.0)^\top$ is first solved. The computed optimal costs, constraints and CPU time for various N are shown in Table 10.3. We can see from Table 10.3 that the obtained optimal cost decreases by increasing the partition number N. It should be noted that the computed optimal costs are all superior to the one reported in [194], which was solved by the dynamic programming together with a penalty method. Also, the terminal states are prominently closer to the origin than those reported in [194]. Then, we solve this problem with several fractional orders when $N = 320$. The obtained numerical results for various values of $\boldsymbol{\alpha}$ are displayed in Table 10.4. It is observed from Table 10.4 that as the fractional orders decrease, the computed optimal costs decrease. The corresponding optimal control policies and optimal states are plotted in Fig. 10.2.

Example 10.4.3 Finally, we consider the following optimal control of crane problem [304]:

$$
\min \quad g_0(\boldsymbol{u}) = \int_0^9 \left[x_3^2(t) + x_6^2(t)\right] dt,
$$

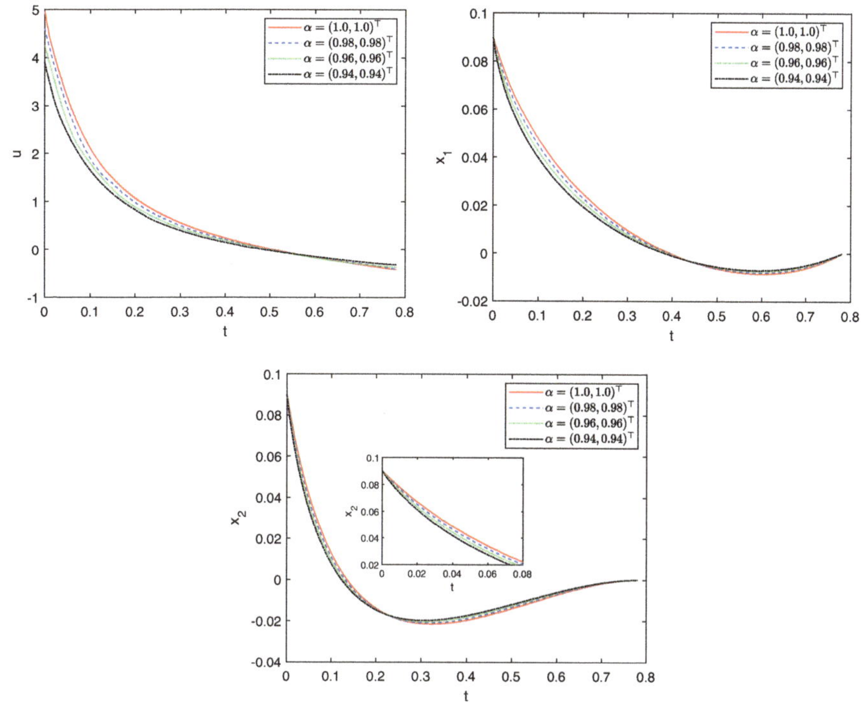

Fig. 10.2 Optimal controls and optimal states of Example 10.4.2

subject to the nonlinear fractional system

$$
\begin{cases}
\frac{d_0^{\alpha_1}}{dt_0^{\alpha_1}} x_1(t) = x_4(t), \\
\frac{d_0^{\alpha_2}}{dt_0^{\alpha_2}} x_2(t) = x_5(t), \\
\frac{d_0^{\alpha_3}}{dt_0^{\alpha_3}} x_3(t) = x_6(t), \\
\frac{d_0^{\alpha_4}}{dt_0^{\alpha_4}} x_4(t) = u_1(t) + 17.2656 x_3(t), \\
\frac{d_0^{\alpha_5}}{dt_0^{\alpha_5}} x_5(t) = u_2(t), \\
\frac{d_0^{\alpha_6}}{dt_0^{\alpha_6}} x_6(t) = -(u_1(t) + 27.0756 x_3(t) + 2 x_5(t) x_6(t))/x_2(t),
\end{cases}
\qquad t \in (0, 9],
$$

with the initial condition

$$\boldsymbol{x}(0) = (0, 22, 0, 0, -1, 0)^\top.$$

Here, x_1 and x_2 represent the horizontal and vertical directions, respectively; x_3 indicates the position of swing; x_4, x_5 and x_6 are the corresponding velocities; and u_1 and u_2 are torques of motors. Furthermore, the following state constraints are imposed:

$$\boldsymbol{x}(9) = (10, 14, 0, 2.5, 0, 0)^\top, \tag{10.32}$$

and

$$|x_4(t)| \leq 2.5, \ |x_5(t)| \leq 1.0, \quad t \in [0, 9]. \tag{10.33}$$

Also, the control function satisfies the following bound constraints:

$$|u_1(t)| \leq 2.83374, \ -0.80865 \leq u_2(t) \leq 0.71265, \quad t \in [0, 9].$$

For constraint (10.32), we define

$$\begin{aligned} g_1(\boldsymbol{u}) = &(x_1(9) - 10)^2 + (x_2(9) - 14)^2 + (x_4(9) - 2.5)^2 \\ &+ x_3^2(9) + x_5^2(9) + x_6^2(9) = 0. \end{aligned}$$

For constraint (10.33), let $h_1(\boldsymbol{x}(t)) = x_4(t) + 2.5$, $h_2(\boldsymbol{x}(t)) = 2.5 - x_4(t)$, $h_3(\boldsymbol{x}(t)) = x_5(t) + 1.0$, and $h_4(\boldsymbol{x}(t)) = 1.0 - x_5(t)$. Then, by using the constraint transcription technique in [303], we can approximate (10.33) by the following integral constraint:

$$g_2(\boldsymbol{u}) = -\gamma - \int_0^9 \sum_{k=1}^4 \mathcal{L}_\epsilon(h_k(\boldsymbol{x}(t)))dt \leq 0.$$

Here, $\epsilon > 0$ and $\gamma > 0$ are two adjustable parameters; and

$$\mathcal{L}_\epsilon(\eta) = \begin{cases} \eta, & \text{if } \eta < -\epsilon, \\ -(\eta - \epsilon)^2/4\epsilon, & \text{if } -\epsilon \leq \eta \leq \epsilon, \\ 0, & \text{if } \eta > \epsilon. \end{cases}$$

By using the control parameterization technique, this example with $\boldsymbol{\alpha} = (1.0, 1.0, 1.0, 1.0, 1.0, 1.0)^\top$ was solved in [304], where the control is parameterized by piecewise linear continuous function with 21 equal spaced knots in the time horizon. As a comparison, this problem with $\boldsymbol{\alpha} = (1.0, 1.0, 1.0, 1.0, 1.0, 1.0)^\top$ for various values of N is solved. The obtained optimal costs and constraints are shown in Table 10.5. From Table 10.5, we see that the computed optimal costs decrease as the values of N increase. In particular, the computed optimal costs are all smaller than the one reported in [304]. This problem with various values of $\boldsymbol{\alpha}$ is then solved when $N = 200$. The numerical results for various values of $\boldsymbol{\alpha}$ are listed in Table 10.6.

Table 10.5 Optimal costs and constraints of Example 10.4.3 for $\alpha = (1.0, 1.0, 1.0, 1.0, 1.0, 1.0)^\top$

N	50	100	150	200	[304]
g_0	0.010415	0.010356	0.010336	0.010319	0.01047
g_1	5.50725e–13	4.95614e–14	1.87901e–13	8.16108e–22	–

Table 10.6 Optimal costs and constraints of Example 10.4.3 with various fractional orders when $N = 200$

α	$(0.99, 0.99, 0.99, 0.99, 0.99, 0.99)^\top$	$(0.98, 0.98, 0.98, 0.98, 0.98, 0.98)^\top$
g_0	0.010541	0.010876
g_1	1.24369e–20	4.95614e–14

From Tables 10.5 and 10.6, it can be seen that the obtained optimal costs under the fractional orders are larger than the computed optimal cost in the case of the integer order. The corresponding optimal control policies and optimal states when $N = 200$ are plotted in Fig. 10.3. From Fig. 10.3, we observe that the obtained optimal states satisfy the constraints (10.32) and (10.33).

10.5 Conclusions

In this chapter, we have focused on numerical solution of a class of nonlinear optimal control problems with conformable fractional derivatives. First, we have formulated the fractional optimal control problems with general equality and inequality constraints. Then, based on the proposed numerical integration, the discretization schemes for the fractional-order system as well as for the cost and constraint functions have been presented, resulting in the discretized approximate problem. To solve the resulting problem, the gradient-based optimization approach has been developed. Finally, three examples have been solved to demonstrate the applicability and effectiveness of the developed approach.

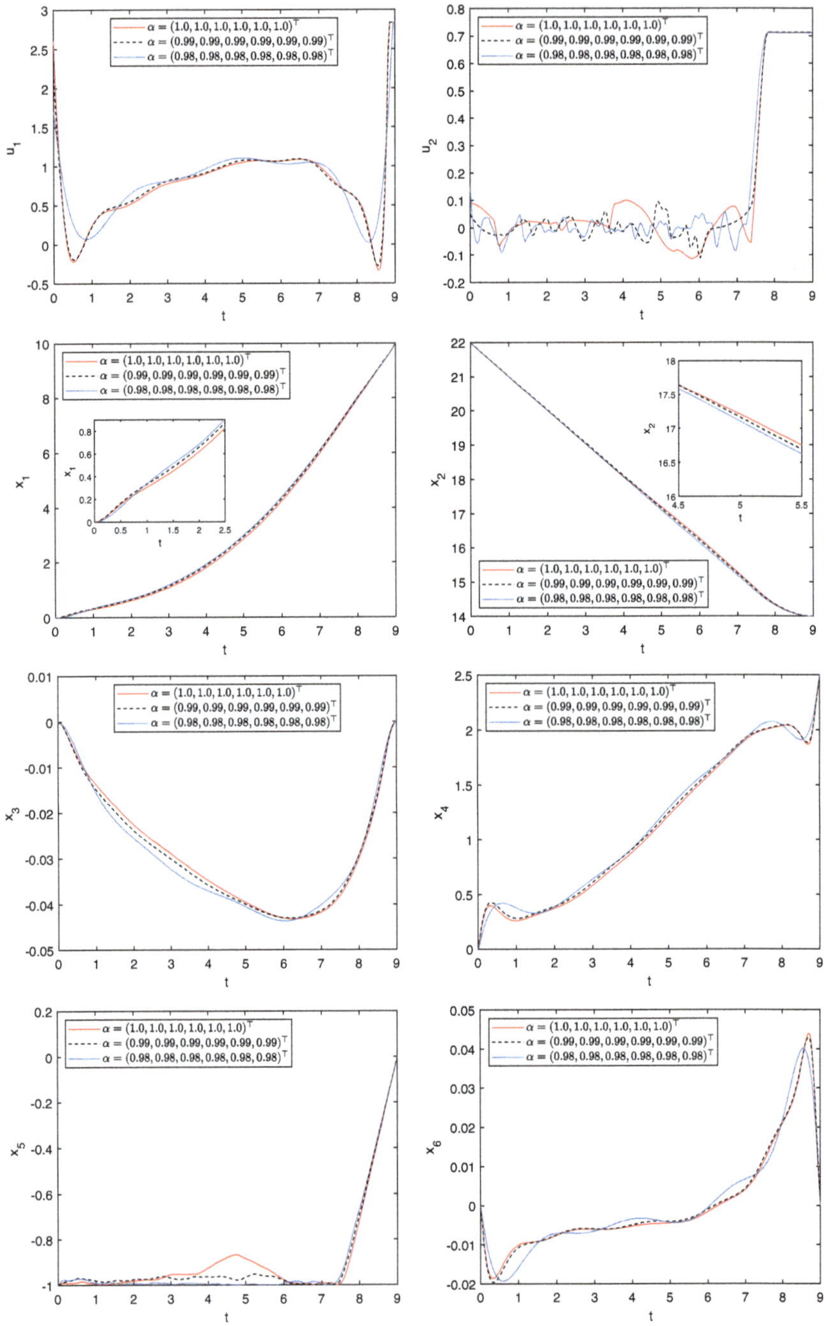

Fig. 10.3 Optimal controls and optimal states of Example 10.4.3

References

1. Abbas, S.: Existence of solution to fractional order ordinary and delay differential equations and applications. Electron. J. Differ. Eq. **2011**, 1–11 (2011)
2. Abdeljawad, T.: On conformable fractional calculus. J. Comput. Appl. Math. **279**, 57–66 (2015)
3. Abd-Elmonmem, A., Banerjee, R., Ahmad, S., Jamshed, W., Nisar, K.S., Eid, M.R., Ibrahim, R.W., El Din, S.M.: A comprehensive review on fractional-order optimal control problem and its solution. Open Math. **21**, 20230105 (2023)
4. Abramowitz, M., Stegun, I.A.: Handbook of Mathematical Functions with Formulas, Graphs, and Mathematical Tables, Series 55. Courier Corporation, Massachusetts (1964)
5. Agarwal, R.P., Baleanu, D., Nieto, J.J., Torres, D.F.M., Zhou, Y.: A survey on fuzzy fractional differential and optimal control nonlocal evolution equations. J. Comput. Appl. Math. **339**, 3–29 (2018)
6. Agrawal, O.P.: A formulation and numerical scheme for fractional optimal control problems. J. Vib. Control **14**, 1291–1299 (2008)
7. Agrawal, O.P.: A general formulation and solution scheme for fractional optimal control problems. Nonlinear Dyn. **38**, 323–337 (2004)
8. Aguila-Camacho, N., Duarte-Mermoud, M.A., Gallegos, J.A.: Lyapunov functions for fractional order systems. Commun. Nonlinear Sci. Numer. Simul. **19**, 2951–2957 (2014)
9. Ahmad, W., Nazir, M.A., Rafiq, M., Butt, A.I.K., Ahmad, N., Hussain, M.: Analyzing optimal control techniques in a nonlinear fractional Rubella model with the Atangana-Baleanu derivative. Comput. Biol. Med. **197**, 110954 (2025)
10. Akman, T., Hosseini, B., Diepolder, J., Grüter, B., Afonso, R., Gerdts, M., Holzapfel, F.: Efficient sensitivity calculation for robust optimal control. Deutscher Luft-und Raumfahrtkongress 2019, Darmstadt, **490248** 1–10 (2020)
11. Alexe, M., Sandu, A.: Forward and adjoint sensitivity analysis with continuous explicit Runge-Kutta methods. Appl. Math. Comput. **208**, 328–346 (2009)
12. Alipour, M., Rostamy, D., Baleanu, D.: Solving multi-dimensional fractional optimal control problems with inequality constraint by Bernstein polynomials operational matrices. J. Vib. Control **19**, 2523–2540 (2013)
13. Alizadeh, A., Effati, S.: An iterative approach for solving fractional optimal control problems. J. Vib. Control **24**, 18–36 (2018)
14. Amin, R., Shah, K., Asif, M., Khan, I.: A computational algorithm for the numerical solution of fractional order delay differential equations. Appl. Math. Comput. **402**, 125863 (2021)
15. Ammi, M.R.S., Torres, D.F.M.: Optimal control of a nonlocal thermistor problem with ABC fractional time derivatives. Comput. Math. Appl. **78**, 1507–1516 (2019)

C. Liu et al., *Nonlinear Fractional Optimal Control*, Springer Optimization and Its Applications 237, https://doi.org/10.1007/978-981-92-0274-4

16. Anapali, A., Öztürk, Y., Gülsu, M.: Numerical approach for solving fractional pantograph equation. Int. J. Comput. Appl. **113**, 45–52 (2015)
17. Andrés-Martínez, O., Biegler, L.T., Flores-Tlacuahuac, A.: An indirect approach for singular optimal control problems. Comput. Chem. Eng. **19**, 106923 (2020)
18. Aoki, M.: Introduction to Optimization Techniques: Fundamentals and Applications of Nonlinear Programming. Macmillan, New York (1971)
19. Atangana, A., Araz, S.İ: New Numerical Scheme with Newton Polynomial: Theory, Methods, and Applications. Elsevier, London (2021)
20. Atangana, A., Baleanu, D.: New fractional derivatives with nonlocal and non-singular kernel: theory and application to heat transfer model. Therm. Sci. **20**, 763–769 (2016)
21. Badri, V., Tavazoei, M.S.: Stability analysis of fractional order time-delay systems: constructing new Lyapunov functions from those of integer order counterparts. IET Control Theory App. **1324**, 2476–2481 (2019)
22. Bai, Z., Chen, Y., Lian, H., Sun, S.: On the existence of blow up solutions for a class of fractional differential equations. Fract. Calc. Appl. Anal. **17**, 1175–1187 (2014)
23. Baleanu, D., António, J., Machado, T., Luo, A.C.J.: Fractional Dynamics and Control. Springer, New York (2012)
24. Baleanu, D., Balas, V.E., Agarwal, P.: Fractional Order Systems and Applications in Engineering. Academic Press, London (2023)
25. Baleanu, D., Fernandez, A.: On some new properties of fractional derivatives with Mittag-Leffler kernel. Commun. Nonlinear Sci. Numer. Simul. **59**, 444–462 (2018)
26. Baleanu, D., Jajari, A., Hajipour, M.: A new formulation of the fractional optimal control problems involving Mittag-Leffler nonsingular kernel. J. Optim. Theory Appl. **175**, 718–737 (2017)
27. Balochian, S., Rajaee, N.: Fractional-order optimal control of fractional-order linear vibration systems with time delay. Int. J. Syst. Dyn. Appl. **7**, 72–93 (2018)
28. Bayour, B., Torres, D.F.M.: Existence of solution to a local fractional nonlinear differential equation. J. Comput. Appl. Math. **312**, 127–133 (2017)
29. Bellman, R.E.: Dynamic Programming. Princeton University Press, Princeton (1957)
30. Benchohra, M., Hamani, S., Ntouyas, S.: Boundary value problems for differential equations with fractional order and nonlocal conditions. Nonlinear Anal. **71**, 2391–2396 (2009)
31. Bertsekas, D.: Constrained Optimization and Lagrange Multiplier Methods. Academic, New York (1982)
32. Betts, J.T.: Practical Methods for Optimal Control and Estimation Using Nonlinear Programming. Society for Industrial and Applied Mathematics, Philadelphia (2010)
33. Bhalekar, S., Daftardar-Gejji, V.: A predictor-corrector scheme for solving nonlinear delay differential equations of fractional order. J. Fract. Calc. Appl. **1**, 1–9 (2011)
34. Bhatti, M.A.: Practical Optimization Methods. Springer, New York (2000)
35. Bhrawy, A.H., Al-Zahrani, A.A., Alhamed, Y.A., Baleanu, D.: A new generalized Laguerre-Gauss collocation scheme for numerical solution of generalized fractional pantograph equations. Rom. J. Phys. **59**, 646–657 (2014)
36. Bhrawy, A.H., Doha, E.H., Machado, J.A., Ezz-Eldien, S.S.: An efficient numerical scheme for solving multi-dimensional fractional optimal control problems with a quadratic performance index. Asian J. Control **17**, 2389–2402 (2015)
37. Bhrawy, A.H., Ezz-Eldien, S.S.: A new Legendre operational technique for delay fractional optimal control problems. Calcolo **53**, 521–543 (2016)
38. Birs, I.R., Muresan, C.I., Nascu, I., Ionescu, C.M.: A survey of recent advances in fractional order control for time delay systems. IEEE Access **7**, 30951–30965 (2019)
39. Boccia, A., Falugi, P., Maurer, H., Vinter, R.B.: Free time optimal control problems with time delays. In: Proceedings of the 52nd IEEE CDC, pp. 520–525. Florence (2013)
40. Bonnet, C., Partington, J.R.: Coprime factorizations and stability of fractional differential systems. Syst. Control Lett. **41**, 167–174 (2000)
41. Bonyah, E.: Fractional optimal control for a corruption model. J. Prime Res. Math. **16**, 11–29 (2020)

42. Bourdin, M.L.: Pontryagin maximum principle for general Caputo fractional optimal control problems with Bolza cost and terminal constraints. ESAIM: Control, Optim. Calc. **26**, 1–38 (2020)
43. Broyden, C.: The convergence of a class of double-rank minimization algorithms. J. Inst. Math. Appl. **6**, 76–90 (1970)
44. Bulirsch, R., Miele, A., Stoer, J., Well, K.H.: Optimal Control: Calculus of Variations, Optimal Control Theory and Numerical Methods. Birkhäuser Verlag, Basel (1993)
45. Cai, R.Y., Zhou, H.C., Kou, C.H.: Kalman rank criterion for the controllability of fractional impulse controlled systems. IET Control Theory Appl. **14**, 1358–1364 (2020)
46. Calderón, A.J., Vinagre, B.M., Feliu, V.: Fractional order control strategies for power electronic buck converters. Signal Process. **86**, 2803–2819 (2006)
47. Caputo, M.: Linear models of dissipation whose Q is almost frequency independent-II. Geophys. J. Int. **107**, 529–539 (1967)
48. Caputo, M.C., Torres, D.F.M.: Duality for the left and right fractional derivatives. Signal Process. **107**, 265–271 (2015)
49. Cesari, L.: An existence theorem in problems of optimal control. SIAM J. Control **3**, 7–22 (1965)
50. Cesari, L.: Existence theorems for optimal solutions in Pontryagin and Lagrange problems. SIAM J. Control **3**, 475–498 (1966)
51. Chai, Q., Wang, W.: A computational method for free terminal time optimal control problem governed by nonlinear time delayed systems. Appl. Math. Model. **53**, 242–250 (2018)
52. Chakraverty, S., Jena, R.M., Jena, S.K.: Computational Fractional Dynamical Systems. Wiley, New Jersey (2023)
53. Chatibi, Y., El Kinani, E.H., Ouhadan, A.: Variational calculus involving nonlocal fractional derivative with Mittag-Leffler kernel. Chaos Soliton. Fract. **118**, 117–121 (2019)
54. Chen, L., Hu, F., Zhu, W.: Stochastic dynamics and fractional optimal control of quasi integrable Hamiltonian systems with fractional derivative damping. Fract. Calc. Appl. Anal. **16**, 189–225 (2013)
55. Chen, T., Liu, W.: An anti-periodic boundary value problem for the fractional differential equation with a p-Laplacian operator. Appl. Math. Lett. **25**, 1671–1675 (2012)
56. Chen, X., Wang, L.: The variational iteration method for solving a neutral functional-differential equation with proportional delays. Comput. Math. Appl. **59**, 2696–2702 (2010)
57. Chen, Y., Lv, Z.: A fractional optimal control model for a simple cash balance problem. Commun. Nonlinear Sci. Numer. Simul. **120**, 107194 (2023)
58. Chen, Y., Han, X., Liu, L.: Numerical solution for a class of linear system of fractional differential equations by the Haar wavelet method and the convergence analysis. Comput. Model. Eng. Sci. **59**, 391–405 (2014)
59. Chen, Y., Wu, Y., Cui, Y., Wang, Z., Jin, D.: Wavelet method for a class of fractional convection-diffusion equation with variable coefficients. J. Comput. Sci. **1**, 146–149 (2010)
60. Chiranjeevi, T., Biswas, R.K.: Application of conformable fractional differential transform method for fractional optimal control problems. IFAC Pap. **55**, 643–648 (2022)
61. Chiranjeevi, T., Biswas, R.K.: Close-form solution of optimal control problem of a fractional order system. J. King Saud Univ. Sci. **31**, 1042–1047 (2019)
62. Chung, W.S.: Fractional Newton mechanics with conformable fractional derivative. J. Comput. Appl. Math. **290**, 150–158 (2015)
63. Clarke, F.H., Ledyaev, S., Stern, R.J.: Nonsmooth Analysis and Control Theory. Springer, New York (1998)
64. Coimbra, C.F.M.: Mechanics with variable-order operators. Ann. Phys. **12**, 692–703 (2003)
65. Cong, N.D., Tuan, H.T.: Existence, uniqueness, and exponential boundedness of global solution to delay fractional differential equations. Mediterr. J. Math. **14**, 193 (2017)
66. Conte D., Farsimadan E., Moradi L., Palmieri F., Paternoster B.: Time-delay fractional optimal control problems: a survey based on methodology. In: Wahab M.A. (ed.) Proceedings of the 8th International Conference on Fracture, Fatigue and Wear, pp. 325–337. Springer Nature, Singapore (2021)

67. Cresson, J.: Fractional Calculus in Analysis, Dynamics and Optimal Control. Nova Science Publishers, New York (2014)
68. Daftardar-Gejji, V., Sukale, Y., Bhalekar, S.: Solving fractional delay differential equations: a new approach. Fract. Calc. Appl. Anal. **18**, 400–418 (2015)
69. Debeljković, D.: Time-Delay Systems. InTech, Rijeka (2011)
70. Dehghan, M., Hamedi, E., Khosravian-Arab, H.: A numerical scheme for the solution of a class of fractional variational and optimal control problems using the modified Jacobi polynomials. J. Vib. Control **31**, 1547–1559 (2016)
71. Deng, W., Li, C.: Numerical schemes for fractional ordinary differential equations. In: Miidla, P. (ed.) Numerical Modelling, pp. 355–374. Rijeka, InTech (2012)
72. Di Pillo, G., Grippo, L.: Exact penalty functions in constrained optimization. SIAM J. Control. Optim. **27**, 1333–1360 (1989)
73. Diethelm, K.: The Analysis of Fractional Differential Equations. Springer, Berlin (2010)
74. Diethelm, K., Ford, N.J., Freed, A.D.: A predictor-corrector approach for the numerical solution of fractional differential equations. Nonlinear Dyn. **29**, 3–22 (2002)
75. Diethelm, K., Ford, N.J., Freed, A.D.: Detailed error analysis for a fractional Adams method. Numer. Algorithms **36**, 31–52 (2004)
76. Dontchev, A.L., Hager, W.W.: The Euler approximation in state constrained optimal control problems. Math. Comput. **70**, 173–203 (2000)
77. Dontchev, A.L., Hager, W.W., Poore, A.B., Yang, B.: Optimality, stability and convergence in nonlinear control. Appl. Math. Optim. **31**, 297–326 (1995)
78. Duarte-Mermoud, M.A., Aguila-Camacho, N., Gallegos, J.A.: Using general quadratic Lyapunov functions to prove Lyapunov uniform stability for fractional order system. Commun. Nonlinear Sci. Numer. Simul. **22**, 650–659 (2015)
79. Elnagar, G.N., Kazemi, M.A.: Pseudospectral Chebyshev optimal control of constrained nonlinear dynamical systems. Comput. Optim. Appl. **11**, 195–217 (1998)
80. Elnagar, G.N., Kazemi, M.A., Razzaghi, M.: The pseudospectral Legendre method for discretizing optimal control problems. IEEE Trans. Autom. Control **40**, 1793–1796 (1995)
81. Elgindy, K.T.: Fourier-Gegenbauer pseudospectral method for solving periodic fractional optimal control problems. Math. Comput. Simul. **225**, 148–164 (2024)
82. Ewoud, J.J., Chu, Q., Croon, G.C.D.: Adaptive incremental nonlinear dynamic inversion for attitude control of micro air vehicles. J. Guid. Control Dyn. **39**, 450–461 (2016)
83. Ezz-Eldien, S.S.: On solving fractional logistic population models with applications. Comput. Appl. Math. **37**, 6392–6409 (2018)
84. Ezz-Eldien, S.S., Doha, E.H., Baleanu, D., Bhrawy, A.H.: A numerical approach based on Legendre orthonormal polynomials for numerical solutions of fractional optimal control problems. J. Vib. Control **23**, 16–30 (2017)
85. Fabien, B.: Indirect solution of inequality constrained and singular optimal control problems via a simple continuation method. J. Dyn. Syst. Meas. Control **136**, 021003 (2013)
86. Fahroo, F., Ross, I.M.: Direct trajectory optimization pseudospectral method. J. Guid. Control Dyn. **25**, 160–166 (2002)
87. Filippov, A.F.: On certain questions in the theory of optimal control. SIAM J. Control Optim. **1**, 76–84 (1962)
88. Fletcher, R.: A new approach to variable metric algorithms. Comput. J. **13**, 317–322 (1970)
89. Fletcher, R.: Practical Methods of Optimization. Wiley-Interscience, New York (1987)
90. Gao, G.H., Sun, Z.Z., Zhang, H.W.: A new fractional numerical differentiation formula to approximate the Caputo fractional derivative and its applications. J. Comput. Phys. **259**, 33–50 (2014)
91. Cao, J., Xu, C.: A high order scheme for the numerical solution of the fractional ordinary differential equations. J. Comput. Phys. **238**, 154–168 (2013)
92. Garg, D., Hager, W.W., Rao, A.V.: Pseudospectral methods for solving infinite-horizon optimal control problems. Automatica **47**, 829–837 (2011)
93. Garg, D., Patterson, M., Hager, W.W., Rao, A.V., Benson, D.A., Huntington, G.T.: A unified framework for the numerical solution of optimal control problems using pseudospectral methods. Automatica **46**, 1843–1851 (2010)

94. Garg, D., Patterson, M., Hager, W., Rao, A., Benson, D.R., Huntington, G.T.: An overview of three pseudospectral methods for the numerical solution of optimal control problems. hal-01615132 (2017)
95. Gasimov, J.J., Asadzade, J.A., Mahmudov, N.I.: Pontryagin maximum principle for fractional delay differential equations and controllled weakly singular Volterra delay intgral equations. Qual. Theor. Dyn. Syst. **23**, 213 (2024)
96. Ge, F.D., Zhou, H.C., Kou, C.H.: Approximate controllability of semilinear evolution equations of fractional order with nonlocal and impulsive conditions via an approximating technique. Appl. Math. Comput. **275**, 107–120 (2016)
97. Ghomanjani, F., Farahi, M.H., Gachpazan, M.: Optimal control of time-varying linear delay systems based on the Bezier curves. Comput. Appl. Math. **33**, 687–715 (2014)
98. Goldfarb, D.: A family of variable-metric methods derived by variational means. Math. Comput. **24**, 23–26 (1970)
99. Goldstine, H.H.: A History of the Calculus of Variations from the 17th to the 19th Century. Springer, New York (1981)
100. Göllmann, L., Kern, D., Maurer, H.: Optimal control problems with delays in state and control and mixed control-state constraints. Optim. Contr. Appl. Met. **30**, 341–365 (2009)
101. Gómez-Aguilar, J.F., Atangana, A., Morales-Delgado, V.F.: Electric circuits RC, LC, and RL described by Atangana-Baleanu fractional derivatives. Int. J. Circ. Theor. App. **45**, 1514–1533 (2017)
102. Gómez-Aguilar, J.F., Yépez-Martínez, H., Calderón-Ramón, C., Cruz-Orduña, I., Escobar-Jiménez, R.F., Olivares-Peregrino, V.H.: Modeling of a mass-spring-damper system by fractional derivatives with and without a singular kernel. Entropy **17**, 6289–6303 (2015)
103. Gong, Z., Liu, C., Teo, K., Wang, S., Wu, Y.: Numerical solution of free final time fractional optimal control problems. Appl. Math. Comput. **405**, 126270 (2021)
104. Gong, Z., Liu, C., Teo, K.L., Wu, Y.: A gradient-based optimization algorithm to solve optimal control problems with conformable fractional-order derivatives. J. Comput. Appl. Math. **454**, 116169 (2025)
105. Gorbunov, V.K.: A method for the parametrization of optimal control problems. USSR Comp. Math. Math. Phys. **19**, 18–30 (1979)
106. Guermah, S., Djennoune, S., Bettayeb, M.: Controllability and observability of linear discrete-time fractional-order systems. Int. J. Appl. Math. Comput. Sci. **18**, 213–222 (2008)
107. Guezane-Lakoud, A., Khaldi, R.: Solvability of a fractional boundary value problem with fractional integral condition. Nonlinear Anal. **75**, 2692–2700 (2012)
108. Guglielmi, N., Hairer, E.: Geometric proofs of numerical stability for delay equations. IMA J. Numer. Anal. **21**, 439–450 (2001)
109. Guo, T.L.: The necessary conditions of fractional optimal control in the sense of Caputo. J. Optim. Theory Appl. **156**, 115–126 (2013)
110. Hadid, S.B., Ta'ani, A.A., Momani, S.M.: Some existence theorems on differential equations of generalized order through a fixed-point theorem. J. Fract. Calc. **9**, 45–49 (1996)
111. Hamdache, A., Saadi, S., Elmouki, I.: Free terminal time optimal control problem for the treatment of HIV infection. Int. J. Optim. Control Theor. Appl. **6**, 33–51 (2016)
112. Hajishafieiha, J., Abbasbandy, S.: Numerical approach for solving the fractional pantograph delay differential equations. Complexity **2022**, Article ID 4134892 (2022)
113. Han, S.P.: Superlinearity convergent variable metric algorithm for general nonlinear programming problems. Math. Program. **11**, 263–282 (1976)
114. Han, S.P.: A globally convergent method for nonlinear programming. J. Optim. Theory Appl. **22**, 297–309 (1977)
115. Harir, A., Malliani, S., Chandli, L.: Solutions of conformable fractional-order SIR epidemic model. Int. J. Differ. Equ. **2021**, 6636686 (2021)
116. Hartley, T.T., Lorenzo, C.F.: Dynamics and control of initialized fractional-order systems. Nonlinear Dyn. **29**, 201–233 (2002)
117. He, B.B., Zhou, H.C., Kou, C.H.: The controllability of fractional damped dynamical systems with control delay. Commun. Nonlinear Sci. Numer. Simul. **32**, 190–198 (2016)

118. Heydari, M.H., Razzaghi, M.: A new wavelet method for fractional integro-differential equations with ψ-Caputo fractional derivative. Math. Comput. Simulat. **217**, 97–108 (2024)
119. Heydari, M.H., Tavakoli, R., Razzaghi, M.: Application of the extended Chebyshev cardinal wavelets in solving fractional optimal control problems with ABC fractional derivative. Int. J. Syst. Sci. **53**, 2694–2708 (2022)
120. Hicks, G.H., Ray, W.H.: Approximation methods for optimal control synthesis. Can. J. Chem. Eng. **49**, 522–528 (1971)
121. Hilfer, R.: Applications of Fractional Calculus in Physics. World Scientific, Singapore (2000)
122. Hosseini, V.R., Chen, W., Avazzadeh, Z.: Numerical solution of fractional telegraph equation by using radial basis functions. Eng. Anal. Bound. Elem. **38**, 31–39 (2014)
123. Hosseini, S., Göttlicher, C., Holzapfel, F.: Optimal input design for flight vehicle system identification via dynamic programming and the direct method for optimal control. AIAA SciTech Forum 1319 (2019)
124. Hosseinpour, S., Nazemi, A.: A collocation method via block-pulse functions for solving delay fractional optimal control problems. IMA J. Math. Control. Inf. **34**, 1215–1237 (2016)
125. Hosseinpour, S., Nazemi, A., Tohidi, E.: Müntz-Legendre spectral collocation method for solving delay fractional optimal control problems. J. Comput. Appl. Math. **351**, 344–363 (2019)
126. Hristov, J.: On the Atangana-Baleanu derivative and its relation to the fading memory concept: the diffusion equation formulation. In: Gómez, J.F., Torres, L., Escobar, R.F. (eds.) Fractional Derivatives with Mittag-Leffler Kernel, pp. 175–193. Switzerland, Springer Nature (2019)
127. Idczak, D., Kamocki, R.: Existence of optimal solutions to Lagrange problem for a fractional nonlinear control system with Riemann-Liouvelle derivative. Math. Control Relat. F. **7**, 449–464 (2017)
128. Iyiolq, O.S., Nwaeze, E.R.: Some new results on the new conformable fractional calculus with application using D'Alambert approach. Progr. Fract. Differ. Appl. **2**, 115–122 (2016)
129. Jaddu, H., Shimemura, E.: Computational methods based on the state parameterization for solving constrained optimal control problems. Int. J. Syst. Sci. **30**, 275–282 (1999)
130. Jafari, H., Ganji, R.M., Sayevand, K., Baleanu, D.: A numerical approach for solving fractional optimal control problems with Mittag-Leffler kernel. J. Vib. Control **28**, 2596–2606 (2022)
131. Jajarmi, A., Ghassabzade, F.A.: Optimal control and general fractional description for a complex biological system. Progr. Fract. Differ. Appl. **9**, 389–396 (2023)
132. Jajarmi, A., Hajipour, M., Mohammadzadeh, E., Baleanu, D.: A new approach for the nonlinear fractional optimal control problems with external persistent disturbances. J. Franklin Inst. **355**, 3938–3967 (2018)
133. Jarad, F., Abdeljawad, T., Hammouch, Z.: On a class of ordinary differential equations in the frame of Atangana-Baleanu fractional derivative. Chaos Soliton. Fract. **117**, 16–20 (2018)
134. Jin, B.T.: Fractional Differential Equations. Springer, Cham (2021)
135. Kafash, B., Delavarkhalafi, A.: Restarted state parameterization method for optimal control problems. J. Math. Comput. Sci. **14**, 151–161 (2015)
136. Kalman, R.E.: Contribution to the theory of optimal control. Bol. Soc. Mat. Mex. **5**, 102–119 (1960)
137. Kamien, M.I., Schwartz, N.L.: Dynamic Optimization: The Calculus of Variations and Optimal Control in Economics and Management. Dover Publications Inc., New York (2012)
138. Kamocki, R.: Pontryagin maximum principle for fractional ordinary optimal control problems. Math. Method. Appl. Sci. **37**, 1668–1686 (2014)
139. Kamocki, R.: On the existence of optimal solutions to fractional optimal control problems. Appl. Math. Comput. **235**, 94–104 (2014)
140. Kamocki, R.: Existence of optimal control for multi-order fractional optimal control problems. Arch. Control Sci. **32**, 279–303 (2022)
141. Kamocki, R.: Pontryagin's maximum principle for a fractional Intgro-differential Lagrange problem. Commun. Nonlinear Sci. Numer. Simul. **128**, 107598 (2024)
142. Kashkari, B.S., Syam, M.I.: Fractional-order Legendre operational matrix of fractional integration for solving the Riccati equation with fractional order. Appl. Math. Comput. **290**, 281–291 (2016)

143. Kaviya, R., Muthukumar, P.: Dynamical analysis and optimal harvesting of conformable fractional prey-predator system with predator immigration. Eur. Phys. J. Plus **136**, 542 (2021)
144. Kaya, C.Y.: Inexact restoration for Runge-Kutta discretization of optimal control problems. SIAM J. Numer. Anal. **48**, 1492–1517 (2010)
145. Kaya, C.Y., Maurer, H.: A numerical method for nonconvex multi-objective optimal control problems. Comput. Optim. Appl. **57**, 685–702 (2014)
146. van Keulen, T.: Indirect solution for optimal control problems with a pure state constraint. IFAC Proc. **47**, 2456–2461 (2014)
147. Khalil, R., Al, H.M., Yousef, A., Sababheh, M.: A new definition of fractional derivative. J. Comput. Appl. Math. **264**, 65–70 (2014)
148. Khodabakhshi, N., Vaezpour, S.M., Baleanu, D.: Numerical solutions of the initial value problem for fractional differential equations by modification of the Adomian decomposition method. Fract. Calc. Appl. Anal. **17**, 382–400 (2014)
149. Khosravian-Arab, H., Dehghan, M., Eslahchi, M.R.: Fractional spectral and pseudo-spectral methods in unbounded domains: theory and applications. J. Comput. Phys. **338**, 527–566 (2017)
150. Kien, B.T., Fedorov, V.E., Phuong, T.D.: Optimal control problems governed by fractional differential equations with control constraints. SIAM J. Control. Optim. **60**, 1732–1762 (2022)
151. Kilbas, A.A., Trujillo, J.J.: Differential equations of fractional order: methods, results and problems-II. Appl. Anal. **81**, 435–493 (2002)
152. Kim, H., Kim, K.H., Lee, S., Jang, B.: New explicit and accelerated techniques for solving fractional order differential equations. Appl. Math. Comput. **379**, 125228 (2020)
153. Kim, M.H., Ri, G.C., Hyong-Chol, O.: Operational method for solving multi-term fractional differential equations with the generalized fractional derivatives. Fract. Calc. Appl. Anal. **17**, 79–95 (2014)
154. Kirk, D.E.: Optimal Control Theory. Prentice-Hall, New Jersey (1970)
155. Kraft, D.: On converting optimal control problems into nonlinear programming problems. In: Schittkowski, K. (ed.) Computational Mathematical Programming, pp. 261–280. Springer, Berlin (1985)
156. Kulczycki, P., Korbicz, J., Kacprzyk, J.: Fractional Dynamical Systems: Methods, Algorithms and Applications. Springer, Cham (2022)
157. Kumar, M.: Optimal design of fractional delay FIR filter using cuckoo search algorithm. Int. J. Circ. Theor. App. **46**, 2364–2379 (2018)
158. Kumar, M.: An efficient numerical scheme for solving a fractional-order system of delay differential equations. Int. J. Appl. Comput. Math. **8**, 262 (2022)
159. Laforgia, A., Natalini, P.: Exponential, gamma and polygamma functions: simple proofs of classical and new inequalities. J. Math. Anal. Appl. **407**, 459–504 (2013)
160. Lakshmikantham, V.: Theory of fractional functional deifferential equations. Nonlinear Anal. **69**, 3337–3343 (2008)
161. Lakshmikantham, V., Leela, S., Vasundhara, D.J.: Theory of Fractional Dynamic Systems. Cambridge Scientific Publishers, Cambridge (2009)
162. Lazarević, M.: Stability and stabilization of fractional order time delay systems. Sci. Tech. Rev. **61**, 31–45 (2011)
163. Lazo, M.J., Torres, D.F.M.: Variational calculus with conformable fractional derivatives. IEEE/CAA J. Automatic. **69**, 340–352 (2017)
164. Lenhart, S., Workman, J.T.: Optimal Control Applied to Biological Models. Chapman & Hall/CRC, New York (2007)
165. Lenz, S.M., Schlöder, J.P., Bock, H.G.: Numerical computation of derivatives in systems of delay differential equations. Math. Comput. Simulat. **96**, 124–156 (2014)
166. Lepik, U.: Solving fractional integral equations by the Haar wavelet method. Appl. Math. Comput. **214**, 468–478 (2009)
167. Li, B., Yu, C., Teo, K.L., Duan, G.R.: An exact penalty function method for continuous inequality constrained optimal control problem. J. Optim. Theory Appl. **151**, 260–291 (2011)

168. Li, C., Zeng, F.: Finite difference methods for fractional differential equations. Int. J. Bifurcat. Chaos **22**, 1230014 (2012)
169. Li, W., Wang, S.: Pricing American options under proportional transaction costs using a penalty approach and a finite difference scheme. J. Ind. Manag. Optim. **25**, 365–389 (2013)
170. Li, W., Wang, S., Rehbock, V.: Numerical solution of fractional optimal control. J. Optim. Theory Appl. **180**, 556–573 (2019)
171. Li, W., Wang, S., Rehbock, V.: A 2nd-order one-point numerical integration scheme for fractional ordinary differential equations. Numer. Algebra Control Optim. **7**, 273–287 (2017)
172. Li, Y., Chen, Y.Q., Podlunby, I.: Mittag-Leffler stability of fractional order nonlinear dynamic systems. Automatica **45**, 1965–1969 (2009)
173. Li, Y., Chen, Y.Q., Podlubny, I.: Stability of fractional-order nonlinear dynamic systems: Lyapunov direct method and generalized Mittag-Leffler stability. Comput. Math. Appl. **59**, 1810–1821 (2010)
174. Lin, J.S., Hwang, C.: Enhancement of the global convergence of using iterative dynamic programming to solve optimal control problems. Ind. Eng. Chem. Res. **37**, 2469–2478 (1998)
175. Lin, Q., Loxton, R., Teo, K.L.: The control parameterization method for nonlinear optimal control: a survey. J. Ind. Manag. Optim. **10**, 275–309 (2014)
176. Liouville, J.: Memoire sue quelques questions de géometrie et de mécanique, et sur un nouveau gentre pour resoudre ces questions. J. Ecole. Polytech. **13**, 1–69 (1832)
177. Liu, C., Gong, Z., Wang, S., Teo, K.L.: Numerical solution of delay fractional optimal control problems with free terminal time. Optim. Lett. **17**, 1359–1378 (2023)
178. Liu, C., Gong, Z., Wu, Y., Wiwatanapataphee, B., Teo, K.L.: Dynamic optimization of nonlinear fractional impulsive switched systems. J. Optim. Theory Appl. **205**, 60 (2025)
179. Liu, C., Gong, Z., Yu, C., Wang, S., Teo, K.L.: Optimal control computation for nonlinear fractional time-delay systems with state inequality constraints. J. Optim. Theory Appl. **191**, 83–117 (2021)
180. Liu, C., Loxton, R., Teo, K.L.: A computational method for solving time-delay optimal control problems with free terminal time. Syst. Control Lett. **72**, 7253–7260 (2014)
181. Liu, C., Yi, X., Gong, Z., Han, M.: The control parametrization technique for numerically solving fractal-fractional optimal control problems involving Caputo-Fabrizio derivatives. J. Comput. Appl. Math. **472**, 116814 (2026)
182. Liu, C., Yu, C., Gong, Z., Cheong, H.T., Teo, K.L.: Numerical computation of optimal control problems with Atangana-Baleanu fractional derivatives. J. Optim. Theory Appl. **197**, 798–816 (2023)
183. Liu, C., Zhou, T., Gong, Z., Yi, X., Teo, K.L., Wang, S.: Robust optimal control of nonlinear fractional systems. Chaos Soliton. Fract. **175**, 113964 (2023)
184. Liu, M., Bai, C.: Optimal harvesting of a stochastic logistic model with time delay. J. Nonlinear Sci. **25**, 277–289 (2015)
185. Liu, S., Yang, R., Zhou, X.F., Jiang, W., Li, X., Zhao, X.W.: Stability analysis of fractional delayed equations and its applications on consensus of multi-agent systems. Commun. Nonlinear Sci. Numer. Simul. **73**, 351–362 (2019)
186. Loewen, P.D., Rockafellar, R.T.: New necessary conditions for the generalized problem of Bolza. SIAM J. Control. Optim. **34**, 1496–1551 (1996)
187. Loewen, P.D., Rockafellar, R.T.: Bolza problems with general time constraints. SIAM J. Control. Optim. **35**, 2050–2069 (1997)
188. Longuski, J.M., Guzmán, J.J., Prussing, J.E.: Optimal Control with Aerospace Applications. Springer, New York (2014)
189. Loxton, R., Lin, Q., Rehbock, V., Teo, K.L.: Control parameterization for optimal control problems with continuous inequality constraints: new convergence results. Numer. Algebra Control Optim. **2**, 571–599 (2012)
190. Loxton, R., Teo, K.L., Rehbock, V.: Robust suboptimal control of nonlinear systems. Appl. Math. Comput. **217**, 6566–6576 (2011)
191. Luenberger, D.G., Ye, Y.: Linear and Nonlinear Programming. Springer, New York (2008)

192. Luus, R.: Optimal control by dynamic programming using systematic reduction in grid size. Int. J. Control **51**, 995–1013 (1990)
193. Luus, R., Galli, M.: Multiplicity of solutions in using dynamic programming for optimal control. Hung. J. Ind. Chem. **19**, 55–62 (1991)
194. Luus, R., Rosen, O.: Application of dynamic programming to final state constrained optimal control problems. Ind. Eng. Chem. Res. **30**, 1525–1530 (1991)
195. Luus, R.: On the application of iterative dynamic programming to singular optimal control problems. IEEE Trans. Auto. Contr. **37**, 1802–1806 (1992)
196. Luus, R.: Optimization of fed-batch fermentors by iterative dynamic programming. Biotech. Bioeng. **41**, 599–602 (1993)
197. Luus, R.: Application of dynamic programming to differential-algebraic process systems. Comp. Chem. Eng. **17**, 373–377 (1993)
198. Luus, R.: Iterative Dynamic Programming. Chapman & Hall/CRC, Boca Raton (2000)
199. Luus, R.: Optimal control of oscillatory systems by iterative dynamic programming. J. Ind. Manag. Optim. **4**, 1–15 (2008)
200. Ma, L.J., Liu, B.: Necessary conditons of fractional optimal control problems with state constraints in the sense of Riemann-Liouville. Asian J. Control **22**, 1494–1512 (2020)
201. Machado, J.A.T., Jesus, I.S., Galhano, A., Cunha, J.B.: Fractional order electromagnetics. Signal Process. **86**, 2637–2644 (2006)
202. Mainardi, F.: Fractional Calculus and Waves in Linear Viscoelasticity: An Introduction to Mathematical Models. Imperial College Press, London (2010)
203. Magin, R.L.: Fractional calculus in bioengineering. Crit. Rev. Biomed. Eng. **32**, 1–104 (2004)
204. Marzban, H.R.: A new fractional orthogonal basis and its application in nonlinear delay fractional optimal control problems. ISA Trans. **114**, 106–119 (2021)
205. Marzban, H.R.: Solution of a specific class of nonlinear fractional optimal control problems including multiple delays. Optim. Contr. Appl. Met. **42**, 2–29 (2021)
206. Marzban, H.R., Malakoutikhah, F.: Solution of delay fractional optimal control problems using a hybrid of block-pulse functions and orthonormal Taylor polynomials. J. Franklin Inst. **356**, 8182–8251 (2019)
207. Marzban, H.R., Pirmoradian, H.: A direct approach for the solution of nonlinear optimal control problems with multiple delays subject to mixed state-control constraints. Appl. Math. Model. **53**, 189–213 (2018)
208. Matignon, D.: Stability result on fractional differential equations with applications to control processing. In: Proceedings of the IMACS-SMC, pp. 963–968. France, Lille (1996)
209. Maurer, H.: Numerical solution of singular control problems using multiple shooting techniques. J. Optim. Theory Appl. **18**, 235–257 (1976)
210. Maurer, H., Büskens, C., Kim, J.H.R., Kaya, C.Y.: Optimization methods for the verification of second-order sufficient conditions for bang-bang controls. Optim. Contr. Appl. Met. **26**, 129–156 (2005)
211. Maurer, H., Gillessen, W.: Application of multiple shooting to the numerical solution of optimal control problems with bounded state variables. Comput. **15**, 105–128 (1975)
212. McCormick, G.: Nonlinear Programming: Theory, Algorithms and Applications. Wiley, New York (1983)
213. Medhin, N.G.: Characterization of optimal pairs for hereditary control problems. J. Optim. Theory Appl. **75**, 355–367 (1992)
214. Mekarapiruk, W., Luus, R.: Optimal control of final state constrained systems. Can. J. Chem. Eng. **25**, 806–811 (1997)
215. Mekarapiruk, W., Luus, R.: Optimal control of inequality state constrained systems. Ind. Eng. Chem. Res. **36**, 1686–1694 (1997)
216. Miller, K.S., Ross, B.: An Introduction to the Fractional Calculus and Fractional Differential Equations. Wiley, New York (1993)
217. Mohammadzadeh, R., Lakestani, M.: Optimal control of linear time-delay systems by a hybrid of block-pulse functions and biorthogonal cubic Hermite spline multiwavelets. Optim. Contr. Appl. Met. **39**, 357–376 (2018)

218. Moghaddam, B.P., Mostaghim, Z.S.: A numerical method based on finite difference for solving fractional delay differential equations. J. Taibah Univ. Sci. **7**, 120–127 (2013)
219. Momani, S., Hadid, S.: Lyapunov stability solutions of fractional integrodifferential equations. Int. J. Math. Math. Sci. **47**, 2503–2507 (2004)
220. Momani, S.M., El-Khazali, R.: Stability analysis of composite fractional systems. Int. J. Appl. Math. **12**, 73–85 (2003)
221. Moon, J.: On the optimality condition for optimal control of Caputo fractional differential equations with state constraints. IFAC Pap. **56**, 216–221 (2023)
222. Moon, J.: The Pontryagin type maximum principle for Caputo fractional optimal control problems with terminal and running state constraints. AIMS Math. **10**, 884–920 (2025)
223. Moradi, L., Mohammadi, F., Baleanu, D.: A direct numerical solution of time-delay fractional optimal control problems by using Chelyshkov wavelets. J. Vib. Control **25**, 310–324 (2019)
224. Mozyrskaa, D., Pawluszewiczb, E., Wyrwas, M.: Local observability and controllability of nonlinear discrete-time fractional order systems based on their linearisation. Int. J. Syst. Sci. **48**, 788–794 (2017)
225. Mu, P., Wang, L., Liu, C.: A control parameterization method to solve the fractional-order optimal control problem. J. Optim. Theory Appl. **187**, 234–247 (2020)
226. Muroya, Y., Ishiwata, E., Brunner, H.: On the attainable order of collocation methods for pantograph Integro-differential equations. J. Comput. Appl. Math. **152**, 347–366 (2003)
227. Muthukumar, P., Ganesh, P.B.: Numerical solution of fractional delay differential equation by shifted Jacobi polynomials. Int. J. Comput. Math. **94**, 471–492 (2017)
228. Nadairou, F., Torres, D.F.M.: Pontryagin maximum principle for incommemnsurate fractional-orders optimal control problems. Math. **11**, 4218 (2023)
229. Naifar, O., Makhlouf, A.B.: Fractional Order Systems-Control Theory and Applications: Fundamentals and Applications. Springer, Cham (2022)
230. Naifar, O., Nagy, A.M., Makhlouf, A.B., Kharrat, M., Hammami, M.A.: Finite-time stability of linear fractional-order time-delay systems. Int. J. Robust Nonlinear Control **29**, 180–187 (2019)
231. Nawaz, M., Jiang, W., Sheng, J., Khan, A.: The controllability of damped fractional differential system with impulses and state delay. Adv. Diff. Equ. **2020**, 1–23 (2020)
232. Nazemi, A., Mansoori, M.: Solving optimal control problems of the time-delayed systems by Haar wavelet. J. Vib. Control **22**, 2657–2670 (2016)
233. Ndoye, I., Zasadzinski, M., Darouach, M., Radhy, N.E.: Observer-based control for fractional-order continuous-time systems. In: Proceedings of the 48th IEEE CDC/28th CCC, pp. 1932–1937. China, Shanghai (2009)
234. Nemati, S., Lima, P., Sedaghat, S.: An effective numerical method for solving fractional pantograph differential equations using modification of hat functions. Appl. Num. Math. **131**, 174–189 (2018)
235. Ngo, H.T.B., Razzaghi, M., Vo, T.N.: Fractional-order Chelyshkov wavelet method for solving variable-order fractional differential equations and an application in variable-order fractional relaxation system. Numer. Algorithms **92**, 1571–1588 (2023)
236. Nocedal, J., Wright, S.: Numerical Optimization. Springer, Berlin (2006)
237. Ockendon, J.R., Tayler, A.B., Temple, G.F.J.: The dynamics of a current collection system for an electric locomotive. Proc. R. Soc. Lond. **322**, 447–468 (1971)
238. Odibat, Z., Erturk, V.S., Kumar, P., Ben Makhlouf, A., Govindaraj, V.: An implementation of the generalized differential transform scheme for simulating impulsive fractional differential equations. Math. Prob. Eng. **2022**, Article ID 8280203 (2022)
239. Odibat, Z.M.: Computational algorithms for computing the fractional derivatives of functions. Math. Comput. Simulat. **79**, 2013–2020 (2009)
240. Padula, F., Visioli, A.: Advances in Robust Fractional Control. Springer, Cham (2014)
241. Papadopoulos, I.P.A., Olver, S.: A sparse spectral method for fractional differential equations in one-spatial dimension. Adv. Comput. Math. **50**, 69 (2024)
242. Podulubny, I.: Fractional Differential Equations. Academic Press, San Diego (1999)

243. Polak, E.: An history survey of computational methods in optimal control. SIAM Rev. **15**, 553–584 (1973)
244. Polak, E.: Optimazation Algorithms and Consistent Approximation. Springer, New York (1997)
245. Pontryagin, L.S., Boltyanski, V.G., Gamkrelidze, V., Mischenko, E.E.: The Mathematical Theory of Optimal Control Process. Fizmatigiz (1961)
246. Pooseh, S., Almieda, R., Torres, D.F.M.: Fractional order optimal control problems with free terminal time. J. Ind. Manag. Optim. **10**, 363–381 (2014)
247. Postavaru, O.: An efficient numerical method based on Fibonacci polynomials to solve fractional differential equations. Math. Comput. Simulat. **212**, 406–422 (2023)
248. Powell, M.J.D.: A fast algorithm for nonlinearly constrained optimization calculations. In: Matson, G.A. (ed.) Numerical Analysis: Lecture Notes in Mathematics, vol. 630, pp. 144–157. Springer, Berlin (1978)
249. Rabiei, K., Ordokhani, Y.: Boubaker hybrid functions and their application to solve fractional optimal control and fractional variational problems. Appl. Math. **63**, 541–567 (2018)
250. Rabiei, K., Ordokhani, Y.: Solving fractional pantograph delay differential equations via fractional-order Boubaker polynomials. Eng. Comput. **35**, 1431–1441 (2019)
251. Radwan, A.G., Soliman, A.M., Elwakil, A.S., Sedeek, A.: On the stability of linear systems with fractional-order elements. Chaos Soliton. Fract. **40**, 2317–2328 (2009)
252. Rahimkhani, P., Ordokhani, Y.: Generalized fractional-order Bernoulli-Legendre functions: an effective tool for solving two-dimensional fractional optimal control problems. IMA J. Math. Control. Inf. **36**, 185–212 (2019)
253. Rahimkhani, P., Ordokhani, Y.: Numerical investigation of distributed-order fractional optimal control problems via Bernstein wavelets. Optim. Contr. Appl. Met. **42**, 355–373 (2021)
254. Rahimkhani, P., Ordokhani, Y., Babolian, E.: An efficient approximate method for solving delay fractional optimal control problems. Nonlinear Dyn. **86**, 1649–1661 (2016)
255. Rahimkhani, P., Ordokhani, Y., Babolian, E.: Numerical solution of fractional pantograph differential equations by using generalized fractional-order Bernoulli wavelet. J. Comput. Appl. Math. **309**, 493–510 (2017)
256. Raja, M.M., Vijayakumar, V., Veluvolu, K.C., Shukla, A., Nisar, K.S.: Existence and optimal control results for Caputo fractional delay Clark's subdifferential inclusions of order $r \in (1, 2)$ with sectorial operators. Optim. Contr. Appl. Met. **45**, 1832–1850 (2024)
257. Rakhshan, S., Effati, S.: Fractional optimal control problems with time-varying delay: a new delay fractional Euler-Lagrange equations. J. Franklin Inst. **357**, 5954–5988 (2020)
258. Rannacher, R.: Finite element solution of diffusion problems with irregular data. Numer. Math. **43**, 309–327 (1984)
259. Rao, A.V.: A survey of numerical methods for optimal control. Adv. Astronaut. Sci. **135**, 1–32 (2010)
260. Rehbock, V., Teo, K.L., Jennings, L.S.: A computational procedure for suboptimal robust controls. Dyn. Control **2**, 331–348 (1992)
261. Richard, J.P.: Time-delay systems: an overview of some recent advances and open problems. Automatica **29**, 1667–1694 (2003)
262. Riemann, B.: Versuch einer allgemeinen Auffassung der Integration und Differentiation, pp. 331–344. Gesammelte Mathematische Werke und Wissenschaftlicler, Teubner, Leipzig (1847)
263. Rodrigues, H.S., Monteiro, M.T.T., Torres, D.F.M.: Optimal control and numerical software: an overview. In: Francisco, M. (ed.) Systems Theory, pp. 93–110. Nova Science Publishers (2014)
264. Ross, B.: The development of fractional calculus 1695–1900. Hist. Math. **4**, 75–89 (1977)
265. Ross, B.: Fractional Calculus and Its Applications. Springer, London (1975)
266. Ross, I.M., Fahroo, F.: Pseudospectral knotting methods for solving optimal control problems. J. Guid. Control. Dyn. **27**, 397–405 (2004)
267. Roxin, E.: The existence of optimal controls. Mich. Math. J. **9**, 109–119 (1962)
268. Sabatier, J., Moze, M., Farges, C.: LMI stability conditions for fractional order systems. Comput. Math. Appl. **59**, 1594–1609 (2010)

269. Sadek, L., Ounamane, S., Abouzaid, B., Sadek, E.M.: The Galerkin Bell method to solve the fractional optimal control problems with inequality constraints. J. Comput. Sci. **77**, 102244 (2024)
270. Saeed, U.: Hermite wavelet method for fractional delay differential equations. J. Differ. Equ 359093 (2014)
271. Saeed, U., Rehman ur, M., Iqbal, M.A.: Modified Chebyshev wavelet methods for fractional delay-type equations. Appl. Math. Comput. **264**, 431–42 (2015)
272. Safaie, E., Farahi, M.H., Ardehaie, M.F.: An approximate method for numerically solving multi-dimensional delay fractional optimal control problems by Bernstein polynomials. Comput. Appl. Math. **34**, 831–846 (2015)
273. Saha, K.K., Sukavanam, N., Pan, S.: Existence and uniqueness of solutions to fractional differential equations with fractional boundary conditions. Alex. Eng. J. **72**, 147–155 (2023)
274. Salati, A.B., Shamsi, M., Torres, D.F.M.: Direct transcription methods based on fractional integral approximation formulas for solving nonlinear fractional optimal control problems. Commun. Nonlinear Sci. Numer. Simul. **67**, 334–350 (2019)
275. Samko, S.G., Kilbas, A.A., Marichev, O.I.: Fractional Integrals and Derivatives: Theory and Applications. Gordon and Breach Science Publishers, Singapore (1993)
276. Sargent, R.W.H.: Optimal control. J. Comput. Appl. Math. **124**, 361–371 (2000)
277. Schittkowski, K.: NLPQL: a Fortran subroutine solving constrained nonlinear programming problems. Ann. Oper. Res. **5**, 485–500 (1986)
278. Sebaa, N., Fellah, Z.E.A., Lauriks, W., Depollier, C.: Application of fractional calculus to ultrasonic wave propagation in human cancellous bone. Signal Process. **86**, 2668–2677 (2006)
279. Selvam, A., Hussin, F., Ibrahim, R., Bingi, K., Nagarajapandian, M.: Optimal Fractional-Order Predictive PI Controllers: For Process Control Applications with Additional Filtering. Springer, Singapore (2022)
280. Seywald, K., Seywald, H.: Desensitized optimal control. AIAA Scitech. Forum 0651 (2019)
281. Shakeri, F., Dehghan, M.: Solution of delay differential equations via a homotopy perturbation method. Math. Comput. Model. **48**, 486–498 (2008)
282. Shanno, D.: Conditioning of quasi-Newton methods for function minimization. Math. Comput. **24**, 647–656 (1970)
283. Shapiro, A., Dentcheva, D., Ruszczyński, A.: Lectures on Stochastic Programming: Modeling and Theory. Society for Industrial and Applied Mathematics, Philadelphia (2009)
284. Sherif, M.N.: Numerical solution of system of fractional delay differential equations using polynomial spline functions. Appl. Math. **7**, 518–526 (2016)
285. Sikora, B., Klamka, J.: Constrained controllability of fractional linear systems with delays in control. Syst. Control Lett. **106**, 9–15 (2017)
286. Singh, A., Kanaujiya, A., Mohapatra, J.: Euler wavelets method for optimal control problems of fractional Integro-differential equations. J. Comput. Appl. Math. **454**, 116178 (2025)
287. Sirisena, H.R., Chou, F.S.: State parameterization approach to the solution of optimal control problems. Optim. Contr. Appl. Met. **2**, 289–298 (1981)
288. Srinivasan, B., Bonvin, D., Visser, E., Palanki, S.: Dynamic optimization of batch processes-II: role of measurements in handling uncertainty. Comput. Chem. Eng. **27**, 27–44 (2003)
289. Stapor, P., Fröhlich, F., Hasenauer, J.: Optimization and profile calculation of ODE models using second order adjoint sensitivity analysis. Bioinformatics **34**, i151–i159 (2018)
290. Stein, O.: How to solve a semi-infinite optimization problem. Eur. J. Oper. Res. **223**, 312–320 (2012)
291. Stoddart, A.W.J.: Existence of optimal controls. Pac. J. Math. **1**, 167–177 (1967)
292. Sun, H.G., Zhang, Y., Baleanu, D., Chen, W., Chen, Y.: A new collection of real world applications of fractional calculus in science and engineering. Commun. Nonlinear Sci. Numer. Simul. **64**, 213–231 (2018)
293. Sun, W., Yuan, Y.: Optimization Theory and Methods: Nonlinear Programming. Springer, New York (2006)
294. Swan, G.W.: Optimal control applications in biomedical engineering-a survey. Optim. Contr. Appl. Met. **2**, 311–334 (1981)

295. Sweilam, N.H., AL-Mekhlafi, S.M.: Optimal control for a time delay multi-strain tuberculosis fractional model: a numerical approach. IMA J. Math. Control Inf. **36**, 317–340 (2019)
296. Sweilam, N.H., Al-Mekhlafi, S.M., Assiri, T., Atangana, A.: Optimal control for cancer treatment mathematical model using Atangana-Baleanu-Caputo fractional derivative. Adv. Differ. Equ. **2020**, 334 (2020)
297. Tajadodi, H., Khan, A., Gómez-Aguilar, J.F., Khan, H.: Optimal control problems with Atangana-Baleanu fractional derivative. Optim. Contr. Appl. Met. **42**, 96–106 (2021)
298. Tang, X., Liu, Z., Wang, X.: Integral fractional pseudospectral methods for solving fractional optimal control problems. Automatica **62**, 304–311 (2015)
299. Tang, X., Shi, Y., Wang, L.L.: A new framework for solving fractional optimal control problems using fractional pseudospectral method. Automatica **78**, 333–340 (2017)
300. Tang, Y., Li, N., Liu, M., Lu, Y., Wang, W.: Identification of fractional-order systems with time delays using block pulse functions. Mech. Syst. Signal Process. **91**, 382–394 (2017)
301. Tavazoei, M., Asemani, M.H.: Stability analysis of time-delay incommensurate fractional-order systems. Commun. Nonlinear Sci. Numer. Simul. **109**, 106270 (2022)
302. Tavazoei, M.S., Haeri, M.: Note on the stability of fractional order systems. Math. Comput. Simulat. **79**, 1566–1576 (2009)
303. Teo, K.L., Li, B., Yu, C., Rehbock, V.: Applied and Computational Optimal Control: A Control Parameterization Approach. Springer, Cham (2021)
304. Teo, K.L., Jennings, L.S., Lee, H.W.J., Rehbock, V.: The control parameterization enhancing transform for constrained optimal control problems. J. Aust. Math. Soc. **40**, 314–335 (1999)
305. Toledo-Hernandez, R., Rico-Ramirez, V., Rico-Martinez, R., Hernandez-Castro, S., Diwekar, U.M.: Numerical solution of fractional optimal control problems: a fractional calculus approach to the dynamic optimization of biological reactive systems. Part II. Chem. Eng. Sci. **117**, 239–247 (2014)
306. Tricaud, C., Chen, Y.: An approximate method for numerically solving fractional order optimal control problems of general form. Comput. Math. Appl. **59**, 1644–1655 (2010)
307. Trélat, E.: Optimal control and applications to aerospace: some results and challenges. J. Optim. Theory Appl. **154**, 713–758 (2012)
308. Tuan, H.T., Trinh, H.: A qualitative theory of time delay nonlinear fractional-order systems. SIAM J. Control. Optim. **58**, 1491–1518 (2020)
309. Tyrtyshnikov, E.E.: A Brief Introduction to Numerical Analysis. Springer, Boston (1997)
310. Vadivoo, B.S., Ramachandran, R., Cao, J., Zhang, H., Li, X.: Controllability analysis of nonlinear neutral-type fractional-order differential systems with state delay and impulsive effects. Int. J. Control Autom. Syst. **16**, 659–669 (2018)
311. Valério, D., Costa, J.: An Introduction to Fractional Control. The Institution of Engineering and Technology, London (2012)
312. Vichitkunakorn, P., Vo, T.N., Razzaghi, M.: A numerical method for fractional pantograph differential equations based on Taylor wavelets. Trans. Inst. Meas. Control **42**, 1334–1344 (2020)
313. Wang, S.: An interior penalty method for a large-scale finite-dimensional nonlinear double obstacle problem. Appl. Math. Model. **58**, 217–228 (2018)
314. Wang, S., Li, W., Liu, C.Y.: On necessary optimality conditions and exact penalization for a constrained fractional optimal control problem. Optim. Contr. Appl. Met. **43**, 1096–1108 (2022)
315. Wang, S., Yang, X.Q., Teo, K.L.: Power penalty method for a linear complementarity problem arising from American option valuation. J. Optim. Theory Appl. **129**, 227–254 (2006)
316. Wang, W.: Stability of solutions of nonlinear neutral differential equations with piecewise constant delay and their discretizations. Appl. Math. Comput. **219**, 4590–4600 (2013)
317. Wang, X., Li, X., Huang, N., O'Regan, D.: Asymptotical consensus of fractional-order multi-agent systems with current and delay states. Appl. Math. Mech. **40**, 1677–1694 (2019)
318. Wang, Z., Huang, X., Shi, G.: Analysis of nonlinear dynamics and chaos in a fractional order financial system with time delay. Comput. Math. Appl. **62**, 1531–1539 (2011)

319. Wang, Z.L., Yang, D.S., Ma, T.D., Sun, N.: Stability analysis for nonlinear fractional-order systems based on comparison principle. Nonlinear Dyn. **75**, 387–402 (2014)
320. Weber, T.A.: Optimal Control Theory with Applications in Economics. The MIT Press, Cambridge (2011)
321. Wei, Z., Li, Q., Che, J.: Initial value problems for fractional differential equations involving Riemann-Liouville sequential fractional derivative. J. Math. Anal. Appl. **367**, 260–272 (2010)
322. Wilson, R.B.: A Simplicial Method for Convex Programming. Ph.D. Thesis, Cambridge, Massachusetts (1963)
323. Xing, A.Q.: The exact penalty function method in constrained optimal control problems. J. Math. Anal. Appl. **186**, 514–522 (1994)
324. Xing, A.Q., Chen, Z.H., Wang, C.L., Yao, Y.Y.: Exact penalty function approach to constrained optimal control problems. Optim. Contr. Appl. Met. **10**, 173–180 (1989)
325. Xiu, D.: Fast numerical methods for stochastic collocation: a review. Commun. Comput. Phys. **5**, 242–272 (2008)
326. Yamamoto, T.: A method for finding sharp error bounds for Newton's method under the Kantorovich assumptions. Numer. Math. **49**, 203–220 (1986)
327. Yan, Y., Pal, K., Ford, N.J.: Higher order numerical methods for solving fractional differential equations. BIT Numer. Math. **54**, 555–584 (2014)
328. Yang, Y., Huang, Y.: Spectral-collocation methods for fractional pantograph delay-integrodifferential equations. Adv. Math. Phys. 821327 (2013)
329. Yang, Z., Cao, J.: Initial value problems for arbitrary order fractional differential equations with delay. Commun. Nonlinear Sci. Numer. Simul. **18**, 2993–3005 (2013)
330. Ye, H., Go, J., Ding, Y.: A generalized Gronwall inequality and its application to a fractional differential equation. J. Math. Anal. Appl. **328**, 1075–1081 (2007)
331. Yi, X., Gong, Z., Liu, C., Cheong, H.T., Teo, K.L., Wang, S.: A control parameterization method for solving combined fractional optimal parameter selection and optimal control problems. Commun. Nonlinear Sci. Numer. Simul. **141**, 108462 (2025)
332. Yi, X., Liu, C., Cheong, H.T., Teo, K.L., Wang, S.: A third-order numerical method for solving fractional ordinary differential equations. AIMS Math. **9**, 21125–21143 (2024)
333. Yousefi, S.A., Lotfi, A.: Legendre multiwavelet collocation method for solving the linear fractional time delay systems. Open Phys. **11**, 1463–1469 (2013)
334. Yu, C., Lin, Q., Loxton, R., Teo, K.L., Wang, G.: A hybrid time-scaling transformation for time-delay optimal control problems. J. Optim. Theory Appl. **169**, 876–901 (2016)
335. Yu, J.M., Hu, H., Zhou, S.B., Lin, X.R.: Generalized Mittag-Leffler stability of multi-variables fractional order nonlinear systems. Automatica **49**, 1798–1803 (2013)
336. Yu, Y.: Optimal control of a nonlinear time-delay system in batch fermentation process. Math. Prob. Eng. 478081 (2014)
337. Zayernouri, M., Cao, W., Zhang, Z., Karniadakis, G.E.: Spectral and discontinuous spectral element methods for fractional delay equations. SIAM J. Sci. Comput. **36**, B904–B929 (2014)
338. Zeid, S.S.: Approximation methods for solving fractional equations. Chaos Soliton. Fract. **125**, 171–193 (2019)
339. Zhang, K., Yang, X., Wang, S., Teo, K.L.: Numerical performance of penalty method for American option pricing. Optim. Meth. Softw. **25**, 737–752 (2010)
340. Zhang, X., Cao, J.: A high order numerical method for solving Caputo nonlinear fractional ordinary differential equations. AIMS Math. **6**, 13187–13209 (2021)
341. Zhao, X., Ge, W.: Unbounded solutions for a fractional boundary value problems on the infinite interval. Acta Appl. Math. **109**, 495–505 (2010)
342. Zhong, W., Wang, L.: Basic theory of initial value problems of conformable fractional differential equations. Adv. Differ. Equ. **2018**, 321 (2018)
343. Zhou, X.F., Jiang, W., Hu, L.G.: Controllability of a fractional linear time-invariant neutral dynamical system. Appl. Math. Lett. **26**, 418–424 (2013)
344. Zhou, Y., Ahmad, B., Alsaedi, A.: Theory of Fractional Evolution Equations. de Gruyter, Berlin (2022)
345. Zhou, Y.: Basic Theory of Fractional Differential Equations. World Scientific, Sigapore (2024)

www.ingramcontent.com/pod-product-compliance
Ingram Content Group UK Ltd.
Pitfield, Milton Keynes, MK11 3LW, UK
UKHW020632260926
475829UK00003BA/48
9789819202737